IET TELECOMMUNICATIONS SERIES 70

Cloud and Fog Computing in 5G Mobile Networks

Other volumes in this series:

Volume 9	**Phase noise in signal sources** W.P. Robins
Volume 12	**Spread spectrum in communications** R. Skaug and J.F. Hjelmstad
Volume 13	**Advanced signal processing** D.J. Creasey (Editor)
Volume 19	**Telecommunications traffic, tariffs and costs** R.E. Farr
Volume 20	**An introduction to satellite communications** D.I. Dalgleish
Volume 26	**Common-channel signalling** R.J. Manterfield
Volume 28	**Very small aperture terminals (VSATs)** J.L. Everett (Editor)
Volume 29	**ATM: the broadband telecommunications solution** L.G. Cuthbert and J.C. Sapanel
Volume 31	**Data communications and networks, 3rd edition** R.L. Brewster (Editor)
Volume 32	**Analogue optical fibre communications** B. Wilson, Z. Ghassemlooy and I.Z. Darwazeh (Editors)
Volume 33	**Modern personal radio systems** R.C.V. Macario (Editor)
Volume 34	**Digital broadcasting** P. Dambacher
Volume 35	**Principles of performance engineering for telecommunication and information systems** M. Ghanbari, C.J. Hughes, M.C. Sinclair and J.P. Eade
Volume 36	**Telecommunication networks, 2nd edition** J.E. Flood (Editor)
Volume 37	**Optical communication receiver design** S.B. Alexander
Volume 38	**Satellite communication systems, 3rd edition** B.G. Evans (Editor)
Volume 40	**Spread spectrum in mobile communication** O. Berg, T. Berg, J.F. Hjelmstad, S. Haavik and R. Skaug
Volume 41	**World telecommunications economics** J.J. Wheatley
Volume 43	**Telecommunications signalling** R.J. Manterfield
Volume 44	**Digital signal filtering, analysis and restoration** J. Jan
Volume 45	**Radio spectrum management, 2nd edition** D.J. Withers
Volume 46	**Intelligent networks: principles and applications** J.R. Anderson
Volume 47	**Local access network technologies** P. France
Volume 48	**Telecommunications quality of service management** A.P. Oodan (Editor)
Volume 49	**Standard codecs: image compression to advanced video coding** M. Ghanbari
Volume 50	**Telecommunications regulation** J. Buckley
Volume 51	**Security for mobility** C. Mitchell (Editor)
Volume 52	**Understanding telecommunications networks** A. Valdar
Volume 53	**Video compression systems: from first principles to concatenated codecs** A. Bock
Volume 54	**Standard codecs: image compression to advanced video coding, 3rd edition** M. Ghanbari
Volume 59	**Dynamic ad hoc networks** H. Rashvand and H. Chao (Editors)
Volume 60	**Understanding telecommunications business** A. Valdar and I. Morfett
Volume 65	**Advances in body-centric wireless communication: applications and state-of-the-art** Q.H. Abbasi, M.U. Rehman, K. Qaraqe and A. Alomainy (Editors)
Volume 67	**Managing the Internet of Things: architectures, theories and applications** J. Huang and K. Hua (Editors)
Volume 68	**Advanced relay technologies in next generation wireless communications** I. Krikidis and G. Zheng
Volume 905	**ISDN applications in education and training** R. Mason and P.D. Bacsich

Cloud and Fog Computing in 5G Mobile Networks

Emerging advances and applications

Edited by
Evangelos Markakis, George Mastorakis,
Constandinos X. Mavromoustakis and Evangelos Pallis

The Institution of Engineering and Technology

Published by The Institution of Engineering and Technology, London, United Kingdom

The Institution of Engineering and Technology is registered as a Charity in England & Wales (no. 211014) and Scotland (no. SC038698).

First published 2017

The Institution of Engineering and Technology
Michael Faraday House
Six Hills Way, Stevenage
Herts, SG1 2AY, United Kingdom

www.theiet.org

British Library Cataloguing in Publication Data
A catalogue record for this product is available from the British Library

ISBN 978-1-78561-083-7 (hardback)
ISBN 978-1-78561-084-4 (PDF)

Typeset in India by MPS Limited
Printed in the UK by CPI Group (UK) Ltd, Croydon

Contents

Editors' biographies

Dr. Evangelos Markakis holds a PhD from the Department of Information & Communication Systems Engineering of the University of the Aegean, Greeve. He has actively participated in more than 15 EU FP5/FP6/FP7/Horizon2020 funded projects (IST/ICT/Health/Security) and over 20 Greek funded research and development projects. He is currently working as a Senior Research Fellow at PASIPHAE Laboratory of the Technological Educational Institute of Crete. His research activities include Interactive Digital Television (DVB), distributed systems and P2P applications, design of large large-scale Heterogeneous Networks, Fog Computing & Networking, as well as Network Management and Virtualisation Techniques, including SDN/NFV Concepts. He is a member of IEEE ComSoc and the Workshop Co-Chair for the IEEE SDN-NFV Conference.

Dr. George Mastorakis is an associate professor in the field of new technologies and marketing at the Technological Educational Institute of Crete and member of the PASIPHAE laboratory. His research interests include network/IT resource management, dynamic resource reservation and trading, interactive broadcasting and generation networks. He has actively participated in more than 10 EC-funded R&D projects and a large number of national research ones. He has more than 160 publications at various international conference proceedings, workshops, scientific journals and book chapters. He is a member of IEEE and Editor of the *Journal of Networks, Information Technology Journal and Journal of Business and Economics*.

Dr. Constandinos X. Mavromoustakis is a professor at the Department of Computer Science at the University of Nicosia, Cyprus. He received his dipl.Eng (BSc/BEng/MEng) in Electronic and Computer Engineering from the Technical University of Crete, Greece, an MSc in Telecommunications from University College of London, UK, and his PhD from the Department of Informatics at Aristotle University of Thessaloniki, Greece. He is leading the Mobile Systems Lab. at the Department of Computer Science at the University of Nicosia, dealing with the design and implementation of hybrid wireless testbed environments, high performance cloud and mobile cloud computing (MCC) systems, modelling and simulation of mobile computing environments and protocol development and deployment for large-scale heterogeneous networks as well as new 'green' mobility-based protocols. He has served as a consultant to many industrial bodies, is a management member of the IEEE Communications Society (ComSoc) Radio Communications Committee (RCC) and a board member of the IEEE-SA

Standards IEEE SCC42 WG2040 where he has served as track Chair and Co-Chair of various IEEE International Conferences. He is the recipient of various grants including the European grant of Early Stage Researcher (ESR), for the excellent research output and research impact (December 2013, EU secretariat/Brussels).

Dr. Evangelos Pallis is an associate professor at the Informatics Engineering Department of the Technological Educational Institute of Crete, and the director of PASIPHAE Laboratory, Greece. His research interests are in the fields of wireless networks and mobile telecommunication systems, linear and interactive broadcasting, multimedia service provisioning and fixed-mobile converged infrastructures. He has more than 100 refereed publications in international journals, conference proceedings and is a member of the IEEE/ComSoc, IEE/IET. He serves in the Editorial Board of the *Information Technology Journal and the Research Journal of Information Technology* and holds the general chair of the international conference on Telecommunications and Multimedia (TEMU). Since 2007, he has acted as a Distinguished Member of the Union of Regional TV Broadcasters of Greece.

Chapter 1
NOMA schemes for 5G green mobile networks

S.M. Riazul Islam, Anish P. Shrestha, Farman Ali and K.S. Kwak

The nonorthogonal multiple access (NOMA) is one of the fledging paradigms that the next generation radio access technologies sprouting toward. The NOMA with superposition coding (SC) in the transmitter and successive interference cancelation (SIC) at the receiver comes with many desirable features and benefits over orthogonal multiple access such as orthogonal frequency division multiple access adopted by long-term evolution. Various studies reveal that the NOMA is a noble spectrum-efficient technique, which can also be designed in the light of energy efficiency. In this chapter, we study the recent progresses of NOMA in fifth-generation (5G) systems. We discuss the basic concepts of NOMA and explain its aspects of importance for future radio access. Then, we provide a survey of the state of the art in NOMA solutions for 5G systems with numerical performances and provide some avenues for future research on NOMA on a set of open issues and challenges.

1.1 Introduction

In order to continue to ensure the sustainability of mobile communication services over the next decade and to meet the business and consumer demands, fifth generation (5G) mobile communication services is expected to be rolled out by 2020. One of the major requirements for 5G networks is the significant spectral efficiency (SE) enhancement compared to fourth generation (4G) as the anticipated exponential increase in the volume of mobile data traffic is huge, for example, at least 1,000-fold in the 2020s compared to 2010. In particular, the peak data rate in 5G should be 10–20 Gbps that is 10–20 times the peak data rate in 4G, and the user experienced data rate should be 1 Gbps (100 times the user experienced data rate in 4G). In addition, the rapid development of Mobile Internet and the Internet of Things (IoT) exponentially accelerates the demands for high data rate applications, including high-quality video streaming, social networking, and machine-to-machine communications.

In cellular network, the design of radio access technology, in general, and multiple access technique, in particular, are one of the most important aspects in improving the system capacity. Multiple access techniques are usually categorized into two orthogonal and nonorthogonal approaches [1]. In orthogonal approaches, signals from different users are orthogonal to each other, that is, their cross correlation is zero (the available resources such as the system bandwidth (BW), and time is divided among users). Nonorthogonal schemes such as code division multiple access (CDMA) allow nonzero cross correlation among the signals from different users. Second and third generation cellular systems such as IS-95, CDMA2000, and wideband-CDMA (WCDMA) have adopted nonorthogonal multiple access (NOMA) techniques. CDMA is usually more robust against fading and cross-cell interference, but is susceptible to intracell interference. With careful cell planning, orthogonal multiple access (OMA) can avoid intracell interference. On that, most of the first and second generation cellular systems adopted orthogonal MA approaches. Even, orthogonal frequency division multiple access–based OMA has been adopted in 4G systems such as long-term evolution (LTE) and LTE-advanced.

Despite a practical advantage of intracell interference avoidance capability, CDMA has limited data rate due to its spread-spectrum nature. OMA is a realistic choice for achieving good performance in terms of system-level throughput. However, due to the aforementioned upcoming wave, 5G networks require further enhancement in the system efficacy. Then again, to get the facilities of on-demand resource processing, delay-aware storage, and high network capacity, the cloud computing–based radio access infrastructure is a possible solution. And, advanced baseband computation and radio frequency communication are required to enable large-scale cooperative signal processing in the physical layer and adapted to new air interfaces in 5G systems. In this regard, researchers over the globe have started investigating NOMA as a promising multiple access scheme for future radio access. NOMA achieves superior spectral efficiencies by combining superposition coding (SC) at the transmitter with successive interference cancelation (SIC) at the receivers [2,3]. On the top of that, the evolution of wireless networks into 5G poses new challenges on energy efficiency (EE), as the entire network will be ultradense. With an extreme increase in number of infrastructure nodes, the total energy consumption may simply surpass an acceptable level. Although the substantial energy is basically consumed by the hardware, the NOMA has an inherent ability to adapt the transmission strategy according to the traffic and users channel state information (CSI). Thus, it can achieve a good operating point where both the spectrum efficiency and EE become optimum. In view of the fact that the IoT is expected to be widely used in our everyday life, the fog computing is growing in popularity. One of the primary objectives of fog networking is minimizing the use of BW. Although fog computing is implemented by handing some application services at edge devices and in a remote data center, some physical and medium access control layer issues can help to achieve its efficient spectrum utilization intention. In this regard, NOMA is important, as its target is also the efficient utilization of available spectrum.

Over the past few years, NOMA has attracted huge attention of researchers to meet the 5G requirements. As a consequence, many research efforts on this field already exist. Research trends in NOMA include diverse topics, for example, performance analysis, cooperative communications, and fairness analysis. However, NOMA in 5G is still in its infancy. At this stage, a comprehensive knowledge on the up-to-date research status of NOMA in 5G systems is extremely useful to researchers to do more research in this area. In this chapter, we appraise the state of the art of

NOMA research trends and disclose various issues that need to be addressed to transform radio access techniques through NOMA innovation.

1.2 Basic concepts of NOMA

First, we present a brief note about SC and SIC as these two techniques play important roles in NOMA and will then describe a typical NOMA scheme.

1.2.1 Superposition coding

The SC which was first proposed by Cover [4] is a technique of communicating information to several receivers by a single source simultaneously. In other words, it allows the transmitter to transmit multiple users' information at the same time. Examples of communications in a superposition fashion include broadcasting TV information to multiple receivers, giving a lecture to a group of different backgrounds, and aptitudes such as a lecture in a class room. To delineate the thought of superposition, we will present a simple example [3] of a speaker who can speak both English and Korean. There are two audience members: one understands only English and the other only Korean. Assume the speaker can transmit $R1 = 20$ bits of information per second to listener 1 by speaking to her continually; in this case, he sends no information to listener 2. Likewise, he can send $R2 = 20$ bits per second (bps) to listener 2 without sending any information to listener 1. Thus, he can accomplish any rate pair with $R1 + R2 = 20$ by simple time-sharing. But, is it possible to send more information? Recall that the English listener, even though he does not understand Korean, can distinguish when the word is Korean. Moreover, the Korean listener can identify when English occurs. The speaker can exploit this to convey information. For example, if the speaker delivers a sequence of 100 words with 50% time-sharing to each listener, there are about $^{100}C_{50}$ ways to order the English and Korean words. Information to both listeners can be sent through one of these orderings. This technique enables the speaker to convey information at a rate of 10 bps to the English listener, 10 bps to the Korean listener, and 1 bps of common information to both of them. Thus, a total rate of 21 bps (more than that achievable by simple time-sharing) is achieved, which is more than that achievable by simple time-sharing. This can be thought as an example of superposition of information. To make SC practical, the transmitter must encode information relevant to each user. For example, for two-user case, the transmitter will have to contain two point-to-point encoders that map their respective inputs to complex-valued sequences of two users' signal. It can be mentioned that the SC is a recognized nonorthogonal scheme that attains the capacity on a scalar Gaussian broadcast channel. Some good strategies for SC and proposes a design technique for SC by using off-the-shelf single-user coding and decoding blocks are available in [5].

1.2.2 Successive interference cancelation

To decode the superposition coded information at each receiver, Cover [4] first proposed the SIC technique. The SIC is conceivable by exploiting the knowledge of the differences in signal strength among the signals of interest. The basic idea of

SIC is that users are successively decoded. After one user is decoded, its signal is subtracted from the combined signal before the next user is decoded. When SIC is applied, one of the users is decoded treating the other user as interferer, but the latter is decoded with the benefit of the signal of the former already removed. However, prior to SIC, users are ordered according to their signal strengths so that the receiver may be able to decode the stronger signal first, subtract it from the combined signal and remove the weaker one from the residue. Note that, each user is decoded treating the other interfering users as noise in using signal reception. To gain a deeper understanding of how SIC performs in wireless communications, in general, and in orthogonal frequency division multiple access (OFDM), and multiple input multiple output (MIMO) systems, in particular, interested readers are referred to [6].

1.2.3 A typical NOMA scheme

Let us consider a single-cell downlink scenario where there is single base station (BS), B, and N users U_i, with $i \in N = \{1, 2, \ldots, N\}$, and all terminals are equipped with single antenna. It can be noted that a similar uplink scenario can also be described and NOMA scheme can equally be utilized there. The BS always sends data to all users simultaneously with the constraint of total power P. We assume the wireless links experience independent and identically distributed (i.i.d.) block Rayleigh fading and additive white Gaussian noise (AWGN). The channels are sorted as $0 < |h_1|^2 \leq |h_2|^2 \leq \cdots \leq |h_i|^2 \cdots \leq |h_N|^2$ which indicates that the user U_i always holds the ith weakest instantaneous channel. The NOMA scheme allows simultaneous serving all users by using the entire system BW to transmit data by with the help of a SC at the BS and SIC techniques at the users. Here, user multiplexing is performed in the power domain. The BS transmits a linear superposition of N users' data by allocating a fraction β_i of the total power to each U_i that is, the power allocated for ith user is $P_i = \beta_i P$. In the receiving side, each user decodes the signals of the weaker users that is, the U_i can decode the signals for each U_m with $m < i$. The signals for weaker users are then subtracted from the received signal to decode the signal of the user U_i itself treating the signals for the stronger users U_m with $m > i$ as interferences. The received signal at the user U_i can be represented as

$$y_i = h_i x + w_i \tag{1.1}$$

here $x = \sum_{i=1}^{N} \sqrt{P\beta_i} S_i$ is the superposition coded signal transmitted by B with S_i be the signal for the user U_i. Also, w_i is the AWGN at the user U_i with zero mean and variance σ_n^2. If signal superposition at B and SIC at U_i are carried out perfectly, the data rate achievable to user U_i for 1 Hz system BW is given by

$$R_i = \log\left(1 + \frac{\beta_i P |h_i|^2}{P|h_i|^2 \sum_{k=i+1}^{N} \beta_k + \sigma_n^2}\right) \tag{1.2}$$

Note that, the data rate of user U_N is $R_N = \log\left(1 + \beta_i P |h_i|^2 / \sigma_n^2\right)$.

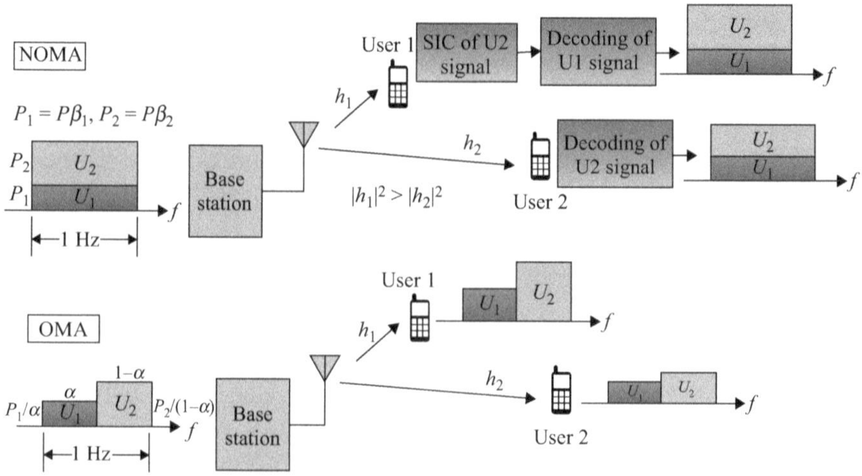

Figure 1.1 NOMA and OMA schemes with spectrum usage comparison for two users' case

Note that, a strong user means it experiences a better channel condition but does not mean that its signal strength is stronger. In fact, the less transmit power is assigned with a strong user, and the weak user is assigned with more power. Thus, the weak user's signal is the strongest one. Therefore, the NOMA does not contradict with the basic concept of SIC that decoding of the strongest signal should be performed first.

Figure 1.1 represents the aforementioned NOMA scheme with two users. This figure also represents the OMA scheme to disclose the particular advantage of NOMA scheme over OMA one. In case of NOMA, the entire 1 Hz BW is simultaneously used by two users. However, in case of OMA, user 1 uses α Hz and the remaining $1 - \alpha$ Hz is assigned to user 2. In NOMA, the user 1 first performs SIC to decode the signal for user 2 as the channel gain of user 1 is higher than that of user 2. The decoded signal is then subtracted from the received signal of user 1. This resultant signal is eventually used for decoding the signal for the user 1 herself. At user 2, no SIC is performed and its signal is directly decoded. Thus, the achievable data rate to users 1 and 2 are given by (1.3) and (1.4), respectively.

$$R_1 = \log\left(1 + \frac{P_1|h_1|^2}{\sigma_n^2}\right) \tag{1.3}$$

$$R_2 = \log\left(1 + \frac{P_2|h_2|^2}{P_1|h_2|^2 + \sigma_n^2}\right) \tag{1.4}$$

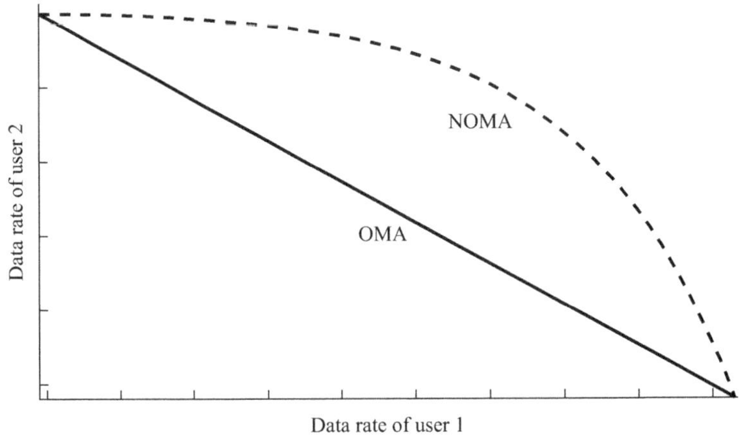

Figure 1.2 Capacity comparison of NOMA and OMA with two users

In case of OMA, the achievable data rate to users 1 and 2 are given by (1.5) and (1.6), respectively.

$$R_1 = \alpha \log\left(1 + \frac{P_1|h_1|^2}{\sigma_n^2}\right) \tag{1.5}$$

$$R_2 = (1-\alpha)\log\left(1 + \frac{P_2|h_2|^2}{\sigma_n^2}\right) \tag{1.6}$$

It is clear from (1.3) and (1.4) that the NOMA scheme controls the throughput of each user by adjusting the power allocation ratio P_1/P_2. Thus, the overall throughput, and user fairness are closely related to the power allocation scheme. If we consider an asymmetric channel (signal-to-noise ratios (SNRs) of the two users are different), we can numerically show that the values of R_1 and R_2 calculated from (1.3) and (1.4), respectively, are considerably much higher than those of R_1 and R_2 calculated from (1.5) and (1.6), respectively. This numerical comparison is basically a special case of the multi-user channel capacity analysis in [3]. Figure 1.2 gives us the idea of a generalized capacity comparison of NOMA and OMA for two users. It shows that the boundary of achievable rate pairs of NOMA is outside of the OMA capacity region in general. Therefore, NOMA is highly effective in terms of system-level throughput when the channels are different for two users. On that, NOMA is being considered as a promising multiple access technique for future radio access.

1.3 Potential NOMA solutions

In this section, we will present some concurrent works on NOMA which can be considered as potential solutions to problems or issues associated with the integration of NOMA in 5G. We will avoid the detail explanations and mathematical derivations of the techniques, since our major focus is to get some primary ideas of the state of art of NOMA research in 5G systems. Interested readers are referred to original articles for this purpose.

1.3.1 NOMA performances in 5G

A substantial number of researches have investigated the performances of NOMA schemes to study the feasibility of adopting this technique as a multiple access scheme for 5G systems. The survey [7] and references therein demonstrate that NOMA can be a promising power domain user multiplexing scheme for future radio access. In a cellular network with randomly deployed users, the performance of NOMA can be evaluated under two situations. In first case, each user has a targeted data rate determined by its assigned quality of service (QoS). Here, the outage probability is an ideal performance metric as it measures the capability of NOMA to supply the users' QoS requirements. In the other case, users' rates are opportunistically allocated according to their channel conditions. In this situation, the achievable ergodic sum rate can be investigated to evaluate the NOMA performances. According to Ding *et al.* [8], if users' data rate and assigned power are chosen properly, NOMA can offer better outage performance than the OMA techniques. This study also shows that NOMA can achieve a superior ergodic sum rate. If the SNR is high, the outage probability of ith user in a typical disk-shaped cell with radius R_D can be given by:

$$P_i^{out} = \frac{\tau_i}{i}\eta^i(\psi_i^*)^i \tag{1.7}$$

where $\tau_i = N!/((i-1)!(N-i)!)$ and $\eta = 1/R_D\sum_{l=1}^{L}\beta_l$ with $\beta_l = \pi/l\sqrt{1-\theta_l^2}$ $((R_D/2)\theta_l + (R_D/2))(1 + ((R_D/2)\theta_l + (R_D/2))^\alpha)$ and $\theta_l = cos((2n-1/2L)\pi)$. In addition, L, α, and ψ_i^* represent complexity trade-off parameter, path-loss factor, and maximum SNR corresponding to data rate of ith user, respectively. Also, with sufficient number of users, N, and adequate transmit SNR, ρ, the NOMA can achieve the following ergodic sum rate:

$$R_{erg} = log(\rho\, log\, log\, N) \tag{1.8}$$

In [9], Xiaohang *et al.* focus on the impact of rank optimization on the performance of NOMA with single user (SU)-MIMO. They show the way of how NOMA combined with SU-MIMO techniques can achieve further system performance improvement by adjusting rank of channel matrix.

Based on (1.7), Figure 1.3 compares the outage performances of NOMA schemes with that of OMA scheme for a cellular network with randomly deployed users with $N = 2, L = 10, \alpha = 2$, and $R_D = 3m$. The users are uniformly located. We use the target data rates of 0.1 bit per channel use (BPCU) and 0.5 BPCU for weak user and strong user (user 2 here) respectively. As the conventional orthogonal scheme has been considered for benchmarking, its target rate is 0.6 BPCU (the addition of two users' data rate). Also, it is to be noted that the numerical results are based on the normalized SNR model. As can be observed from this figure, the NOMA outperforms

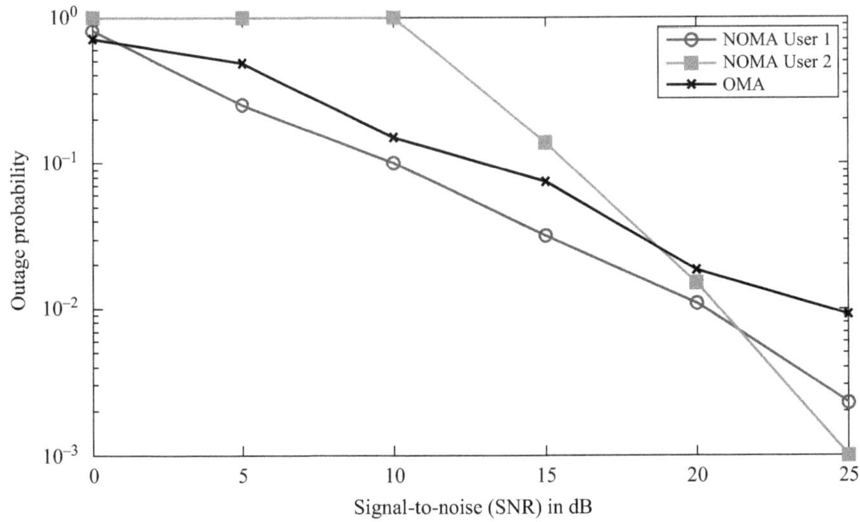

Figure 1.3 Outage performance of NOMA in 5G systems with random users

the comparable scheme, and the diversity order of the users is a function of their channel conditions. Note that, in this case, the ratio of the power assigned to strong user to the power assigned to weak user is 1:4. The outage probability given by (1.7) is basically valid at high SNR condition. On that, in order to recognize the comparative outage performance, we need to focus on high SNR regions where both users outperform the OMA scheme. As the assigned power to the strong user is proportionally lower, the outage performance at low SNR region is poor. However, as the SNR becomes high enough, the power-domain multiplexing becomes dominant and thereby shows the best performance with superior diversity order.

1.3.2 Cooperative NOMA

In wireless networks, cooperative communications have gained huge attention due to its ability to offer spatial diversity for mitigating fading, while resolving the difficulties of mounting multiple antennas on small communication terminals [10]. In cooperative communication, several relay nodes are assigned to assist a source in forwarding its information to the respective destination. Therefore, the integration of cooperative communication with NOMA can further improve the system efficiency in terms of capacity and reliability. The cooperative NOMA (C-NOMA) scheme proposed in [11] exploits prior information available in NOMA systems. In this scheme, users with better channel conditions decode the messages for the others, and therefore, these users act as relays to improve the reception reliability for the users with poor connections to the B. The cooperative communication from the users with better channel conditions to the ones with poor channel conditions can be done by using short range communication techniques, such as ultra-wide-band and BT. It is demonstrated that C-NOMA can achieve the maximum diversity gain for all the users. The overall outage probability of cooperative is defined in the following equation [11]:

$$P^{out} \triangleq 1 - \prod_{i=1}^{N} \left(1 - P_i^{out}\right) \tag{1.9}$$

The cooperative NOMA scheme ensures that the ith best user experiences a diversity of order of N conditioned on a specific power allocation ratio. However, C-NOMA is expensive in terms of additional time-slots, as its cooperative phase requires messages retransmission from each user acting as relay in a serial manner. To reduce system complexity, C-NOMA performs user pairing based on distinctive channel gains. The performance of C-NOMA can be further enhanced by adopting optimal power allocation schemes [12,13]. The direct derivation of theoretical achievable rate in NOMA is quite difficult. However, if we compare the rates of conventional time division multiple access (TDMA) with that of noncooperative NOMA, we can observe that the performance difference is not a function of power allocation coefficients but rather depends on how disparate two users' channels are. And the similar observation can also be noted to C-NOMA.

With the same number of users and power allocation ratio as we used in case of Figure 1.3, Figure 1.4 presents the outage probability, based on (1.9), achieved by the noncooperative NOMA, and cooperative NOMA as a function of SNR. It shows that cooperative NOMA transcends the comparable scheme as it ensures that the maximum diversity gain is achievable to all the users. This high diversity gain can be explained as below. Under C-NOMA, a user with the worst channel condition gets assistance from the other $N - 1$ users along with its own direct link to the source, whereas noncooperative NOMA can attain only a diversity order of i for the

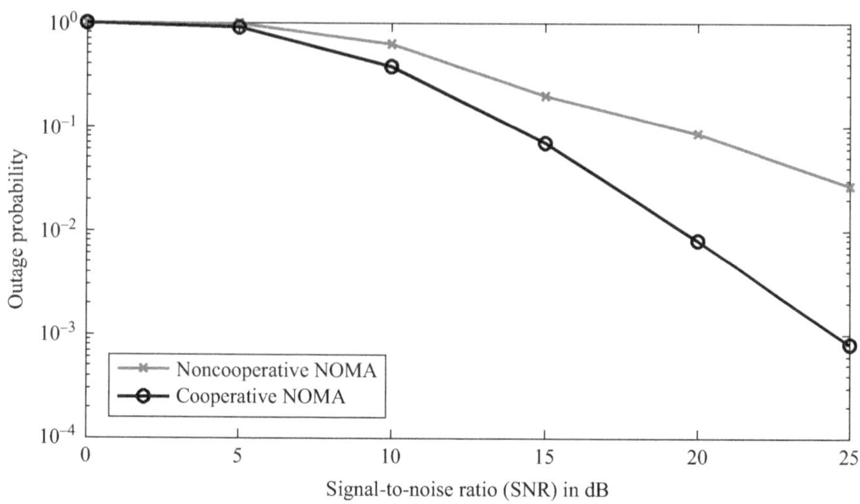

Figure 1.4 Outage performance of cooperative NOMA

ith ordered user, C-NOMA ensures that a diversity order of N is achievable by all users by exploiting user cooperation.

1.3.3 Fairness in NOMA

The investigation of the impact of power allocation on the fairness performance of the NOMA scheme has been studied in [14]. Therein, authors study the power allocation problem from a fairness viewpoint under two assumptions: (i) BS has perfect CSI, and hence, users' data rates are adopted to the channel conditions and (ii) when users have fixed targeted data rates under an average CSI. They provide low-complexity algorithms that yield globally optimal solutions. This study confirms that the NOMA scheme outperforms conventional MA approaches by significantly improving the performances of the users with worst channel conditions. If instantaneous CSIs are available at BS, fairness among users can be ensured by maximizing the minimum achievable user rate, that is,

$$\max_{\beta} \min_{i \epsilon N} R_i(\beta) \tag{1.10a}$$

$$\text{s.t.} \sum_{j=1}^{N} \beta_j = 1 \tag{1.10b}$$

$$0 \leq \beta_j, \quad \text{for } j \epsilon N \tag{1.10c}$$

As the problem (1.10a)–(1.10c) is not convex, it needs to be converted into a sequence of linear programming first. Eventually, the optimal solution to (1.10a)–(1.10c) can be given by the following equation:

$$\beta_i = \frac{2^t - 1}{P|h_i|^2} \left(P|h_i|^2 \sum_{k=i+1}^{N} \beta_k + \sigma_n^2 \right) \qquad i = N, N-1, \ldots, 1. \tag{1.11}$$

where t represents the minimum data rate. If instantaneous CSIs are not available, we should optimize the outage probability with the knowledge of average CSI. In this case, the fairness among users can be ensured by minimizing the maximum outage probability as $\min_{\beta} \max_{i \epsilon N} P_i^{out}(\beta)$ conditioned on (1.10b) and (1.10c). Unfortunately, this fairness study suffers from the inadequate performance comparison. It does not graphically demonstrate the achievable maximum fairness rate and does not provide a visual comparison between NOMA and TDMA. Also, this chapter does not explicitly explain what it means by fixed NOMA.

1.3.4 NOMA with beamforming

As a representative chapter for NOMA with multiuser beamforming (NOMA-BF), we focus on the research work reported in [13]. The proposed NOMA-BF technique allows two users to share a single beamforming vector. To reduce the interbeam interferences (from users of other beams) and intrabeam interferences (from users sharing the same beamforming vector), the NOMA-BF comes with a clustering and power allocation algorithm based on correlation among users and channel gain difference, respectively. The NOMA-BF system improves the sum capacity, compared to the conventional multiuser beamforming system. The NOMA-BF also guarantees the weak users' capacity to ensure user fairness. A power allocation scheme for nth cluster of two-users NOMA-BF consisting of N clusters with 2 users in each cluster that maximizes the sum capacity while keeping the weak user's capacity at least equal to that of the conventional multiuser beamforming system can be formulated as below conditioned on (1.10b) and (1.10c):

$$\beta_1^n = \arg\max_{\beta_1^n} (R_1 + R_2) \tag{1.12a}$$

$$\text{s.t. } R_2 \geq \frac{1}{2} R_{2,conv-\mathrm{BF}} \tag{1.12b}$$

where R_1 and R_2 are the capacities of the strong and the weak users, respectively. $R_{2,conv-\mathrm{BF}}$ is the capacity of the weak user if the weak user would be supported by conventional beamforming. β_1^n and $1 - \beta_1^n = \beta_2^n$ are the power fractions of strong, and weak user, respectably in the nth cluster. The optimal solution to (1.12a) and (1.12b) can be obtained by using the Karush–Kuhn–Tucker condition as below:

$$\beta_1^n = \frac{1}{\sqrt{\left(1 + |h_{2,n}|^2 \rho\right)}} - \frac{\left\{\sqrt{\left(1 + |h_{2,n}|^2 \rho\right)} - 1\right\}\left\{\sum_{i=1, i\neq n}^{N} |h_{2,i} w_i|^2 \rho + 1\right\}}{\rho |h_{2,i} w_i|^2 \sqrt{\left(1 + |h_{2,n}|^2 \rho\right)}} \tag{1.13}$$

Figure 1.5 shows the sum capacities of the NOMA-BF and the conventional multiuser beamforming with correlation threshold $\rho = 0.75$, system BW 4.32 MHz, maximum transmission power per cluster 43 dBm, and noise density –169 dBm/Hz.

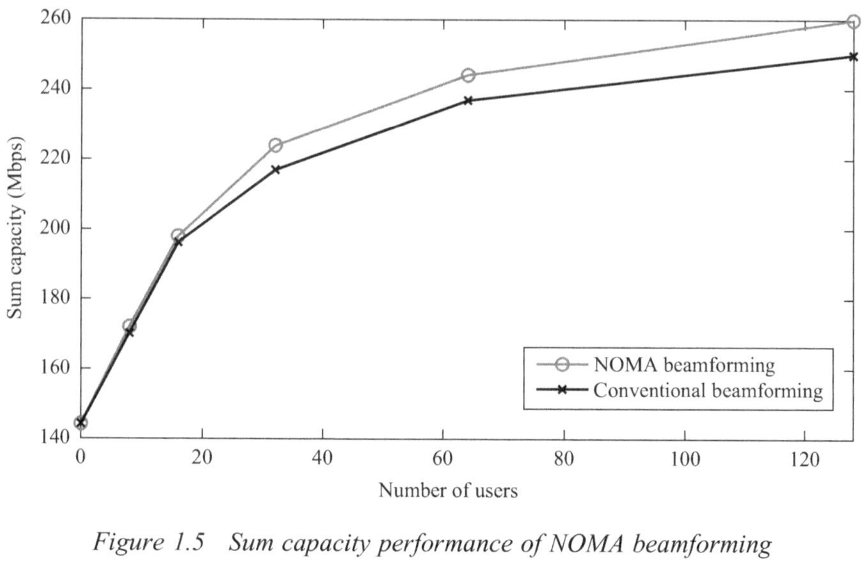

Figure 1.5 Sum capacity performance of NOMA beamforming

As can be observed, the NOMA-BF improves the sum capacity. Here, the users are randomly located with uniform distribution in a cell of radius 500 m. The NOMA-BF is better in terms of sum capacity compare to conventional multiuser beamforming as correlation-based clustering with effective power allocation reduces the interbeam and intrabeam interferences. As two users share a single beamforming vector, the number of supportable users can easily be increased by utilizing the NOMA-BF.

1.3.5 NOMA in coordinated system

In cellular systems, a cell-edge user usually experiences lower data rate compare to that experienced by a user near to a BS. The coordinated multipoint (CoMP) transmission (and reception) techniques, where multiple BSs support cell-edge users together are usually employed to increase transmission rates to cell-edge users. And the associated BSs for CoMP need to allocate the same channel to a cell-edge user. As a result, the SE of the system becomes worse as the number of cell-edge users increases. To avoid this problem, Choi [15] employs NOMA and thus proposes coordinated SC (CSC)-based NOMA scheme by considering SC for downlink transmissions to a group of cell-edge user and user near to a BS simultaneously with a common access channel [16]. In other words, BSs transmit Alamouti (space-time) [17] coded signals to user c (a cell-edge user), while each BS also transmits signals to a user near to the BS. The CSC-NOMA scheme with the Alamouti code provides a cell-edge user with reasonable transmission rate without demeaning the rates to near users and increases the SE. If R_{c1}, R_{c2}, and R_c are the rates to user (U_1) near to BS 1, user (U_2) near to BS 2, and coordinated user (U_c), the sum rate becomes $R_{c1} + R_{c2} + R_c$ with

$$R_{c1} = E\left[log_2\left(1 + \frac{|h_{1,1}|^2 P_1}{E\left[|h_{1,2}|^2\right] P_2 + \sigma_n^2}\right)\right] \tag{1.14}$$

$$R_{c2} = E\left[log_2\left(1 + \frac{|h_{2,2}|^2 P_2}{E\left[|h_{2,1}|^2\right] P_1 + \sigma_n^2}\right)\right] \tag{1.15}$$

$$R_c = \min\{Z_1, Z_2, Z_c\} \tag{1.16}$$

where $Z_1 = E[log_2(1 + SINR_1)]$, $Z_2 = E[log_2(1 + SINR_2)]$, and $Z_c = E[log_2(1 + SINR_c)]$. And $SINR_i$ be the signal-to-interference-plus-noise ratio at user U_i in decoding the signal of U_c. Note that, $h_{i,j}$ denotes the channel coefficient from BS j to user i. Because of the use of Alamouti code for CoMP communications, it does not require the exchanges of instantaneous CSI. This is a significant advantage over coherent transmission schemes that require instantaneous CSI exchange, which results in an excessive backhaul overhead for high mobility cell-edge users.

Figure 1.6 compares the sum rate performances of CSC-based NOMA with that of non-CSC-based NOMA under symmetric channel conditions with the path

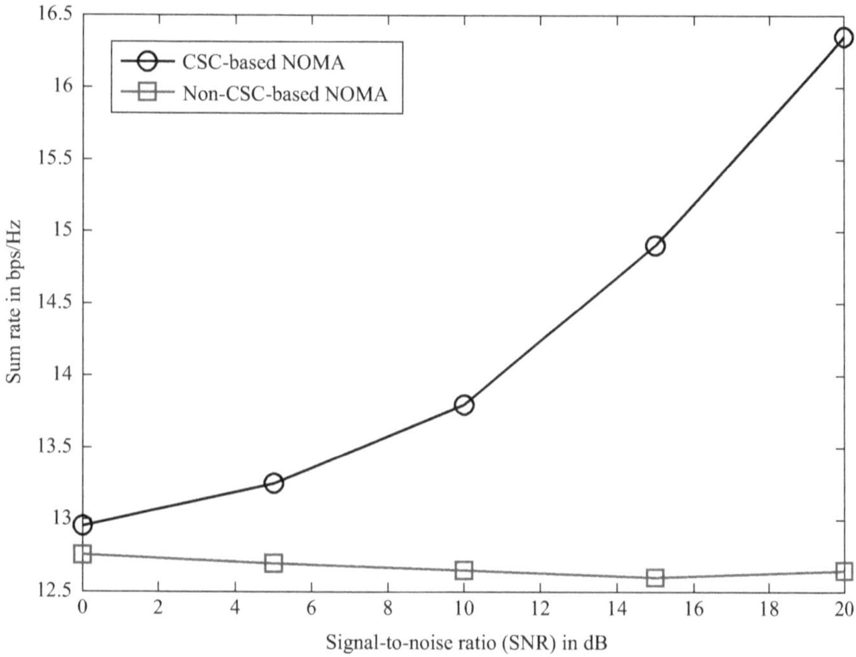

Figure 1.6 Sum rate performance of CSC-based NOMA

loss exponent of 3. Note that a non-CSC-based NOMA considers only one BS, either one, to employ SC to serve a pair of cell-edge and near users simultaneously. We observe that the sum rate of CSC-based system exponentially increases with SNR and is higher than that of non-CSC-based system.

1.3.6 Network NOMA

Let us consider a simple two-cell scenario of a cellular system (Figure 1.7), where U_3 and U_4 are served by BS 1, whereas U_1 and U_2 are served by BS 2. Also, we assume that a two-user NOMA scheme is adopted so that U_3 is paired with U_4, and U_1 is paired with U_2. In this situation, cell edge user U_4 at cell 1 and cell edge user U_1 at cell 2 may experience strong interferences from BS 2 and BS 1, respectively, as the power allocation by each transmitter may be biased to the distant user. To deal with the problems, for example, intercell interference associated with the employment of NOMA in multi-cell scenario, the straightforward application of single-cell NOMA solutions will not be appropriate; the single-cell NOMA need to be extended to the network NOMA. One possible solution to mitigate the intercell interference in Network NOMA is to utilize joint procoding of users' signals across the neighboring cells. However, the design of an optimal precoder is difficult as each BS should know all users' data and CSI. The correlation-based precoder design needs dynamic user selection for each NOMA pair [13]. Moreover, the multi-user precoding applicable for a single-cell NOMA may not be realistic in

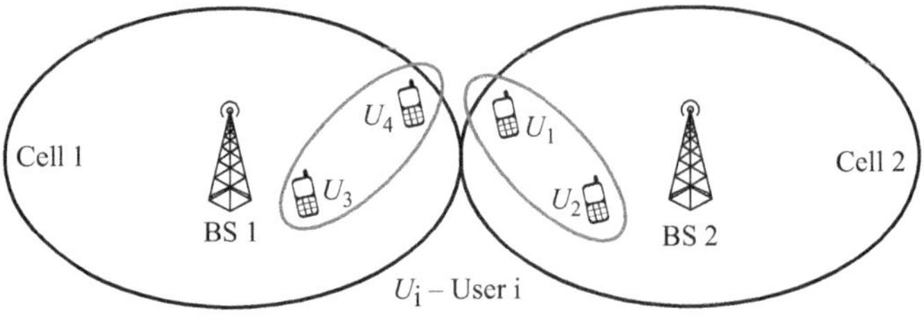

Figure 1.7 NOMA in a two-cell scenario

network NOMA as a beam generated via geographically separated BS does not support more than one spatially separated user for intrabeam NOMA. A low complexity precoding scheme for network NOMA has been proposed in [18] based on the fact that large-scale fading would be very disparate between the links to different cells. Here, the joint precoder is applied only to cell edge users (e.g., U_4 and U_1 in Figure 1.7) and the resulting $SINR$ of each user U_i of power P_i can be found in the following equation:

$$SINR_1 = \frac{\left[\left\{\left(\boldsymbol{H}_{41}(\boldsymbol{H}_{41})^{\boldsymbol{H}}\right)^{-1}\right\}_{1,1}\right]^{-1} P_1}{|h_{11}|^2 P_0 + |h_{12}|^2 P_2 + N_0 B} \tag{1.17a}$$

$$SINR_2 = \frac{|h_{22}|^2 P_2}{|h_{21}|^2\left(P_0 + |w_{0,0}|^2 P_1 + |w_{0,1}|^2 P_3\right) + N_0 B} \tag{1.17b}$$

$$SINR_3 = \frac{|h_{30}|^2 P_3}{|h_{32}|^2\left(P_2 + |w_{1,0}|^2 P_1 + |w_{1,1}|^2 P_3\right) + N_0 B} \tag{1.17c}$$

$$SINR_4 = \frac{\left[\left\{\left(\boldsymbol{H}_{41}(\boldsymbol{H}_{41})^{\boldsymbol{H}}\right)^{-1}\right\}_{0,0}\right]^{-1} P_4}{|h_{41}|^2 P_0 + |h_{42}|^2 P_2 + N_0 B} \tag{1.17d}$$

where $\boldsymbol{H}_{41} = [\boldsymbol{h}_4, \boldsymbol{h}_1]^T$ with the channel vector of ith user, $\boldsymbol{h}_i = [h_{i1}, h_{i2}]$ and h_{ij} ($i \in \{1, 2, 3, 4\}$ and $j \in \{1, 2\}$) being the channel response between the jth BS and ith user. The zero-forcing precoder, W, is the normalized pseudo-inverse of $\boldsymbol{H}_{41}$, that is, $W = (\boldsymbol{H}_{41})^H(\boldsymbol{H}_{41}(\boldsymbol{H}_{41})^{\boldsymbol{H}})^{-1}$, B is the system BW and N_0 is the noise density.

1.3.7 NOMA in MIMO systems

As we discussed in Section 1.3.4, the random opportunistic beamforming is first proposed in [13] for the MIMO NOMA systems under the assumption of perfect CSI at transmitter. It becomes evident that, with relatively large number of users, the combination of NOMA and MIMO can achieve a sufficient throughput gain [19]. In case of unavailability of perfect CSI at transmitter due to limited feedback, statistical CSI can be utilized for long-term power allocation to maximize the ergodic capacity of MIMO NOMA systems. Both optimal and low complexity suboptimal power allocation schemes are proposed in [20] to maximize the ergodic capacity with total transmit power constraint. The proposed MIMO NOMA system outperforms the conventional OMA scheme. It is also intuitive that the extension of NOMA in massive MIMO systems can further enhance the SE.

It is well known that relaying in wireless communications is very effective in terms of extended service coverage and increased system capacity. NOMA for multiple-antenna relaying network is studied in [21]. This chapter analyzes the outage behavior of the mobile users and derives the closed-form expressions for the exact outage probability. If NOMA is combined with multiple-antenna amplify-and-forward relaying network, where the BS and the mobile users are equipped with multiple antennas, the relay locations have a substantial impact on the outage performance. When the relay location is close to the BS, NOMA outperforms conventional OMA. However, conventional OMA attains better outage performance when the relay location is close to the users. In either case, NOMA offers better performances in terms of SE and user fairness.

1.3.8 Energy-efficient NOMA

NOMA employs some controllable interference by nonorthogonal resource allocation and realizes overloading at the cost of slightly increased receiver complexity. Consequently, higher SE can be achieved by NOMA for 5G. Although SE shows how efficiently a limited spectrum resource is utilized, it fails to provide any insight on how efficiently energy is utilized. With the rise of green communication in the recent years, reducing energy consumption has become a prime importance for researchers. 5G has also targeted EE as one of the major parameters to be achieved. Nonetheless, Shannon's capacity theorem illustrates that the two objectives of minimizing the consumed energy and maximizing the SE are not achievable simultaneously and calls for a trade-off. It can be noted that with circuit power under consideration, there always exists an optimal point in EE–SE curve. An energy efficient two-user single-cell NOMA is studied in [18]. Under fixed total power consumption, the EE–SE relationship is found to be linear with positive slope. Appropriate power allocation between two users allows achieving any point in the EE–SE curve. For given SE for each user, the maximal EE performance can be achieved. The degree of efficiency can be adjusted by varying the total power using power control schemes. If the sum rate capacity of the cell is R_{sum} with the total power consumption P_{cell}, the EE can be written as $\eta_E = R_{sum}/P_{cell} = B\eta_S/P_{cell}$, where η_S is the spectrum efficiency.

1.3.9 Other NOMA solutions

1.3.9.1 NOMA in light communication

One of the major downsides of visible light communication (VLC) systems is the narrow modulation BW of the light sources, which results in a barrier to attain the competent data rates. Like wireless communications, optical wireless communications also consider various signal processing techniques and multicarrier and multi-antenna systems for achieving higher data-rates in VLC systems. As the NOMA is now a potential candidate for next generation wireless communications, the feasibility of NOMA in VLC can also be a subject of interest. In [22], Marshoud *et al.* apply NOMA scheme to enhance the achievable throughput in high-rate VLC. This study reveals that NOMA is a promising MA scheme for the downlink of VLC networks.

1.3.9.2 NOMA with Raptors codes

For a given integer, k, and a real ϵ, Raptor codes, which was first proposed in [23], encode a message of k symbols into a potentially limitless sequence of encoding symbols such that any subset of $k(1+\epsilon)$ encoding symbols allows the message to be recovered with high probability. Raptor codes have recently been found effective in several cooperative communication scenarios. The integration of Raptor codes with NOMA has been studied in [24], where an interfering channel with Raptor code has been added to an existing main nonorthogonal wireless channel. It is demonstrated that the coded interference does not affect the performance of the main channel, whereas the interfering signal itself can successfully be decoded with high probability.

1.3.9.3 NOMA with network coding

Random linear network coding (RLNC) is a good encoding scheme which allows data retransmission. In RLNC scheme, the source does not need to be aware of the packets lost by intended receivers. To date, various RLNC techniques have been proposed to improve the transmission efficiency in the case of both multicast and broadcast services. The performances of multicast services in downlink networks can be furthered enhanced by integrating RLNC with the NOMA. The NOMA with RLNC has been studied in [25]. In conventional NOMA, the power domain multiplexing of multiple receivers is considered for unicast services, whereas NOMA-RLNC utilizes power domain multiplexing of multiple reception groups of receivers for multicast services. It is found that the NOMA-RLNC improves the packet success probability to provide multicast services where a source superposes multiple-coded packets before transmitting.

1.3.9.4 Coexistence of NOMA and OMA

In terms of capacity enhancement, which is a major goal of 5G, NOMA is a potential candidate for future radio access. Conversely, this does not mean that OMA schemes will be entirely replaced by NOMA. For example, OMA might be preferred over NOMA in case of small cells if the number of users is small and the near-far effect is not important. It can be concluded that both OMA and NOMA will coexist to fulfill varied requirements of different services and applications in future 5G. As a matter of fact, the long-term coexistence of different radio access technologies is, in general, an import feature of 5G networks. In [26], Dai *et al.* talk on some NOMA schemes for 5G and analyze their basic principles, key features, and receiver complexity. They also conclude that the concept of software defined multiple access can offer us various services and applications with different requirements.

1.4 NOMA challenges

We are now familiar with the fact that there exist many research efforts to design and implement NOMA scheme. In addition to these research concerns, there are several other challenges and open issues which should also be addressed with utmost efforts. In this section, we will briefly provide some research directions to the researchers interested to investigate NOMA in a larger scale.

1.4.1 Distortion analysis

The transmission of source information, for example, voice and video, over communication channels is generally considered lossy. The transmitted data always experience distortion while it propagates to receiver. To deal with this lossy transmission, considerable theoretical attention in assessing source fidelity over fading channels has been paid up-to-date. Different source coding and channel coding diversities have been framed for minimizing the end-to-end distortion. However, source coding diversity and channel coding diversity provide conflicting situations over preferring amount of distortion, cost, and complexity. Choudhury and Gibson [27] compare the source distortion for two definitions of channel capacity, namely, ergodic capacity and outage capacity. Both information capacity and distortion depends on outage probability. It is evident that outage probability that maximizes outage rate may not provide the minimum expected distortion. An investigation can be carried out to optimize the outage probability for which NOMA scheme can provide the maximum outage rate with acceptable distortion.

1.4.2 Interference analysis

Although interference analysis is a generic term in wireless communications, we focus on cooperative NOMA suggested in [11]. This chapter proposes Bluetooth (BT)-like short-range communication in cooperative phase. However, the uses of BT radio in cellular communication will face an extreme interference scenario from the existing wireless personal area network operations. The BT interference decreases the coverage, and throughput; causes intermittent or complete loss of the connectivity; and results in difficult paring during user's discovery phase. In fact, interference of deployed environment, payload size and distance between cooperative users affects the deployment of channel allocation. Also, self-organizing scatternet to manage BT nodes need to be reformulated to make it functional with NOMA, as the users in NOMA are paired according to their CSIs. In addition, a robust scatternet should offer valid routes between nodes with high probability, even though users' mobility causes the complete loss of some of the wireless links. Furthermore, due to the mobility of users, the interference becomes dynamic. Therefore, the performance analysis of a cooperative NOMA scheme in this dynamic interfering environment will be an interesting task.

1.4.3 Resource allocation

In order to accommodate a diverse set of traffic requirements, 5G systems should be capable of supporting high data rates at very low latency and in reliable ways.

However, this is very difficult job, as the resources are limited. So, resource management has to get involved to assist with effective utilization. Wireless resource management is a series of processes required to determine the timing and amount of related resources to be allocated to each user [28]. It also depends on the type of resources. According to Shannon's Information-theoretical capacity, BW is one of the wireless resources. As a part of effective management of system BW in a communication system, the total BW is first divided into several chunks. Each chunk is then assigned to a particular user or a group of users as in case of NOMA. Also, number of packets in each user varies over time. Therefore, user-pairing and optimum power allocation among users in NOMA requires a sophisticated algorithm to provide best performances with the usages of minimum resources.

1.4.4 Heterogeneous networks

A heterogeneous network (HetNet) is a wireless network consisting of nodes with diverse transmission powers and coverage sizes. The HetNet is potential enough for next generation wireless network in terms of capacity and coverage with reduced energy consumption. The infrastructure featuring a high density deployment of low power nodes can also significantly increase EE compared to the one with a low density deployment of fewer high power nodes. There are several research works in HetNets, for example, node cooperation, optimal load balancing, and enhanced intercell interference coordination [29]. A system framework of cooperative HetNet for 5G has recently been studied in [30] with the aim of both spectrum efficiency and EE. As the objective of NOMA coincides with that of HetNet, the specific utilization of NOMA in a particular HetNet can offer extended benefits. Also, the nonuniform spatial distribution of mobile users will preassembly affect the performance of NOMA. Therefore, investigation of outage performance, ergodic capacity, and user fairness of NOMA schemes with spatial user distribution can be a worth work.

1.4.5 Beamforming outage

We learned that NOMA-BF system improves the sum capacity, compared to the conventional multiuser BF system [13]. When NOMA comes with beamforming, outage probability of users will be changed. On that, outage performance analysis of NOMA-BF can be investigated.

1.4.6 Practical channel model

To support the ever-growing consumer data, next generation wireless networks requires not only an efficient radio access technique but also the spectrum availability. For this time being, it is obvious that the 5G will use spectrum allocations at unused millimeter wave (mmW) frequency bands. Also, the backbone networks of 5G are expected to move from copper and fiber to mmW wireless connections, allowing rapid deployment and mesh-like connectivity. The mmW frequencies between 30 and 300 GHz are a new frontier for cellular networks that offers huge amount of BWs. The understanding of the challenges of mmW cellular communications, in general, and channel behavior, in particular, is therefore extremely important and is a fundamental requirement to develop 5G mobile systems as well as backhaul techniques [31]. The existing studies on NOMA assume the wireless links between transmitter and receiver exhibits Rayleigh fading channel with AWGN. A more realistic analysis would be revealed if we could consider the measured path loss and delay spread values [32] to reflect exact radio channel for 5G cellular.

1.4.7 Uniform fairness

In mmW cellular, at locations at distance greater than 175 m, most locations experience a signal outage [31]. As outage is highly dependent on environment, actual outage may be more significant if there were more local obstacles. Deriving a NOMA scheme which provides users (especially located at distance greater than 150 m up to cell boundary in case of mmW cellular), uniform outages experiences would be an excellent work.

1.4.8 Other challenges

There are also some other challenges need to be addressed before NOMA becomes a part of 5G in future. In a downlink scenario, for example, the transmitter allocates the transmit power to the users based on the respective CSIs. Therefore, a proper mechanism for CSI feedback, a suitable channel estimation scheme with proper reference signal design is important for achieving the robust performances. In multicarrier communications, the peak to average power ratio (PAPR) can cause the transmitter's power amplifier (PA) to run within a nonlinear operating region. This causes significant signal distortion at the output of the PA. The effect of PAPR is thus critical to determine what techniques to use for achieving the best NOMA performances. To adopt NOMA in 5G, NOMA should also be made robust in terms of system scalability, since the 5G must support heterogeneous traffic and diverse radio environments.

1.5 NOMA implementation issues

In this section, we discuss a number of implementation issues regarding NOMA, including computational complexity, and error propagation.

1.5.1 Decoding complexity

The signal decoding by using SIC requires additional implementation complexity compare to orthogonal scheme as the receiver has to decode other users' information prior to decoding its own information [3]. Also, this complexity increases as the number of users in the cell of interest increases. However, the users can be clustered into a number of groups, where each cluster contains a small number of users with bad channels. The SC/SIC can then be performed within each group. This group-wise SC and SIC operation does basically provide a trade-off between performance gain and implementation complexity.

1.5.2 Error propagation

It is intuitive that once an error happens for a user, all the other users' information subsequently will likely be decoded erroneously. However, this error can easily be compensated by using a slightly stronger code. Specially, it is evident that error propagation has almost no impact on NOMA performance [2] as a user with bad channel gain is assigned with another user with good channel gain during NOMA scheduling. In case of the degradation of the performances of some users, some nonlinear detection techniques can be considered to suppress the error propagation.

1.5.3 Quantization error

When the received signals strengths of the users are very disparate, the analog-to-digital converter needs to support a very large full-scale input voltage range and requires high resolution to accurately quantize the weak signal as the more levels the ADC uses for quantization, the lower is its quantization noise power. However, there is a limitation placed on arbitrarily high resolution ADC due to its cost, conversation time, and hardware complexity. This constraint eventually leads to a trade-off between the quantization error, and SIC gain.

1.5.4 Power allocation complexity

The achievable throughput of a user is affected by the transmit power allocation to that user. This particular power allocation also affects the achievable capacity of other users, since the basis of NOMA is power-domain user multiplexing. To achieve the best throughput performances of NOMA, a brute-force searching over the possible user pairs with dynamic power allocation is required. However, this kind of exhaustive searching is computationally expensive.

1.5.5 Signaling and processing overhead

There are several sources of additional signaling and processing overhead in NOMA compare to orthogonal counterparts. For example, to collect the CSIs from different receivers and to inform the receivers the SIC order, some time slots need to be elapsed. This causes rate degradation in NOMA. Also, with dynamic power allocation and encoding and decoding for SC and SIC, the NOMA signal processing requires additional energy overhead.

References

[1] P. Wang, J. Xiao, and L. Ping, "Comparison of orthogonal and non-orthogonal approaches to future wireless cellular systems," IEEE Vehicular Technology Magazine, vol. 1, no. 3, pp. 4–11, 2006.

[2] A. Benjebbour, Y. Saito, Y. Kishiyama, L. Anxin, A. Harada, and T. Nakamura, "Concept and practical considerations of non-orthogonal multiple access (NOMA) for future radio access," International Symposium on

Intelligent Signal Processing and Communications Systems (ISPACS), pp. 770–774, Nov. 2013.

[3] D. Tse and P. Viswanath, Fundamentals of Wireless Communication, Cambridge University Press, New York, 2005.

[4] T. Cover, "Broadcast channels," IEEE Transactions on Information Theory, vol. 18, no. 1, pp. 2–14, 1972.

[5] S. Vanka, S. Srinivasa, Z. Gong, P. Vizi, K. Stamatiou, and M. Haenggi, "Superposition coding strategies: design and experimental evaluation," IEEE Transactions on Wireless Communications, vol. 11, no. 7, pp. 2628–2639, 2012.

[6] N.I. Miridakis and D.D. Vergados, "A survey on the successive interference cancellation performance for single-antenna and multiple-antenna OFDM systems," IEEE Communications Surveys & Tutorials, vol. 15, no. 1, pp. 312–335, 2013.

[7] K. Higuchi and A. Benjebbour, "Non-orthogonal multiple access (NOMA) with successive interference cancellation," IEICE Transactions on Communications, vol. E98-B, no. 3, pp. 403–414, 2015.

[8] Z. Ding, Z. Yang, P. Fan, and H.V. Poor, "On the performance of non-orthogonal multiple access in 5G systems with randomly deployed users," IEEE Signal Processing Letters, vol. 21, no. 12, pp. 1501–1505, 2014.

[9] C. Xiaohang, A. Benjebbour, L. Yang, L. Anxin, and J. Huiling, "Impact of rank optimization on downlink non-orthogonal multiple access (NOMA) with SU-MIMO," IEEE International Conference on Communication Systems (ICCS), pp. 233–237, Nov. 2014.

[10] A.S. Ibrahim, A.K. Sadek, W. Su, and K.J.R. Liu, "Cooperative communications with relay-selection: when to cooperate and whom to cooperate with," IEEE Transactions on Wireless Communications, vol. 7, no. 7, pp. 2814–2827, Jul. 2008.

[11] Z. Ding, M. Peng, and H.V. Poor, "Cooperative non-orthogonal multiple access in 5G systems," IEEE Communications Letters, vol. 19, no. 8, pp. 1462–1465, 2015.

[12] Y. Hayashi, Y. Kishiyama, and K. Higuchi, "Investigations on power allocation among beams in non-orthogonal access with random beamforming and intra-beam SIC for cellular MIMO downlink," in Proc. IEEE Vehicular Technology Conference, Las Vegas, NV, USA, Sep. 2013.

[13] B. Kim, S. Lim, H. Kim, *et al.*, "Non-orthogonal multiple access in a downlink multiuser beamforming system," in Proc. IEEE Military Communications Conference, San Diego, CA, USA, Nov. 2013.

[14] S. Timotheou and I. Krikidis, "Fairness for non-orthogonal multiple access in 5G systems," IEEE Signal Processing Letters, vol. 22, no. 10, pp. 1647–1651, 2015.

[15] J. Choi, "Non-orthogonal multiple access in downlink coordinated two-point systems," IEEE Communications Letters, vol. 18, no. 2, pp. 313–316, 2014.

[16] S. Vanka, S. Srinivasa, Z. Gong, P. Vizi, K. Stamatiou, and M. Haenggi, "Superposition coding strategies: design and experimental evaluation," IEEE Transactions on Wireless Communications, vol. 11, no. 7, pp. 2628–2639, 2012.

[17] S. Alamouti, "A simple transmit diversity technique for wireless communications," IEEE Journal on Selected Areas in Communications, vol. 16, no. 8, pp. 1451–1458, 1998.
[18] S. Han, C.-L. I, Z. Xu, and Q. Sun, "Energy efficiency and spectrum efficiency co-design: from NOMA to network NOMA," IEEE Communications Society MMTC E-Letter, vol. 9, no. 5, pp. 21–24, 2014.
[19] K. Higuchi and Y. Kishiyama, "Non-orthogonal access with random beamforming and intra-beam SIC for cellular MIMO downlink," IEEE Vehicular Technology Conference, pp. 1–5, Sep. 2013.
[20] Q. Sun, S. Han, I.C. Lin, and Z. Pan, "On the ergodic capacity of MIMO NOMA systems," IEEE Wireless Communications Letters, vol. 4, no. 4, pp. 405–408, 2015.
[21] J. Men and J. Ge, "Non-orthogonal multiple access for multiple-antenna relaying networks," IEEE Communications Letters, vol. 19, no. 10, pp. 1686–1689, 2015.
[22] H. Marshoud, V.M. Kapinas, G.K. Karagiannidis, and S. Muhaidat, "Non-orthogonal multiple access for visible light communications," IEEE Photonics Technology Letters, vol. 28, no. 1, pp. 51–54, Jan. 2016.
[23] A. Shokrollahi, "Raptor codes," IEEE Transactions on Information Theory, vol. 52, no. 6, pp. 2551–2567, 2006.
[24] M.J. Hagh and M.R. Soleymani, "Raptor coding for non-orthogonal multiple access channels," IEEE International Conference on Communications (ICC), pp. 1–6, Jun. 2011.
[25] S. Park and D. Cho, "Random linear network coding based on non-orthogonal multiple access in wireless networks," IEEE Communications Letters, vol. 19, no. 7, pp. 1273–1276, 2015.
[26] L. Dai, B. Wang, Y. Yuan, S. Han, C.-L. I, and Z. Wang, "Non-orthogonal multiple access for 5G: solutions, challenges, opportunities, and future research trends," IEEE Communications Magazine, vol. 53, no. 9, pp. 74–81, 2015.
[27] S. Choudhury and J.D. Gibson, "Information transmission over fading channels," IEEE Global Telecommunications Conference (GLOBECOM), pp. 3316–3321, Nov. 2007.
[28] B. Lee, D. Park, and H. Seo. Wireless Communications Resource Management. Wiley, Singapore, 2009.
[29] R.Q. Hu and Y. Qian, Heterogeneous Cellular Networks, Wiley, Chichester, 2013.
[30] R.Q. Hu and Y. Qian, "An energy efficient and spectrum efficient wireless heterogeneous network framework for 5G systems," IEEE Communications Magazine, vol. 52, no. 5, pp. 94–101, 2014.
[31] S. Rangan, T.S. Rappaport, and E. Erkip, "Millimeter-wave cellular wireless networks: potentials and challenges," Proceedings of the IEEE, vol. 102, no. 3, pp. 366–385, 2014.
[32] T.S. Rappaport, S. Shu, R. Mayzus, *et al.*, "Millimeter wave mobile communications for 5G cellular: it will work!" IEEE Access, vol. 1, pp. 335–349, 2013.

Chapter 2

Fog computing in 5G networks: an application perspective

Harshit Gupta[1], *Sandip Chakraborty*[2], *Soumya K. Ghosh*[2] *and Rajkumar Buyya*[3]

Abstract

Fifth generation (5G) cellular network promises to offer to its users sub-millisecond latency and 1 Gbit/s transmission speed. However, the current cloud-based computation and data delivery model do not allow these quality of service guarantees to be efficiently harnessed, due to the number of hops of wired networks between the 5G-base stations and the cloud, that leads to a significant increase in latency. Forwarding all the data generated by devices directly to the cloud may devour the bandwidth and lead to congestion. Therefore, it is necessary that processing be hosted near the devices, close to the source of the data, so that the high speed transmission of 5G can be utilized and data can be processed and filtered out by the time it reaches the cloud. This bringing down of computation, storage, and networking services to the network edge opens up many new research areas of applying fog computing over cellular network architecture. This chapter discusses the advantages of extending the cloud services to the edge by presenting use-cases that can be realized by fog computing over 5G networks.

2.1 An introduction to fog computing

The Internet has been evolving from the time it was conceived, and is now going beyond traditional desktop computers. The proliferation of the Internet of Things (IoT) has brought about a transformation in the way the world interacts on the Internet. The World Wide Web connected computers together, smartphones brought humans into the fold of the Internet, and now IoT is poised to connect devices, people, environments, virtual objects, and machines in ways that the world has never known. IoT deployments like smart cities, smart homes, and the

[1]School of Computer Science, Georgia Institute of Technology, Atlanta, GA 30332, USA
[2]Department of Computer Science and Engineering, IIT Kharagpur, Kharagpur, West Bengal, India
[3]CLOUDS Laboratory, The University of Melbourne, Melbourne, VIC, Australia

like – things that were more of fiction – are now becoming a reality, and are projected to affect as many aspects of human life as possible.

Internet of things

- The number of 'things' connected to the Internet surpassed people in 2008. By 2020, the population of Internet-connected things will reach 50 billion, garnering profit and cost savings worth $19 trillion over the next decade [4].
- General Electric predicts that amalgamation of machines, data, and analytics will become a global industry worth $200 billion in a period of 3 years [6].
- A whopping 94% of all businesses have seen a return on their IoT investments [7].

Typical IoT systems consist of a myriad of devices, ranging from sensors embedded in roads to mobile devices like cars and trains. With such a large number of *things* involved in an IoT deployment, the number of devices connected to the Internet is growing by leaps and bounds. At present, the number of endpoints (typically smart phones and laptops) has been estimated to be around 3–4 billion and is expected to grow to a trillion in a few years. Such a lot of devices will generate gigantic volumes of data, in a phenomenon that has been attributed the term *data tsunami*. The applications and the network infrastructure will have to adapt accordingly to such a massive increase in the amount of data that they will have to handle given the constraint of the amount of bandwidth available.

The IoT brings a data Tsunami

Development in IoT has brought about the proliferation of cheap, distributed sensors resulting in a huge volume of data in a short amount of time. Virgin Atlantic's new fleet of highly connected planes is expected to create over half a terabyte of data per flight [1]. According to Cisco Systems most recent visual networking index, mobile data traffic will grow 10-fold globally between 2014 and 2019, reaching 24.3 exabytes per month worldwide in 2019 [5].

Development in engineering has always aimed at designing systems that can function with as low human intervention as possible. The IoTs is a perfect platform for designing such applications, particularly because connecting every device to the Internet gives every device the power to make decisions on its own, thus reducing the need of human intervention. Research on such autonomous systems has revealed that they heavily rely on low response time of the application. IoT systems like smart grids, collaborative object detection and others require latency of sub-millisecond order – requirements that the Internet will have to provide for the application to work in the desired manner, failing to do that may defeat the entire purpose of the application.

This change in the nature of devices connected to the Internet and the concomitant increase in the amount of data generated demands an evolution in the network infrastructure as well. The present cloud-model of execution will prove to be inefficient, if at all feasible, for the futuristic applications that development in IoT brings to vision.

2.1.1 Limitations of the current computation paradigm

The current computation paradigm has cloud datacentres as the only point for execution after the basic processing available at the devices. However, such a large number of IoT devices continuously sending data to the cloud for analysis would lead to scalability issues in the core network. Levels of congestion in the backbone network will increase manifold and may lead to aggravated packet loss and delay, spoiling the user experience. Furthermore, sending lots of data to the cloud for processing may lead to the cloud becoming a bottleneck, again leading to increase in response time.

A lot of IoT applications, typically those that run in industrial settings like smart grids, need their devices to react very quickly to an impulse. In such a case, sending the data related to the impulse to the cloud and then getting the response back may not be desired due to the high communication latency involved in the network in between. This latency is unavoidable due to the large number of hops that a packet has to travel through to reach the cloud. Therefore, the *do-it-on-cloud* paradigm of computation will become disruptive with the advent of latency-critical applications for IoT systems. Such a scenario poses the requirement of distributed computation, storage, and networking services that are close to the source of data, or, in other words, *fog computing*.

2.1.2 Fog computing

Fog computing [10] is a term coined by professor Salvatore J. Stolfo [32], that has recently been picked up by Cisco [3]. Fog computing is a paradigm that extends cloud computing and services to the edge of the network allowing applications to run in close proximity of users, be highly geo-distributed and support user mobility. Due to such characteristics, fog cuts down latency of service requests, and improves quality of service (QoS), resulting in superior user-experience. Fog computing is a necessity for emerging Internet of Everything applications (like industrial automation, transportation, etc.) that demand real-time/predictable latency. Owing to its wide and dense geographical distribution, the fog paradigm is well-positioned for real-time big data analytics. The data collection points in fog computing are densely distributed, hence adding a fourth axis – *geo-distribution* – to the often mentioned big data dimensions (volume, variety, velocity, and veracity).

Fog computing

Fog computing is a non-trivial extension of cloud computing – by providing compute, storage and networking services near the edge of an enterprise's network. The peculiar characteristics of the fog are its proximity to end-users, its dense geographical distribution, and its support for mobility.

Fog provides the same services as the cloud (compute, storage and networking) and shares the same mechanisms (virtualization, multi-tenancy, etc.). These common attributes of the cloud and the fog makes it possible for developers to build applications that utilize the interplay between the fog and the cloud. According to Bonomi *et al.* [12], fog computing was conceived to support applications whose requirements don't quite match the QoS guarantees provided by the cloud. Such applications include (as illustrated in Figure 2.1) the following:

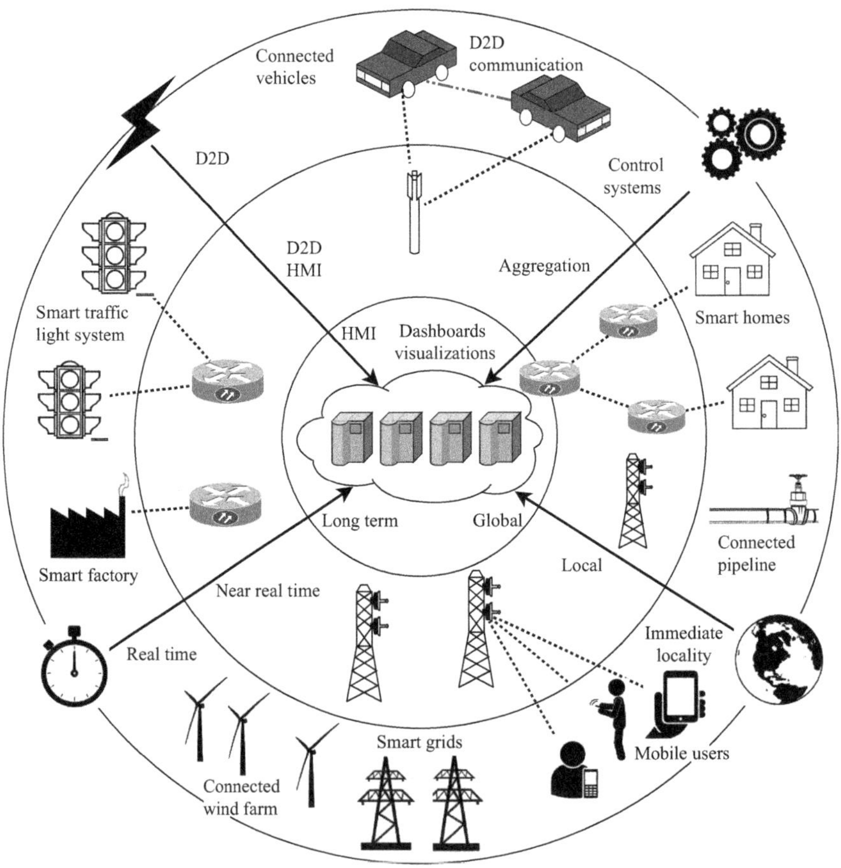

Figure 2.1 Applications supported by fog computing

- Applications having stringent latency requirements, for example mobile gaming, video conferencing and others. Running these applications on the cloud can mar user experience due to the unreliability of QoS offered by the cloud.
- Geo-distributed applications where the data collection points are distributed over a wide area, for instance, pipeline monitoring or sensor networks to monitor the environment.
- Fast mobile applications involving high mobile users smart connected vehicle (SCV), connected rail.
- Large-scale distributed control systems consisting of a vast number of sensors and actuators working in a coordinated manner to improve user experience. For example smart grid, connected rail and smart traffic light systems (STLSs).

It is important to note that the fog is not a substitute for the existing cloud computing paradigm, instead, fog is an extension to the cloud, and application built for the fog should be able to exploit both the flexibility and power of the cloud and the real-time capabilities of the fog.

2.2 Fog computing on 5G networks

Fog computing and fifth-generation (5G) networks are two concepts having different origins but will soon converge as the promises made by the vision of 5G networks makes it necessary to bring processing down to the edge.

2.2.1 Fog computing – a requirement of 5G networks

5G mobile networks, though not a reality at present, is expected to hit the market by 2020. Communication in 5G networks will be based on high-frequency signals – in the millimetre-wave frequency band – that can allocate more bandwidth to deliver faster, higher-quality video, and multimedia content. 5G networks promise to provide millisecond and sub-millisecond latency while offering a data rate of more than 1 Gbit/s [30]. This latency is so small that it eliminates the possibility of the radio interface being the bottleneck. Next generation mobile networks are designed in a way that can handle communications not restricted to humans (where one can possibly mask the latency) – they are built to support reliable and fast machine-to-machine communication as well, a use-case that needs low latency to be effective.

For 5G to be successful, it has to support fog computing; otherwise, the low latency radio interfaces will be of no avail. A typical 5G network have mobile users connecting to a base station, which would in turn be connected to the core network through wired links. Requests to a cloud-based application would go through the base station and the core network to finally reach the cloud servers. In such a deployment, even though the low latency radio interfaces enable sub-millisecond communication between the mobile device and base station, but sending the request from the base station to the cloud will lead to a delay increase in orders of magnitude.

The true value of 5G cannot be harnessed by running applications having the cloud as the only processing unit, and it is required to enable the deployment of application code at devices in close proximity to the users [9].

It is imperative for the 5G networks to be more than just a communication infrastructure. Computation and storage services, if supplied by the network, close to the devices, will allow applications to take benefit of low latency radio to provide very fast end-to-end response time. This will highly benefit both the customers (by giving timely responses) and the provider (by alleviating the load on the backbone network). This descent of processing from the cloud to the edge forms the definition of fog computing, and it would not be wrong to say that 5G networks cannot fulfil its promises without fog computing. Fog computing is not a feature, as most view it, but a necessary requirement for 5G networks to be able to succeed.

A key element of 5G networks that enables fog computing is small cell (pico and femto cells), also known as micro-cells. Small cells can alleviate the burden on roof-top base stations (macro-cells) by allowing end points to connect to them. A device can connect either to the macro-cell or to a micro-cell. This makes the architecture of 5G networks a hierarchical one – with the core network (cloud) at the apex, followed by macro-cell base stations and micro-cell base stations, and finally end devices. Hence, from the perspective of fog computing, both macro- and micro-cell base stations form the fog nodes, that is networking nodes providing computation and storage as well. Packets sent uplink by the devices will be analysed at the micro-cell or macro-cell base stations before reaching the core network.

Another major advancement in communications that 5G brings along is efficient device-to-device communication. Application data sent will be sent from the sender device directly to the receiver device, with the base station handling only control information of this transfer. This allows inter-device communication to take place without burdening the base station, thus beatifying fog systems with scalability of handling numerous devices interacting with each other. This will be categorically useful for applications that involve numerous connected points and continuous communication between these points, for example smart homes.

The rest of this section discusses the network architecture of 5G networks and how they will realize fog computing. In addition to this, the architecture of fog applications is also described – a segregation of application logic into components that can harness the services provided by fog computing.

2.2.2 Physical network architecture

The physical network architecture of a fog network over 5G will extend the architecture of the state-of-the-art heterogeneous cloud radio access networks (HCRANs) [28]. In the traditional HCRAN architecture, all application processing tasks are performed on the cloud inside the core network, which requires billions of end devices to communicate their data to the core network. Such a massive amount of communication may vitiate the fronthaul capacity and may overburden the core

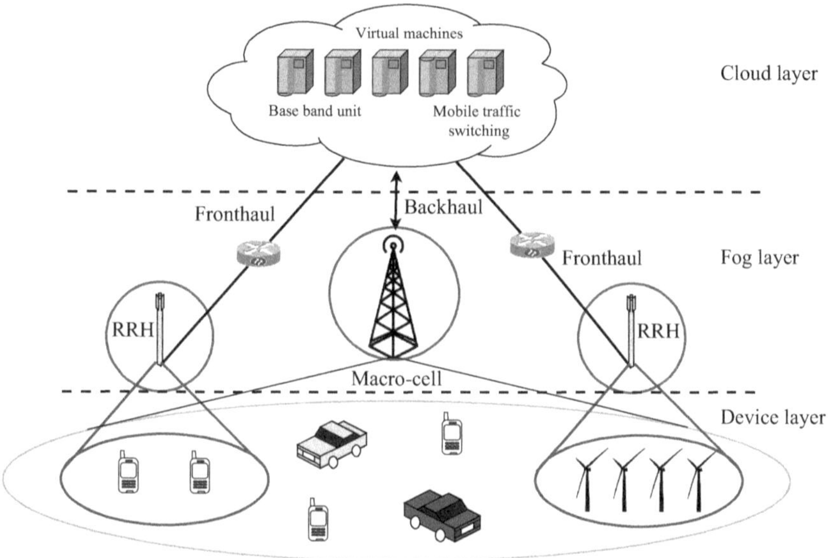

Figure 2.2 Architecture of 5G network with fog computing – a three-layered architecture

network, which will have a detrimental impact on the QoS experienced by the end-users.

An intuitive solution to this problem is to bring down computation and storage capabilities from the cloud near the edge, so that the need to send all the data generated by end-devices to the cloud is done way with, hence alleviating the fronthaul and the core network of the immense traffic surge. Figure 2.2 depicts the various locations where this offload of computation and storage can be done. The fog network architecture consists of three logical layers that are shown in Figure 2.2. The devices in each layer are capable of hosting computation and providing storage, hence making it possible for creating complex processing off-load policies.

- *Device layer*: The device layer subsumes all the end-devices connected to the fog network. The devices include IoT devices like sensors, gateways and others and also mobile devices like smartphones, tablets and others. These devices may be exchanging data directly with the network, or may be performing peer-to-peer communication among themselves. Being the source of all data entering the network and the prime actuators performing tasks, these devices are the lowest tier of fog devices. The device layer hosts computation either by embedded coding (for low-end devices like sensors) or as a software running on the operating system of the device.
- *Fog layer*: The fog layer consists of intermediate network devices located between the end-devices in the device layer and the cloud layer. The first point

of offload in this layer are the remote radio heads (RRHs) and small cells that are connected by fibre fronthaul to the core network. Processing incoming data here will considerably reduce the burden on fronthaul. Macro cells also form a point of offloading processing that send the processed data to the core network through backhaul links. Both fronthaul and backhaul is realized by Ethernet links and the intermediate devices like router and switches in the path from the radio heads to the core also form potential places where computation and storage tasks can be offloaded.

Deploying applications on these devices is made possible by advances in virtualization technology. Each application is packaged in the form of a virtual machine and is launched on the appropriate device. The application virtual machines run alongside the host OS virtual machine (which performs the original network operations) over a hypervisor on the fog device.

- *Cloud layer*: This layer forms the apex of the hierarchical architecture, with cloud virtual machines being the computation offload points. The theoretically infinite scalability and high-end infrastructure of the cloud makes it possible to handle processing that requires intensive computation and large storage – which cannot be done at the edge devices. In addition to application layer processing, the cloud layer contains baseband units which process data coming from RRHs and small cells via fronthauls and route processed data to application servers.

2.2.3 Application architecture

For an application to be called fog-ready, it must be designed to harness the full potential of the fog. Typically, an application built for execution on fog infrastructure would have three components—device, fog and cloud components—as shown in Figure 2.3 [12].

- *Device component*: The device component is bound to the end devices. It performs device level operations, mostly, power management, redundancy elimination and others. At times, when the end-device is not just a light client, it also hosts application logic demanding very low latency responses as this component is executed on the device itself. However, due to the resource constraints of the underlying device, this component should not contain heavy processing tasks.
- *Fog component*: The fog component of an application performs tasks that are critical in terms of latency and require such processing power that cannot be provided by end-devices. Furthermore, as the fog component is meant to run on fog devices close to the edge, the coverage of this component is not global. Thus, this component should host logic that requires only local state information to execute.

  The fog component is not bound to a particular kind of device. It is free to reside in any kind of device between the edge (consisting of end-devices) and the cloud. The mapping of the fog components to devices depends on the points of offload in the path from the edge to the cloud. Depending on the geographical coverage and latency requirements of the application, the fog

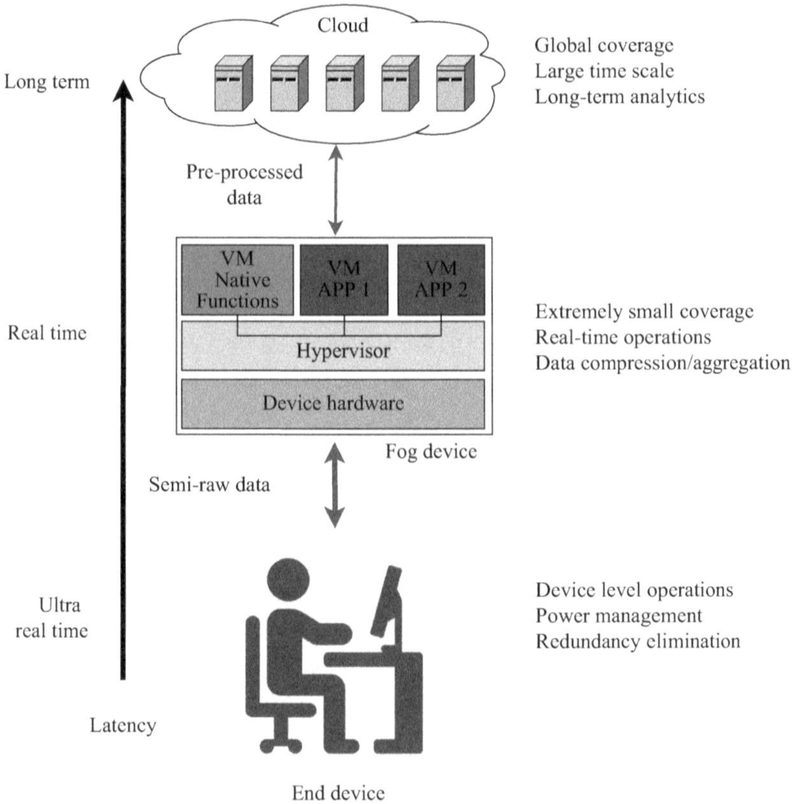

Figure 2.3 Application architecture

component can be hosted on any of these points of offload. In fact, placement of fog component on appropriate fog nodes forms an interesting and important area for research.

- *Cloud component*: Cloud component is bounded to the cloud servers in the core network. It contains logic for long-term analytics of the data collected from the lower layers and for operations that don't have any sort of latency constraints per se. Application tasks requiring large processing power and storage are suitable to be placed in the cloud component, so that they can harness the infinite resources of the cloud. Moreover, as the cloud layer is located at the apex of the network, it receives information from all devices and hence has a global knowledge of the entire system. Thus, application logic requiring knowledge of the global state of the system should be placed in the cloud component of the application.

Coding logic into the various layers of the a fog-ready application determines the performance of the application. Incorrect placement of logic can cripple an application and makes it unable to use the benefits that fog computing has to offer.

The following sections discuss several use-cases whose requirements can be satisfied by the unique QoS provided by fog computing when deployed over a 5G cellular network. For each use-case, we also present a suitable mapping of application logic to application layers for each use-case.

2.3 Smart traffic light system [use case 1]

A STLS is a network of connected traffic lights which intelligently, and in a coordinated fashion, takes decisions that prevent accidents, reduce traffic congestion, minimize noise and fuel consumption and gives the drivers a better experience by long-term monitoring. The STLS is but a component of the larger vision of SCV and Advanced Transportation Systems, but it is rich enough to drive some key requirements for fog computing.

2.3.1 Requirements

An STLS needs to take full control of the traffic in an area and perform a broad spectrum of tasks – more than what a traffic policeman would have to – right from accident prevention to flow control. The various use-cases of an STLS have been listed in the following sections.

2.3.1.1 Accident prevention

The most important concern of any automated system directly affecting humans is user safety. Given the number of traffic accidents that occur daily, accident prevention is one of the key requirements of an STLS, failing to do which can have serious repercussions involving loss of life and property. The STLS should be able to detect vehicles not following traffic rules – for instance, not stopping at a red signal – and should inform vehicles that can potentially be affected by this rogue vehicle (typically those on an orthogonal street). This information can be conveyed by communication between traffic lights on adjacent streets. The orthogonal streets can ask their vehicles too to stop for some time. Also, in the event of a pedestrian crossing a road when he/she should not, and there is a vehicle coming her way, the STLS will calculate, from factors like speed of approaching vehicle and pedestrian, whether an accident may take place and take suitable action. In addition to this, over-speeding vehicles can be asked to stop by these traffic lights in order to avoid accidents [40]. The traffic lights may also determine whether the over-speeding vehicle is an emergency vehicle, like an ambulance and accordingly decide whether to make it stop or let it go.

2.3.1.2 Re-synchronization and flow control

Activation of accident prevention mechanism causes the traffic light cycles in the affected area to go out of synchronization. To dampen this perturbation in traffic light cycles, few neighbouring traffic lights need to re-adjust their cycle. This task is not very critical in terms of latency, since at the most it may lead to prolonged red lights on few lanes causing vehicles to stop more than required. Moreover, this use-case is of a slightly global nature, that is, the actors involved are spread across a few streets.

Flow control is essential to ensure a smooth movement of traffic without having to make the drivers stop too often. An STLS can collect information about the level of traffic in each lane of the city from sensors and based on a routing policy route vehicles to reduce congestion. Traffic lights can coordinate and maintain a *green wave*, reducing the number of times a vehicle would have to stop at traffic signals. By doing this, the STLS can reduce noise and fuel consumption, since vehicles would not have to accelerate often. This will be especially useful for emergency vehicles like ambulances or fire engines, for which the STLS can create green waves on demand, so that they do not have to stop at traffic signals and can reach their destination as soon as possible.

2.3.1.3 *Long-term monitoring*

This use-case is required for monitoring the entire traffic light system over a large time scale and looking at ways to enhance the performance of the system. The STLS can improve its congestion-aware traffic routing policy continuously based on analytics on data collected over a long time. Through long-term analysis on observed pedestrian movement, the STLS would be able to decide the optimal time for which pedestrians should be allowed to cross roads. Policymakers will be able to make decisions such as whether creation of alternate routes is required with the help of long-term analysis of traffic congestion data. The main actors involved in this use-case are policymakers that analyse the road traffic over a long-time period and come up with changes to improve driver experience.

Design requirements

The use-cases entailed by a STLS highlight the following design requirements of the application:

- *Low-latency response*: Accident prevention requires a very low-response time to alert the involved person in a timely manner, failing to do which will mar the very purpose of the accident prevention mechanism. Furthermore, detecting a rogue vehicle (based on his movement) and alerting the rest of the drivers also requires a quick response, so that the chances of a mishap can be minimized.
- *Handling large volume of data*: An STLS contains a large number of sensors deployed on roads throughout the city – generating data at a high rate. Due to the large volume of data that needs to be analysed, the network should be scalable and robust enough to handle large traffic. Poor network architecture can be a victim of bandwidth over-utilization and become congested, leading to further delay in responses.
- *Heavy processing power and global coverage*: The tuning of traffic routing algorithm and analysis for policymaking requires processing a large amount of data, that too on a large time-scale, which is a computationally intensive task. Moreover, the analysis has to be done on a city level – and thus requires to be done on a device with global coverage.

2.3.2 Deployment details

It is worth noting that the requirements of an STLS showcase a variety of requirements, both in terms of response time and geographical area affected. We now discuss a model deployment of an STLS on fog infrastructure – that utilizes services provided both by the edge and the cloud.

2.3.2.1 Physical deployment

The data collection points in an STLS are primarily sensors that are deployed on the roads, like induction loop sensors which can detect a crossing vehicle and speed detection sensors. CCTV cameras installed at intersections also fall under the data collection points of the STLS. Traffic lights are the actuators of the system, as any action performed by the STLS is reflected by a change of traffic lights. Consider a traffic intersection, having a set of traffic lights and two intersecting roads as shown in Figure 2.4.

Each intersection will be equipped with a 5G small cell which would connect the devices on that intersection together, allowing real-time device-to-device communication among them. The small cell is equipped with compute and storage facilities which will be utilized by the STLS application. The small cell is in turn connected by a high bandwidth connection to the cloud through intermediate network devices (typically belonging to the Internet Service Provider (ISP)). These network devices in the STLS deployment are also fog-enabled, meaning that they too are points for offloading application logic. These intermediate devices will be used for communicating between devices belonging to neighbouring intersections.

2.3.2.2 Application architecture

The application logic of STLS has been broken down into components that can be mapped to the three-layered architecture of a fog application. This partitioning of application tasks has been described in the following sections.

Device component

As the devices in this system include only sensors, CCTV cameras and traffic lights, the device component of the application is not complicated. For sensors, they need to be able to send updates to the small cell over a 5G network. The logic running in the CCTV cameras should process the recorded video stream in real time to detect events of interest, such as a human crossing the road or an approaching emergency vehicle, and convey this to the small cell in such a case. As concerns traffic lights, they are the only actuators in the system, as all control decisions taken by the system are ultimately realized via changes in traffic light sequences. The application component running on a traffic light should be able to receive messages from the small cell on the intersection and change light sequence accordingly.

Fog component

The fog component of the application runs on the small cell at each intersection as well as on the intermediate network devices connecting the small cells to the Internet. The application logic running in these devices handles most of the requirements of an STLS.

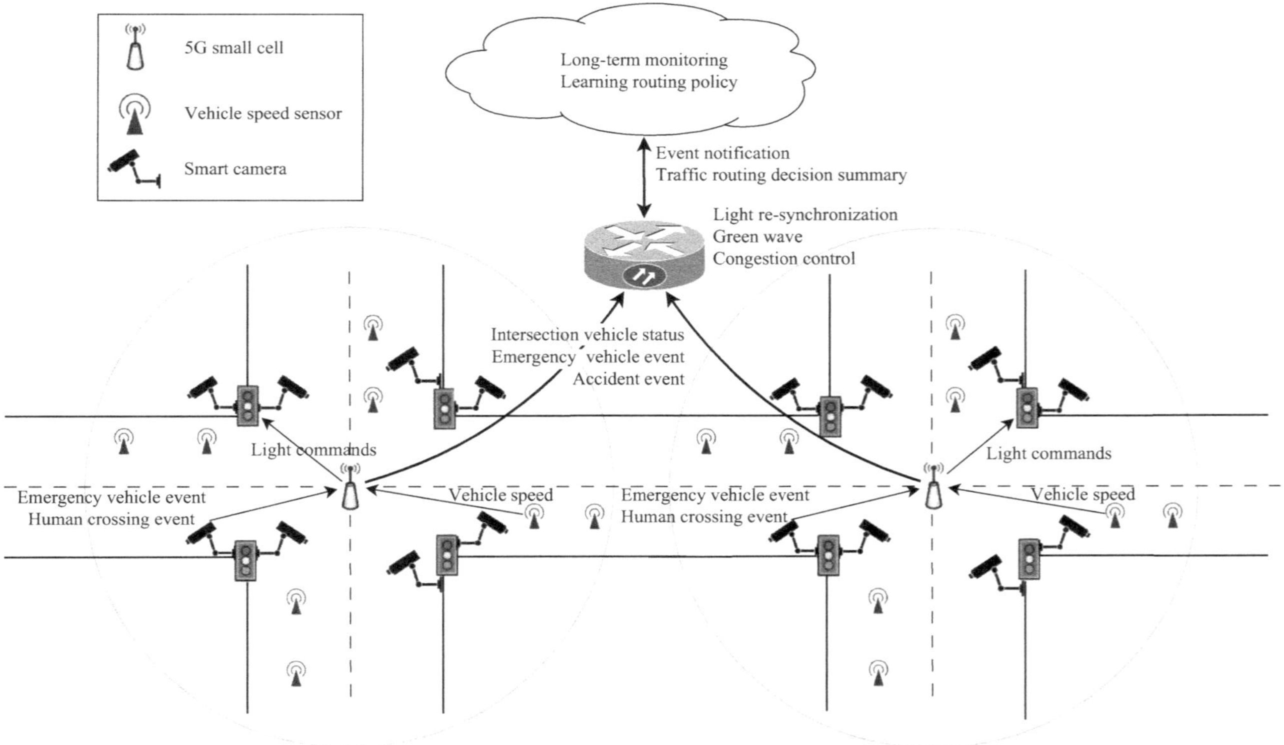

Figure 2.4 Deployment of STLS on fog infrastructure over 5G mobile network

For accident prevention, the fog component needs to handle the event of *human-crossing-road* sent by those CCTV cameras that surveil the lane through which traffic is flowing. In case an accident is possible, the application will send a message to the traffic lights on that lane to change to red immediately so that traffic stops and also blow a horn to alert the human. This task should take place in real-time; hence, it is handled on small cell itself – that is the place which directly receives the human crossing event from the CCTV cameras. This proximal placement of application logic, coupled with the sub-millisecond latency of 5G transmission, allows the accident prevention mechanism to happen with a very small delay, hence minimizing the risk of a human getting injured.

For re-synchronization of traffic lights to dampen a sudden change in light sequence due to activation of the accident prevention mechanism, there has to be a communication between nearby traffic lights so that they may run a distributed algorithm and re-synchronize their light cycles. The STLS allows smooth traffic flow and maintains green waves by coordinating between multiple traffic lights and maintaining an appropriate traffic light sequence. Research works like [21,24,41] have explored the possibilities of improving the traffic flow and minimizing congestion by running a distributed algorithm on multiple traffic lights based on information collected by sensors. The fog component of the STLS also receives events of an approaching emergency vehicle from the CCTV cameras, to which the system will respond by triggering the traffic light on the vehicle's path green and inform the next neighbouring intersection of the approaching emergency vehicle so that it may take necessary actions. This component requires swift communication between neighbouring traffic lights, however, the latency requirement is not as critical as the accident prevention use-case. Moreover, this use-case requires the knowledge of the state of traffic lights at more than one intersection, a coverage which is more global in nature than accident prevention. Hence, this component is hosted on the intermediate network devices connecting the small cells at intersections to the Internet. Being hosted on a device just a few hops away from the small cells, the small cells are able to send messages to each other with a very low delay, and hence are able to control traffic lights in other intersections. In addition to the low delay, running this component in the fog reduces the amount of raw sensor data sent to the cloud, thus alleviating the core network of the risks of congestion and minimizing the consumption of bandwidth.

Cloud component

The cloud component receives data from the small cells about the traffic conditions and events at regular intervals. Small cells aggregate information over a period of time and send it to the cloud, which reduces the volume of data sent. The cloud component of the STLS performs long-term analysis on the incoming data, based on the results of which, experts can infer whether to create a new route for reducing load on existing roads, or whether the crossing time for pedestrians needs to change. Several studies have been conducted on such analysis of traffic data [27,38,39]. Through long-term analysis on congestions levels in the city, the STLS can improve the traffic routing policy that runs in the fog component to reduce traffic congestion. This analysis may require heavy processing since it needs to analyse data related to a large time-scale. Moreover, this component needs to know the traffic state on a global scale, and does not require any guarantee on response time. Due to these characteristics of the application logic in the cloud component, it is appropriate for hosting it on the cloud.

2.4 Mobile gaming [use case 2]

Cloud gaming, sometimes called gaming on-demand, is a new kind of gaming platform made possible by the proliferation of cloud technologies, allowing physically distant users to play together. Cloud gaming is an efficient and cost-effective way to deliver high-quality gaming experience and has opened up a lot of business opportunities. In a cloud gaming system, computer games run on powerful game servers on the cloud, while gamers interact with the games using thin clients connected to the Internet. The thin clients are light-weight applications and can be hosted on resource-constrained devices, such as mobile devices. Cloud gaming is ubiquitous, allowing gamers to play a game from anywhere and at any point of time, while the game developers can optimize their games for a particular machine configuration.

A cloud gaming system essentially renders a gaming application on cloud servers and streams the scenes of the application as a video sequence back to the player. A player of the game interacts with the game through a thin client, which is responsible for displaying the video received from the cloud server as well as sending the interactions of the player with the game to the cloud. Cloud gaming is one of those applications requiring a strict latency guarantee, failing to provide which will lead to detrimental impact on the user experience. In addition, cloud gamers are also particular about the video quality that is rendered on their light clients. Thus the implementation of a cloud gaming system needs to take resource allocation, scalability, and fault tolerance into account as well apart from meeting the gamers' needs.

The traditional implementation of mobile gaming involves hosting all the computation and storage in the cloud, hence making *mobile gaming* synonymous to *cloud gaming*. However, communicating with the cloud for every request may not always be the best practice, especially when latency requirements are stringent. Choy *et al.* [14] have shown through a large-scale empirical study that contemporary cloud infrastructure cannot meet the stringent latency requirements necessary for acceptable game play for many end-users, thus imposing a limit on the number of potential users for an on-demand gaming service. Based on empirical results, they have concluded that augmenting the cloud infrastructure with edge-servers can significantly increase the feasibility of on-demand gaming or cloud gaming. Hence, it makes sense to offload some computation involved in the cloud-based game to the edge. They have described three computation approaches: *cloud-only*, *edge-only* and a *hybrid* approach in [15]. Experiments show that the percentage of users served increased from 70% in an only-cloud deployment to 90% in a hybrid-deployment that used both cloud and edge-servers.

The future of cloud games

'Lets say our industry had never done consoles or consumer clients. Even if we just started out with cloud gaming, you'd actually go in the direction of pushing intelligence out to the edge of the network, simply because its a great way of caching and saving you on network resources.' – Gabe Newell, co-founder and managing director of video game development and online distribution company Valve Corporation [8].

These studies give ample support to the fact that fog computing is an efficient platform for deploying on-demand games, and in this section, we discuss the deployment of a cloud-game on fog infrastructure.

2.4.1 Requirements

Cloud gaming is a highly interactive application posing stringent requirements in terms of latency and video quality, failing to do which can directly affect user experience. The typical requirements of an on-demand game are discussed in the following sections.

2.4.1.1 Interaction delay

The authors of [33] have performed a categorical analysis of state-of-the-art cloud gaming platforms, and brought out the novelty in their framework design. They have highlighted *interaction latency* and *streaming quality* as the two QoS requirements of cloud gaming. As for the interaction latency, Table 2.1 lists out the maximum delay allowed for different types of traditional games before the user experience begins to degrade.

However, the latency requirements in cloud gaming are more stringent. Traditional online games can perform the rendering on the local machine and then update the game state in the game server in some time. Hence, the player of a traditional online game does not feel the effects of interaction delay. But in case of cloud gaming, the rendering is offloaded to the cloud, thus the thin client does not have the ability to hide the interaction delay from the user. This makes cloud gaming less delay tolerant than traditional online gaming systems. The maximum interaction delay for all cloud-based games should be at most 200 ms. Other games, specifically such action-based games as first person shooting games, likely require

Table 2.1 Delay tolerance in traditional gaming

Example game type	Perspective	Delay threshold (ms)
First person shooter (FPS)	First person	100
Role playing game (RPG)	Third person	500
Real-time strategy (RTS)	Omnipresent	1,000

less than 100 ms interaction delay so that players' quality of experience is not affected.

2.4.1.2 *Video streaming and encoding*

When a player of a cloud-based game issues a command, the command has to traverse the Internet to the game server in the cloud, be processed by the gaming logic, rendered by the processing unit, compressed by video encoder and streamed back to the player. This encoding/compression and distribution to end users has to take place in a very timely manner in order to prevent degradation of users' quality of experience (QoE). In addition to timeliness in encoding, the quality of the video being streamed is also an important factor in determining user-experience.

Design requirements

- *Low-latency response*: User experience will be hampered in case of high response time, hence making low-latency response a critical requirement of mobile gaming. For guaranteeing low-response time, the infrastructure should be strong enough that the user inputs reach the game server, be processed by the game logic, and the audio/video be captured, encoded and sent in a timely manner.
- *High bandwidth*: Transferring video streams constitutes most of the data exchanged in a cloud-game. For transferring such a huge amount of data, that too in real time, requires a high bandwidth connection between the game server and client.
- *Global coverage*: To be able to support users from multiple geographical regions, the cloud-game application needs to be accessible from anywhere. Hence, it is imperative for such an application to have a global coverage.

2.4.2 *Deployment details*

Bharambe *et al.*, in [11], have presented *Colyseus*, a distributed architecture for hosting interactive multiplayer games on the internet. Colyseus distributes dynamic game-play state and computation to multiple nodes across the Internet, adhering to stringent latency constraints and maintaining communication costs at the same time.

Properties of multiplayer games

- Games can tolerate weak consistency in the game state. Present client-server implementations cut down interaction delay by presenting the player with a weakly consistent view of the game world.
- Game-play is generally driven by a rule-set that makes it easy to predict reads/writes of the shared game state. For instance, most reads and writes of a player relate to objects which are located physically close to the player.

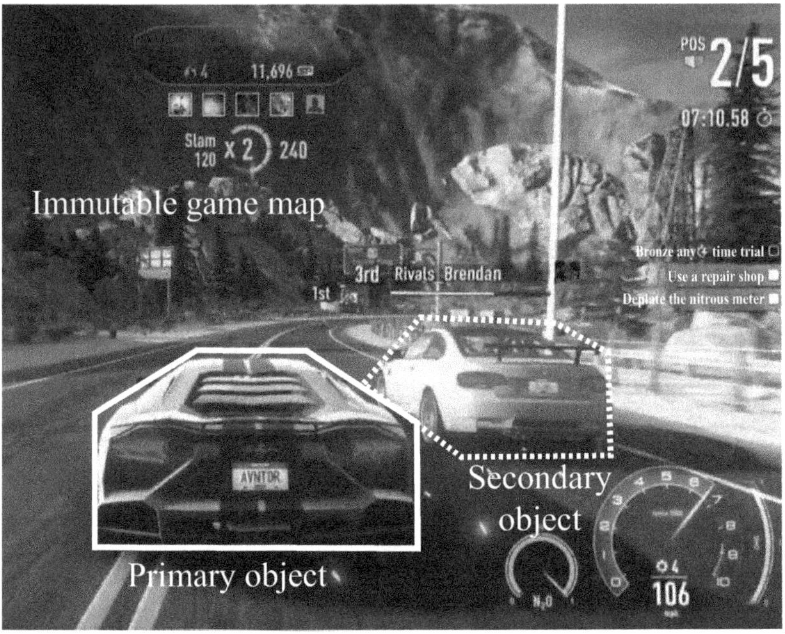

Figure 2.5 Game play showing primary and secondary objects. The game player owns the left car and interacts with the right car. The device closest to the owner of the left car would contain the primary copy of the left car object and a replica of the right car object (courtesy: http://www.metacritic.com).

Using Colyseus [11], the game state concerning a player can be located in a node very close to her, so that the interaction delay is minimized for a smooth gaming experience. Each game is described as a collection of game objects – where each object can be a player's avatar or the user's representation in the game (e.g. a car in a racing game as shown in Figure 2.5). Colyseus maintains a primary copy and several replicas of each object, with each device holding primary copies of user objects that are directly connected to it, and replicas of other objects which primary objects interact with. Figure 2.5 elucidates the concept of primary and secondary objects. Distributed objects in Colyseus follow a single-copy consistency model, that is all writes to an object are serialized through exactly one node in the system – the one containing the primary copy of it. This allows low-latency reads and writes at the cost of weak consistency, since most of the communication is made to the player's own object (which is present right at the edge).

Furthermore, Colyseus utilizes the locality and predictability in the movement patterns of players to pre-fetch objects needed for driving game logic computation. This pre-fetching of objects hides the delay in communicating with the node containing the primary copy of required object, hence giving a smooth user experience without any lags.

For a successful implementation of a cloud-game on fog infrastructure, the game service provider will have to incorporate a Colyseus-like system that distributes game state across multiple nodes based on proximity to user. In the absence of such a mechanism, communication with the cloud for every request will severely hamper user experience, especially for highly interactive games. In the next subsections, we discuss a model deployment of a cloud-game on fog over a 5G network.

2.4.2.1 Physical deployment

Gaming clients running on mobile devices host the device component of the game application. They are connected to 5G base stations, which have the ability to host computation and storage. These base stations host the fog component of the gaming application as virtual machines. Base stations are connected to the cloud via high speed Ethernet links. Virtual machines in the cloud are responsible for carrying out the logic described in the cloud component of the application.

2.4.2.2 Application architecture

To discuss effectively the deployment of a cloud-based game on a fog computing infrastructure, we need to host the aforementioned components of a typical cloud-based game on the three different kinds of computation offload points available. Figure 2.6 shows the mapping of application components to fog infrastructure.

Device component

The device component, that is the application logic running on game clients, hosts the real time streaming protocol (RTSP) reception module for receiving incoming video and audio frames. In addition to this, the component needs to have input handler modules for capturing inputs from users' consoles and sending them to the server.

Fog component

The fog component of the application holds most of the computation and storage involved in the distributed game system. This computation hosting is realized by virtualization technology on fog-enabled edge devices. The application component of a particular game runs on the fog devices inside a virtual machine.

The gaming virtual machine contains, as any cloud-gaming server would, an input handling module for receiving events from users and applying them to the gaming logic, and an output module that captures the rendered audio and video, encodes them and sends them to the clients via RTSP. The purpose of these modules is to allow a basic game to function (the way it did on a cloud-based deployment) and is agnostic to the gaming application. In addition to these modules, a fog game server hosts a distributed game state mechanism (Figure 2.6 shows Colyseus), so that game-play state and computation is distributed among all such fog nodes in the network. This module makes the gaming experience look transparent to the number and geographical distribution of users, by pre-fetching game objects in the area of interest, making users get the impression of a single gaming

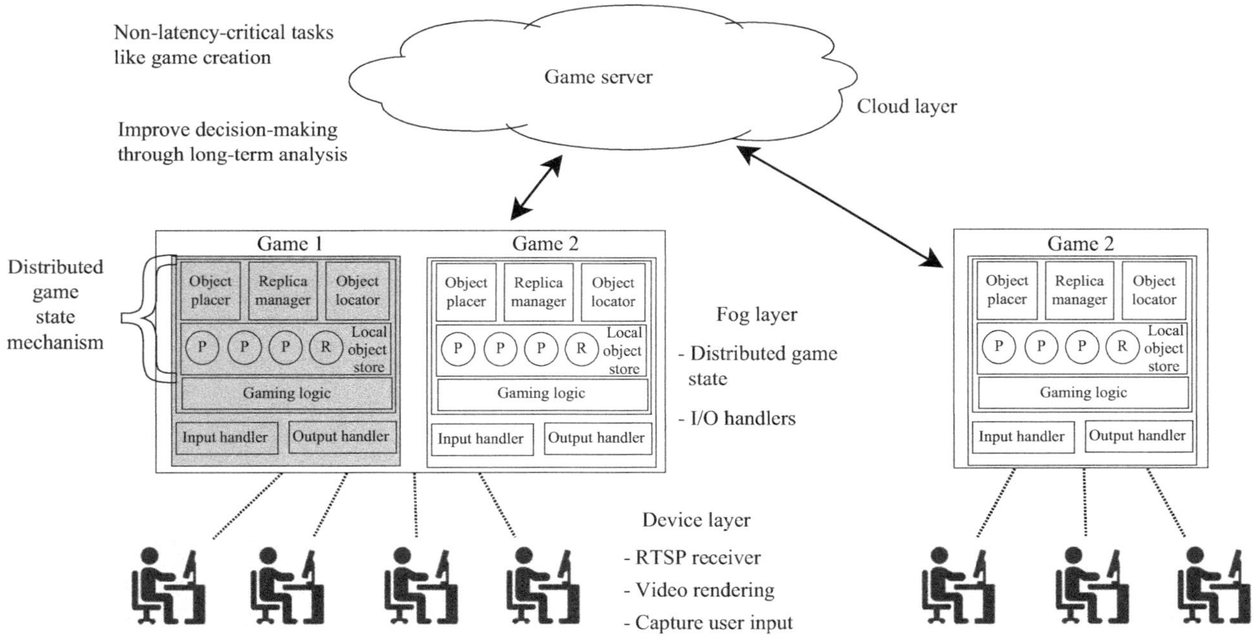

Figure 2.6 Deployment of cloud gaming on fog infrastructure over 5G mobile network

server while being able to overcome the concomitant latency and scalability issues of a single-server implementation.

The majority of data transfer in the gaming application takes place between the fog component and the device component, since the fog needs to stream the video to the game client. To be able to effectively harness the gigabit speed and sub-millisecond latency of 5G transmission, it is necessary that these components are separated by no more than one hop in 5G. Hence, hosting the fog component on the base stations – to which a game client directly connects – is a requirement. Other communications are not heavy, and mostly pertain to game events – which can be sent on fibre links without considerable delay and bandwidth consumption.

Cloud component

The cloud layer hosts logic that are not required to work under strict latency constraints – like game initiation and hosting static game-maps. The cloud can also assist the gaming virtual machines running on fog devices by learning the optimal control strategy through long-term analytics on the decisions taken by them in the past. Furthermore, to assist communication between edge servers running the fog component of the gaming application, the cloud component will provide a message-passing interface – like a publish–subscribe protocol.

2.5 Smart homes [use case 3]

The proliferation of the IoT has given a great boost to smart home automation systems. The smart home market is presaged to cross $44 billion in 5 years from now [23], bringing with it new opportunities for mobile network operators and the rest of the mobile ecosystem. The omnipresence of mobile networks makes them indispensable for connecting smart home devices and home energy management gateways, just as mobile phones are emerging as the main interface for home energy management applications.

2.5.1 Requirements

Smart home is an amalgamation of various technologies which collectively improve the lifestyle and experience of the user through coordinated functionalities. A typical smart home applications should fulfil the following requirements. Because of the large number of devices participating in a smart home and the concomitant large volume of data generated, it poses a number of requirements that any deployment will have to cater to, the most common of which are discussed in the following sections.

2.5.1.1 Energy efficiency

A smart home environment contains a lot of different kinds of devices apart from the appliances normally found in homes. Such devices consume a considerable amount of energy, thus making energy minimization one of the key objectives of an efficient smart home design.

Average American home electrical usage [22]

Heating and cooling accounts for 54% of a household's electricity bill while lighting consumes 25% of the total use. Standby power leaked by devices accounts for 10% of total electrical use.

'Little changes can make a big difference in a home's energy consumption. And with modern technology, it's so easy to automate energy use, so you don't even have to think about it,' – Adam Justice, founder of **ConnectSense**, a wireless, cloud-based home automation device.

Occupancy sensors installed in smart homes can detect the absence of any activity in an area and turn the lights of that location off in order to save electricity. The same can be done for air conditioners, geysers, and room heaters to cut down the expenses incurred on heating and others. Another way to save power is by eliminating phantom energy loss from devices like microwaves even though they are switched off but plugged into the socket. Studies like [25,31] have shown the potential of reducing power consumption of a house by cutting down standby power loss. Data about power consumption of a device – detected by power sensors – coupled with the knowledge of whether the device is on or off can be used to detect the phantom energy loss and the outlet powering the device can be switched off. This requirement is not complex in terms of implementation and can save a large amount of electricity, thus making it one of the most popular requirements of a smart home.

2.5.1.2 Safety

User safety is one of the key concerns of any system in general, and smart homes in particular. A smart home application should be able to detect intruders or any suspicious activity happening around the house. CCTV cameras installed outside the house can detect suspicious activity and send an alert message to the smart home application, which can take action by activating an alarm and turning on the lights of the area. Glass-breakage systems can detect an intruder, whereas motion sensors can detect movement in the house when the owner is away and inform the owner and also call the police in case the situation demands it. Products offering these services like *Canary* [2] make use of the computation on both the cloud and edge-devices, but these products use a separate hardware and form a different ecosystem that is difficult to tie with the complete smart home fog ecosystem.

2.5.1.3 Maintaining home environment

The most important purpose of a smart home is to improve the experience of the house owner by maintaining optimal physical conditions like temperature and humidity inside the home and providing assistance for daily tasks, like preparing coffee on waking up, maintaining optimal lighting by drawing curtains based on time of day and weather. Such application logic works by processing streams of data generated by sensors that sense the physical conditions and detect the activities of the user. The volume of data from a single smart home can be huge, left alone the volume from a colony or a city of smart homes. Smart home applications need to handle voluminous data and yet respond in a timely manner so as not to mar user experience.

2.5.1.4 Mobile dashboard and long-term analysis

A smart home application should have a mobile dashboard using which a user can check the state of his house even from a remote location. The user can use the dashboard to even control objects in the house, like opening/closing doors, talking to people who visit in his absence, or informing the police in case of an emergency. This use-case requires analysis of data generated by the smart home and a user-friendly presentation of information extracted from this analysis. It also demands global coverage since the smart home application may have to communicate with a user at a remote location.

Design requirements

- *High-speed communication*: Due to the number of devices connected in a smart home ecosystem, there is a need for an efficient machine-to-machine (M2M) communication mechanism that incurs a very small delay. For a smooth user experience, it is necessary for the devices coordinate and perform in real time – thus requiring high speed M2M communication.
- *Handling high data volume*: Smart homes generate massive volumes of data, particularly due to the number of devices connected. Hence, the device processing the data as well as the connecting network should be able to handle such an immense volume of data.

2.5.2 Deployment details

The requirements of a smart-home system are peculiarly handled by fog computing owing to its *near-the-edge* processing and resultant low-latency. A schematic of the smart home use-case on fog computing infrastructure has been shown in Figure 2.7. The deployment of a smart-home system on fog infrastructure over a 5G network is described in the following sections.

2.5.2.1 Physical deployment

A smart home system consists of a myriad of connected devices serving a variety of functions, ranging from sensors measuring temperature, humidity or detecting presence or fire to high-level appliances like smart air-conditioners and CCTV cameras. These devices need to communicate with each other and perform coordinated functions to serve the requirements of a smart home, requiring an efficient and reliable M2M communication. The M2M communication facility provided by 5G mobile networks and its ability to support a huge number of connected devices makes it an enabling technology for smart home automation systems. Smart devices are connected to a *small cell* hosted inside the house, that in turn is connected to the core network via a high-speed broadband connection. This small cell acts as the *smart home gateway* for the devices in the smart home and serves as a point for offloading computation and storage of smart home applications (Figure 2.7).

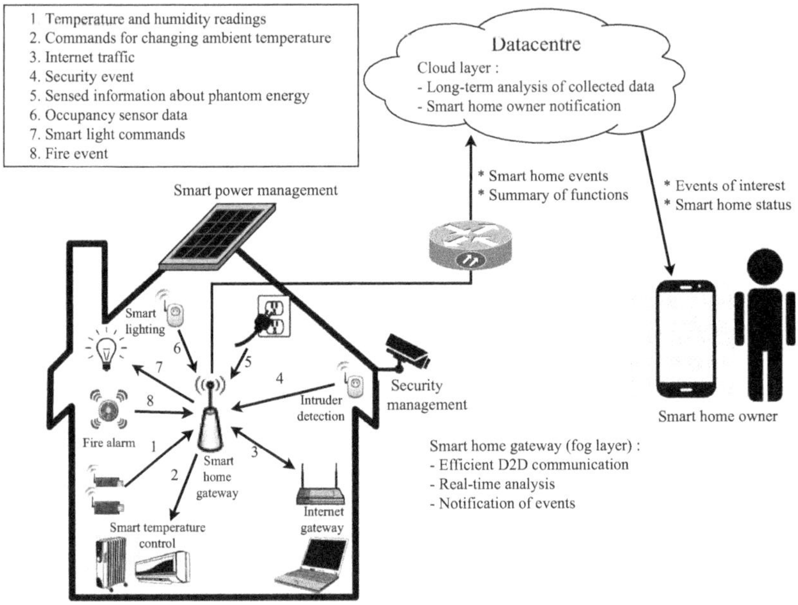

Figure 2.7 Deployment of a smart home application on fog infrastructure

The intermediate network devices (typically belonging to the ISP) connecting a group of smart homes to the Internet also serve as offload points for the smart home application. Owners of smart homes connect to the smart home application through high-speed 5G mobile network using their smartphone application and can access the smart home dashboard to view/modify the status of the smart home.

2.5.2.2 Application architecture

Efficient use of the services provided by fog computing is possible only when the smart-home application logic is partitioned into the three components of a fog application. This partitioning is elucidated in the following sections.

Device component

The data collection points of a smart home are mainly sensors, which need to send the sensed data to the smart home gateway in a timely manner. The application logic deployed in CCTV cameras should process the captured video and detect events of interest in real time and inform the smart home gateway.

The application component running on actuators – like air-conditioners, fire alarm and others – in a smart home should be able to receive commands from the smart home gateway and implement those in real time.

Fog component: smart home gateway

The *smart home gateway* is the seat of control of the smart home and is responsible for running applications that coordinate the activities of various smart devices to create a holistic smart home experience. In order to fulfil the requirements of a smart home system (as described earlier), the smart home gateway needs to provide the following services: such applications leverage the following services provided by the *smart home gateway*.

- *Efficient M2M communication*: The smart home gateway needs to provide interface to an M2M communication interface for the smart home devices to communicate with each other. Energy management requires communication between the sensors and electrical appliances, so that they may be turned off when sensors detect absence of activity. For ensuring safety of the home, CCTV cameras and motion sensors should be able to communicate with the smart home gateway in a timely manner, thus requiring efficient communication between smart home devices. The devices in a smart home are generally resource constrained, and using standard protocols like hyper text transfer protocol (HTTP) for message passing will be inefficient.

  In recent years, a lot of effort has been made towards developing protocols for M2M communication between resource constrained devices, some of that have yielded popular outcomes like Message Queue Telemetry Transport (MQTT), Constrained Application Protocol (COAP) and Session Initiation Protocol (SIP).

MQ Telemetry Transport (MQTT)

MQTT works on an asynchronous publish–subscribe architecture and is realized by sending control packets. MQTT packet headers are kept as small as possible, making this protocol apt for IoT by lowering the amount of data transmitted. Hence, this protocol is suitable for constrained networks (low bandwidth, high latency and fragile connections).

Daş *et al.* [19] have demonstrated an example implementation of a smart home application using M2M communication between resource-constrained sensors and home appliances. In their implementation, communication between devices and sensors was enabled by a SIP server in the smart home. Drawing parallels from the proposed implementation, the *smart home gateway* will support a number of such protocols which can be used by application developers to build useful smart home applications.

- *Real-time data analysis*: The smart home application needs to process events, especially those related to safety and energy management, in real time. Works like [13,43] have shown how real-time analysis of data can benefit smart homes in terms of energy management and security respectively.

  Processing offload could not get any closer to the appliances than the *smart home gateway*. Such a close vicinity to the device reduces the communication delay of between the gateway and the sensors and appliances, allowing smart home applications to make real-time decisions. In cloud-based smart homes, the control system used to reside in the cloud, giving rise to a high latency between the devices and the control system.

- *Network traffic reduction*: Devices in a smart home generate a lot of data, typically because of continuously sensing the environment. In fact, a large portion of the data is redundant or not useful. Smart home applications analyse this little big data to extract valuable information. However, the volume of this data, from a group if not from a single house, can be too huge to be continuously sent to the cloud for processing. The presence of a control system right at the gateway solves this problem by performing the data cleaning and analysis to generate control signals for the appliances. Being located at the very edge, the analysis happens in real time.

Cloud component

A typical smart home application should allow its users to view and control the status of the smart home from any remote location. Such a use-case requires a global coverage of the smart home application, thus justifying the deployment of this logic in the cloud. This component of the application will receive aggregated summaries of the smart home's status at regular intervals as well as notifications of events requiring immediate attention from the smart home gateway and inform the owner of the house.

This component will also perform long-term analytics on data collected from a smart home as well as data from multiple smart homes to detect any kind of usage patterns that it may leverage to improve the services provided in smart homes.

2.6 Distributed camera networks [use case 4]

Distributed system of cameras surveilling an area has garnered a lot of attention in recent years particularly by enabling a broad spectrum of interdisciplinary applications in areas of the likes of public safety and security, manufacturing, transportation and healthcare. The widespread use of these systems has been made possible by the proliferation of economical cameras and the availability of high-speed wired and wireless networks. Such a large number of cameras makes these systems generate data at very high rates. Monitoring these video streams manually is not practical, if at all feasible, thus engendering the need for tools that automatically analyse data coming from cameras and summarize the results in a way that is beneficial to the end-user.

Centralized tools for analysing camera-generated data are not desirable in a lot of cases primarily because of the huge amount of data that needs to be sent to the central processing machine. This would not only lead to a high latency in the system, but would also devour the bandwidth. Hence, processing the video streams in a decentralized fashion is a more advisable method of analysis. A number of research works have explored distributed camera networks [20,29]. The requirements of such a system have been listed in the following sections.

2.6.1 Requirements

Distributed camera network involves communication between devices only, and thus poses unique requirements, that have been discussed in [34].

2.6.1.1 Real-time consensus among cameras

Decisions taken by cameras need to be coordinated in order to attain a consensus about the task they are performing (e.g. activity recognition). Hence, there is a need for communication, that too one with a low delay, between cameras covering overlapping or adjacent regions. Furthermore, arriving at a consensus in real-time demands that the processing of video streams be done in a latency-critical manner. Sending all video streams to the cloud will be inefficient in this case due to the high delay incurred in communicating with the cloud.

2.6.1.2 Real-time PTZ tuning

In case of fixed cameras, video analysis becomes difficult because the fixed resolution or viewpoint may not be able to capture the target. The distributed camera network should support active sensing allowing cameras' parameters such as pan-tilt-zoom (PTZ) and resolution to be controlled by the video analysis system. This tuning of camera parameters has to be done in real time in order to effectively capture the target. Apart from functioning in real time, the parameter tuning of cameras needs to be adaptive, being able to learn from previous decisions and improve.

2.6.1.3 Event notification

The distributed camera network should inform the security personnel monitoring the area about the occurrence of an event. This use-case requires a global coverage, since the user may be present at a remote location.

Design requirements

- *Low-latency communication*: For effective object coverage, the PTZ parameters of multiple cameras need to be tuned in real-time based on the captured image. This requires ultra-low latency communication between the cameras and the seat of camera control strategy.
- *Handling voluminous data*: Video cameras continuously send captured video frames for processing, which amounts to a huge traffic, especially when all cameras in a system are taken into account. It is necessary to handle such a large amount of data without burdening the network into a state of congestion.
- *Heavy long-term processing*: The camera control strategy needs to be updated constantly so that it learns the optimal PTZ parameter calculation strategy. This requires analysis of the decisions taken by the control strategy over a long-period of time, which makes this analysis computationally intensive.

2.6.2 Deployment details

There have been several studies like [20,29] that cover distributed sensing in camera networks. Of particular relevance to fog computing is the work by Peng *et al.* [29] in which they have proposed the use of camera servers physically close to the cameras for processing real-time queries on networks of distributed camera networks. Based on the concept presented by Peng *et al.*, we discuss a typical deployment of distributed camera analysis system on fog infrastructure in the following sections.

2.6.2.1 Physical deployment

The data generation points in a distributed camera network are the numerous surveillance cameras that generate data in the form of captured frames at a constant rate. Cameras are also the prime actuators in this system as they need to constantly change the PTZ parameters in order to get the best coverage of the target. Cameras in a DCN are connected via high bandwidth 5G connection to a small cell located in physical proximity of the cameras. The small cell connects the DCN to the cloud via high-speed Ethernet.

2.6.2.2 Application architecture

Figure 2.8 shows the necessary components in a smart distributed surveillance system and the interactions between them. We now discuss the application architecture, that is the placement of application logic into components that can be deployed at different offload points in the fog network, as shown in Figure 2.9.

Device component

The device component of a distributed camera network application runs on a camera and contains the code for handling it. It essentially consists of two modules – *video sender* and *command receiver*. The video sender module sends the recorded frames to the associated small cell at a constant rate. The command receiver module receives instructions to change the camera parameters from the small cell and applies them to get a better coverage of the target. The encoding of video and sending it should take place in real time, as well as the PTZ change commands received from small cells should be applied in real time so as to bring down the response time of the DCN to real-time domain.

Fog component

The fog component of the application is responsible for detecting events based on spatio-temporal relations between objects across video streams coming from different cameras. The application logic first filters out objects of interest from the live camera feeds by using image processing techniques. It then uses the spatio-temporal relations between the detected objects to detect if an event has occurred. Works like [18,29,42] have proposed techniques performing event detection in real time from live camera feeds. In case an event is detected, the fog component

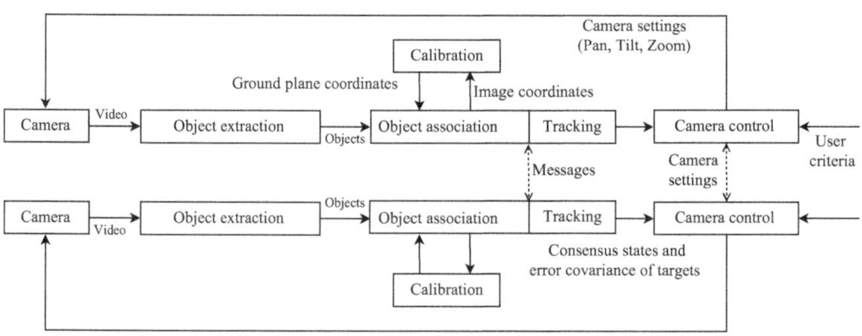

Figure 2.8 Schematic diagram of a distributed camera analysis system

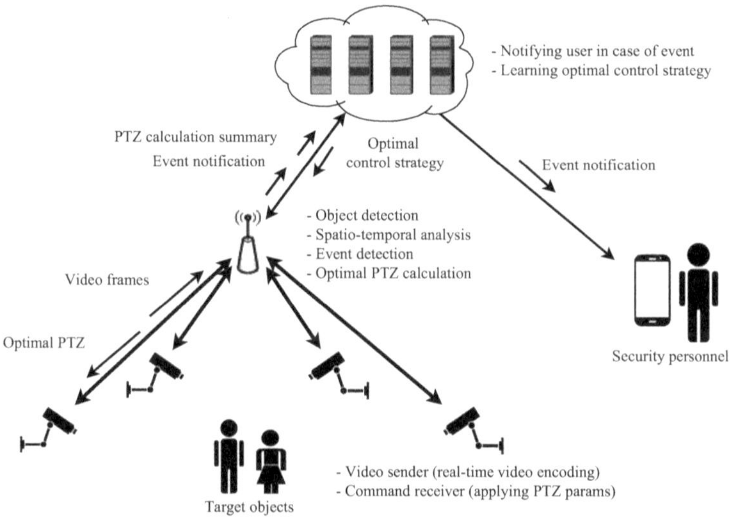

Figure 2.9 Deployment of a distributed camera network on fog over 5G network

informs the cloud component of the application so that users of the system, even at remote locations, can get notified of the occurrence of the event.

The fog component of a DCN application is also responsible for the camera control strategy, that is tuning the parameters of the cameras in order to optimize the scene acquisition capabilities of the cameras. Based on the camera feeds, the application calculates the optimal PTZ parameters for each camera. In addition to the PTZ parameters, the application also responds to scene complexity by determining the optimal resolution for the camera to capture, for example to capture a higher resolution video when the scene is relatively empty and lower the resolution as the number of objects in the scene increases. Starzyk *et al.* [35,36] has elucidated on optimal camera parameter calculation in, the calculation being based on images captured by the cameras. These optimal parameters are sent to the cameras in real-time which apply them to improve the quality of the captured scene. The control strategy hosted by the fog component is adaptive and tries to improve itself based on previous decisions. For this, the fog component sends an aggregate of camera control decisions taken in the past to the cloud component for determining the optimal control strategy, which is then communicated back to the fog component.

In centralized systems, video streams had to be sent to the cloud for processing, and cameras would receive instructions for tuning camera parameters from the cloud, both these communications exhibiting an unpredictable delay that could mar the purpose of the system. Furthermore, sending live video streams to the cloud at all times would devour the bandwidth and may lead to congestion, further delaying frame delivery. In a fog setting, it is apt to place the fog component (which takes video streams as input) on the small cells connecting a group of cameras. Placing this component at the very edge, close to the source of data and actuators, greatly cuts down the delay. The only communication between the small cells and the cloud takes place when an object of interest is detected or when the control strategy of the fog component needs to be updated – which is relatively inexpensive in terms of bandwidth consumed.

Cloud component

Research works like [35] suggest that the control strategies of cameras can be learnt using online learning algorithms. Since the latency requirement of updating the control strategies is not that stringent, the cloud component can perform the learning based on the information about previous decisions taken by the fog component. The cloud component uses advanced learning tools to determine the optimal control strategy at every time and updates the control strategy currently running on the fog component. This task is computationally intensive, hence making it a good fit to be run on the cloud.

The cloud component of a DCN application enables security personnel present at remote locations to monitor the activities in the area surveilled by the DCN by sending notifications pertaining to events of interest to the security personnel, who can respond accordingly. In special cases, the cloud component may also stream video related to the event so that security personnel may have a look at the actual situation. This part of the application demands a wide coverage as the users monitoring the activity in the surveilled region may not be located in vicinity of the cameras or small cells. Hence, it makes a lot of sense to deploy this application logic to the cloud which has a global coverage.

2.7 Open challenges and future trends

Fog computing and 5G networks are the enabling technologies for futuristic applications, especially in the realm of the IoT. 5G mobile networks need to have inherent support for fog computing in order to be efficient and successful. The amalgamation of these two concepts will enable the developers to come up with applications that solve large problems faced by the masses. However, large-scale successful deployment of fog computing systems on 5G mobile networks is bound by research in a number of domains. Fog computing in 5G networks is as much of a vision today as 5G networks itself and a plethora of challenges need to be addressed to make it a reality. These challenges are described as follows:

- *Computation offloading in network base stations*: Fog applications run in the form of virtual machines (or containers) on virtualized fog-devices. This would require shifting the network functions – originally implemented in dedicated hardware – to software (a concept called network function virtualization (NFV)). However, implementing NFV on such a heterogeneous network as a fog-enabled 5G network is still not lucid. Works like [16,17] have explored into the implementation of network virtualization on mobile networks. Further advancements need to be made in this domain for efficiently realizing fog computing on 5G networks.
- *Energy efficiency*: Fog computing on 5G network requires the base stations to be enabled with virtualization for running applications. Running applications on a hypervisor a higher energy requirement due to the heavier processing involved in virtualization. This carbon footprint would be amplified to a great extent by the numerous fog-devices in a network with a dense geographical

distribution. Minimizing energy consumption is a key challenge that needs to be addressed for successful commercialization of fog computing on 5G.

- *Pricing policies*: A fog ecosystem will consist of two kinds of stakeholders: (1) internet service providers who construct the fog infrastructure and (2) the application service providers who want to extend their applications to the edge. Thus for enabling *pay-as-you-go* pricing model, it is necessary to decide the price for resources and the division of payment for different parties. This will be difficult given the widely distributed network of fog devices. For realizing this utilization-based pricing policy, accounting and management systems working at a very fine granularity of the network are required.
- *Resource management*: Efficient resource provisioning and management has been a strong reason for the success of cloud computing – and will continue to be so for fog computing as well. However, the problem of resource management is even tougher, because of the added dimension of network latency involved. Besides, the vast number of heterogeneous devices in the network further complicates resource management. Ottenwälder *et al.* [26] have proposed MigCEP and operator migration policy for fog infrastructure so as to optimize on end-to-end latencies as well as bandwidth consumption and their work is one of the few contributions to resource management on fog.
- *Privacy and security*: Fog computing virtualizes the network and decouples network functionality from the hardware provider. Hence, fog applications process application data on third-party hardware, which poses strong concerns about visibility of data to the third-party. 5G networks handle voice and data packets in the same manner which may lead to leakage of sensitive voice data. This makes privacy measures even more necessary for fog computing on 5G networks. Stojmenovic *et al.* [37] have discussed the security issues in fog computing in their work.

Addressing these challenges are necessary to make fog computing on 5G networks commercially viable. One can then envision the development of services like IaaS, PaaS and SaaS on the fog environment as well, which would be a major milestone for the road to a future with fog-enabled applications.

2.8 Conclusion

Fog computing is a recently emerging computing paradigm that offers facilities like low latency and dense geographical distribution, which is essential for a number of applications. This chapter looks at a few peculiar use-cases apt for fog computing on 5G mobile networks, each of the use-cases being contingent on a specific offering of the Fog. The necessity of an inherent support for fog computing in 5G networks has been presented and the deployment of use-cases (applications on the Fog) – on a model 5G network have been shown. Each use-case has been broken down into components meant for execution in the devices, the fog and the cloud, and the interplay between these components has been shown. Both 5G networking and fog computing technologies are compatible with each other and their amalgamation will be the enabler for future Internet applications. Due to its close affinity for applications on IoT, fog computing facilitate in developing green and sustainable future IoT applications.

References

[1] "Boeing 787s to create half a terabyte of data per flight, says virgin Atlantic", 6 March 2013. http://www.computerworlduk.com/news/data/boeing-787s-create-half-terabyte-of-data-per-flight-says-virgin-atlantic-3433595/, visited 2015-12-06.

[2] "Canary: The first smart home security device for everyone", 29 January 2014. http://canary.is/, visited 2015-12-06.

[3] "Cisco delivers vision of fog computing to accelerate value from billions of connected devices". http://newsroom.cisco.com/press-release-content?type=webcontent&articleId=1334100, visited 2015-12-08.

[4] Cisco IBSG, April 2011. http://www.cisco.com/c/dam/en_us/about/ac79/docs/innov/IoT_IBSG_0411FINAL.pdf, visited 2015-12-06.

[5] "Cisco visual networking index: Global mobile data traffic forecast, 2014–2019", 3 February 2016. http://www.cisco.com/c/en/us/solutions/collateral/service-provider/visual-networking-index-vni/vni-forecast-qa.pdf, visited 2015-12-06.

[6] "Internet of things market statistics – 2015", 5 March 2015. http://www.ironpaper.com/webintel/articles/internet-things-market-statistics-2015/, visited 2015-12-06.

[7] "Machine-to-machine (M2M): Profiting from the internet of things", August 2013. http://info.csgi.com/m2m-internet-of-things, visited 2015-12-06.

[8] "Valve on why cloud gaming will never be mainstream", 8 February 2013. http://www.geeky-gadgets.com/valve-on-why-cloud-gaming-is-not-viable-02-08-2013/, visited 2015-12-06.

[9] "Why edge computing must be a companion to NFV and 5G?", 20 November 2015. http://www.telecomtv.com/articles/mec/why-edge-computing-must-be-a-companion-to-nfv-and-5g-13043/, visited 2015-12-06.

[10] Jonathan Bar-Magen. "Fog computing: Introduction to a new cloud evolution". In *Escrituras Silenciadas: Paisaje Como Historiografía*, pages 111–126. Servicio de Publicaciones, 2013.

[11] Ashwin R Bharambe, Jeff Pang, and Srinivasan Seshan. *A distributed architecture for interactive multiplayer games*. Citeseer, Doctoral Dissertation, INRIA Paris, 2005.

[12] Flavio Bonomi, Rodolfo Milito, Preethi Natarajan, and Jiang Zhu. "Fog computing: A platform for internet of things and analytics". In *Big Data and Internet of Things: A Roadmap for Smart Environments*, pages 169–186. Springer, 2014.

[13] Wen-Tsuen Chen, Po-Yu Chen, Wei-Shun Lee, and Chi-Fu Huang. "Design and implementation of a real time video surveillance system with wireless sensor networks". In *Vehicular Technology Conference, 2008. VTC Spring 2008. IEEE*, pages 218–222. IEEE, Calgary, AB, Canada, 2008.

[14] Sharon Choy, Bernard Wong, Gwendal Simon, and Catherine Rosenberg. "The brewing storm in cloud gaming: A measurement study on cloud to end-user latency". In *Proceedings of the 11th Annual Workshop on Network and Systems Support for Games*, page 2. IEEE Press, Venice, Italy, 2012.

[15] Sharon Choy, Bernard Wong, Gwendal Simon, and Catherine Rosenberg. "A hybrid edge-cloud architecture for reducing on-demand gaming latency". *Multimedia Systems*, 20(5):503–519, 2014.

[16] Xavier Costa-Pérez, Jorg Swetina, Tao Guo, Rajesh Mahindra, and Sampath Rangarajan. "Radio access network virtualization for future mobile carrier networks". *Communications Magazine, IEEE*, 51(7):27–35, 2013.

[17] Jose Costa-Requena, Jesus Llorente Santos, Vicent Ferrer Guasch, *et al.* "SDN and NFV integration in generalized mobile network architecture". In *European Conference on Networks and Communications (EuCNC), 2015*, pages 154–158. IEEE, Paris, France, 2015.

[18] Abir Das, Anirban Chakraborty, and Amit K Roy-Chowdhury. "Consistent re-identification in a camera network". In *Computer Vision–ECCV 2014*, pages 330–345. Springer, Zurich, Switzerland, 2014.

[19] Das, R., and G. Tuna. "Machine-to-machine communications for smart homes." *International Journal of Computer Networks and Applications (IJCNA)*, EverScience Publications 2.4 (2015): 196–202.

[20] Chong Ding, Bi Song, Akshay Morye, *et al.* "Collaborative sensing in a distributed PTZ camera network". *IEEE Transactions on Image Processing*, 21(7):3282–3295, 2012.

[21] Sébastien Faye, Claude Chaudet, and Isabelle Demeure. "A distributed algorithm for multiple intersections adaptive traffic lights control using a wireless sensor networks". In *Proceedings of the First Workshop on Urban Networking*, pages 13–18. ACM, Nice, France, 2012.

[22] Ilana Greene, 20 December 2013. http://www.huffingtonpost.com/ilana-greene/smart-houses-help-reduce-_b_4472919.html, visited 2015-11-26.

[23] GSMA. "Vision of smart home: The role of mobile in the home of the future", September 2011. http://www.gsma.com/connectedliving/wp-content/uploads/2012/03/vision20of20smart20home20report.pdf, visited 2015-11-20.

[24] Stefan Lämmer and Dirk Helbing. "Self-control of traffic lights and vehicle flows in urban road networks". *Journal of Statistical Mechanics: Theory and Experiment*, 2008(04):P04019, 2008.

[25] Ing Nils F Nissen. "EuP preparatory study lot 6 'standby and off-mode losses'". Compiled by: Fraunhofer Institute for Reliability and Micro-integration, IZM, Berlin, 2007.

[26] Beate Ottenwälder, Boris Koldehofe, Kurt Rothermel, and Umakishore Ramachandran. "Migcep: Operator migration for mobility driven distributed complex event processing". In *Proceedings of the Seventh ACM International Conference on Distributed Event-based Systems,* pages 183–194. ACM, Arlington, TX, USA, 2013.

[27] FT Park. *Traffic data collection and analysis*. The National Academies of Sciences, Engineering, and Medicine, Washington DC, USA, 1980.

[28] Mugen Peng, Yuan Li, Jiamo Jiang, Jian Li, and Chonggang Wang. "Heterogeneous cloud radio access networks: A new perspective for enhancing spectral and energy efficiencies". *Wireless Communications, IEEE*, 21(6):126–135, 2014.

[29] Rui Peng, Alex J Aved, and Kien A Hua. "Real-time query processing on live videos in networks of distributed cameras". In *Research, Practice, and Educational Advancements in Telecommunications and Networking*, IGI Global, page 27, 2012.

[30] Zhouyue Pi, Farooq Khan, and Jianzhong Zhang. *Techniques for millimeter wave mobile communication*, October 29, 2010. Google Patents, US Patent App. 12/916,019.

[31] K Roth, K McKenney, R Ponoum, and C Paetsch. "Residential miscellaneous electric loads: Energy consumption characterization and savings potential in 2006 and scenario-based projections for 2020". In *Final Report by TIAX LLC to US Department of Energy*, 121, National Technical Reports Library, USA, 2008.

[32] Noah Shachtman. "Feds look to fight leaks with 'fog of disinformation'", 3 July 2012. http://www.wired.com/2012/07/fog-computing/, visited 2015-12-12.

[33] Ryan Shea, Jiangchuan Liu, Edith Ngai, and Yong Cui. "Cloud gaming: Architecture and performance". *Network, IEEE*, 27(4):16–21, 2013.

[34] Bi Song, Chong Ding, Ahmed T Kamal, *et al.* "Distributed camera networks". *Signal Processing Magazine, IEEE*, 28(3):20–31, 2011.

[35] Wiktor Starzyk and Faisal Z Qureshi. "Learning proactive control strategies for PTZ cameras". In *Fifth ACM/IEEE International Conference on Distributed Smart Cameras (ICDSC), 2011*, pages 1–6. IEEE, Het Pand, Ghent Belgium, 2011.

[36] Wiktor Starzyk and Faisal Z Qureshi. "Multi-tasking smart cameras for intelligent video surveillance systems". In *8th IEEE International Conference on Advanced Video and Signal-Based Surveillance (AVSS), 2011*, pages 154–159. IEEE, Klagenfurt University, Austria, 2011.

[37] Ivan Stojmenovic and Sheng Wen. "The fog computing paradigm: Scenarios and security issues". In *Federated Conference on Computer Science and Information Systems (FedCSIS), 2014*, pages 1–8. IEEE, Warsaw, Poland, 2014.

[38] Michael AP Taylor and William Young. *Traffic Analysis: New technology and new solutions*. Hargreen Publishing, 1988.

[39] Michael AP Taylor, William Young, and Peter W Bonsall. *Understanding traffic systems: Data, analysis and presentation*. Ashgate Publishing Company, 1996.

[40] The Telegraph. "Smart traffic lights to stop speeders", 19 May 2011. http://www.telegraph.co.uk/motoring/news/8521769/Smart-traffic-lights-to-stop-speeders.html, visited 2015-12-12.

[41] Marco Wiering, Jelle Van Veenen, Jilles Vreeken, and Arne Koopman. "Intelligent traffic light control". *Institute of Information and Computing Sciences*, Utrecht University, Utrecht University, the Netherlands, 2004.

[42] Shu Zhang, Yingying Zhu, and Amit Roy-Chowdhury. "Tracking multiple interacting targets in a camera network". *Computer Vision and Image Understanding*, 134(C):64–73, 2015.

[43] Suyang Zhou, Zhi Wu, Jianing Li, and Xiao-ping Zhang. "Real-time energy control approach for smart home energy management system". *Electric Power Components and Systems*, 42(3–4):315–326, 2014.

Chapter 3

The in-band full duplexing wireless exploiting self-interference cancellation techniques: algorithms, methods and emerging applications

Geili T. A. El Sanousi[1] and Mohammed A. H. Abbas[2]

Abstract

The most significant recent evolution in the wireless communication theory (WCT) is Full Duplexing of channel; that is, communicating a transmission simultaneously with reception. This has full impact on the communication system and the overall constituents of the WCT. This chapter aims at highlighting the developments in the in-band full duplexing (IBFD) access technique and its impact on the whole system linking the information source to termination channel.

In this chapter, all techniques, algorithms and emerging applications are related to the reader to provide an updated starting point for the fresher and a comprehensive review of the current state of the art, for the experienced professional. The flow of the chapter started by relating the origins of the concept through to different evolved forms and finalizing with the emerging applications. The technique variants are presented in a categorized approach, providing the foundations of the concepts, how these progressed and how the older techniques have been incorporated into the newer context. The categorization was system based – that is relating the system block where the associated technique is exploited- and network based, that is relating how these fit into different communication networks' topologies and applications. The categorization was well-related integrability and hybridization of the techniques, as well as presenting the reader with useful reviews and referencing to further readings.

The impact of the IBFD is significant, and the pace of associated developments is extremely huge and fast; this art here presents a useful guide which is only relating the current state of art in the immediate temporal zone. IBFD field impacts every detail of the wireless communication system, so it is fair to anticipate the forthcoming decade to lay emphasis on the technique and its exploitations. Pertaining to the current temporal frame, this chapter covered mostly every foundation point in the concept.

[1]Assistant Professor, Self Based Researcher.
[2]Associate Professor, Faculty of Engineering, University of Khartoum.

Keywords

In-band/single channel full duplexing (IBFD/SCFD), self-interference cancellation (SIC), antenna cancellation technique (ACT), radio frequency (RF) cancellation, RF suppression, analogue SIC, digital SIC, channel state information, relaying protocols, 5G, IoT, Fog and Cloud

3.1 Introduction

The wireless communication theory (WCT) is witnessing tremendous growth in this era. New concepts are envisioned every day and new ways of solving problems are devised. What was at a time a demerit is treated in newer systems as a merit for example multi input multi output (MIMO) theory where the multipath became the merit which provides for bandwidth and signal power gains. And as new merits are exploited new demerits are discovered and new challenges are defined. The theory is hence always evolving around a compromise between merits and demerits. Of recent the most evolving area of the communication system has been the *link* [1], particularly the wireless link, whereas remaining blocks of the system are passing through a revolution to accommodate developments in the link theory rather than an evolution in most of its aspects.

Variants of exploitations of communications theory and systems are many and part of every aspect of modern life; but the most thriving field with the major impact on our human lives is so far the networking field, in particular the personal communication networks and the associated parallel data networks; the most recent development of which is the 5G networks and the Internet of Things (IoTs). These networks naturally depend on the communication theory and directly exploit its' variants. As a matter of fact, networking entities are gradually integrating to cooperate as a unified large scale system units rather than separated communication entities, for example, Cooperation of Multiple Points (CoMPs) and cooperative relaying. Thus the focus in this chapter is subjected to in band link techniques as applied in system level but which as well have huge impact on the networking frames, architectures and topologies.

This chapter starts by reviewing the theoretical background of the technique, relating current theory to older existing literature, explaining the variants of the theory and pointing out the challenges and technical limitations it faces. The variants of the technique are all RF wireless link (passband) located but are complemented with system (baseband) techniques; these will also be elucidated as well. Based on the current state of the art, as would be highlighted; the technique has progressed well enough that it is sufficiently sensible to envision insights on the opportunities the technique can provide and possibilities to incorporate the in-band full duplexing (IBFD) in the forthcoming systems, designs and protocols. This would be the ensuing and wrapping up work to this chapter.

3.2 The in-band full duplexing communications: the concept and the background

IBFD communications refer to a link system whereby the communication system transmits and receives in the same frequency simultaneously. This definition might also be extended to distinguish and include both wireless links that propagate in the same source/destination spatial channel (same physical route); and/or those wireless channels that link using more than one different source/destination spatial channels (or diverse routes).

The heritage of WCT avoided full duplexing (FD) in the same band. It has been considered inapplicable throughout decades because of the strong self-Interference (SI). Usually the far-field signal travels a long way and reaches the receiver after being annihilated by convolution with channel and unwanted noise. The most common metrics used for describing these channel effects are the signal-to-noise ratio (SNR) and signal-to-interference plus noise ratio (SINR). At the receiver, transmitting in-band simultaneously with reception provides a very strong loop of self-interfering signal compared to the received weak far-field signal and the SINR drops severely as a result of this strong SI. Receiver front end amplifiers are thus driven into saturation resulting in receiver desensitization. Ambiguity between sent and received signals becomes very difficult or impossible to resolve, even for receivers with very high sensitivity. The WCT therefore resorted to means to separate the sent/received signals in domains.

Conventional WCT favoured time division duplexing (TDD) whereby the timing of the bursts differs for receive and transmit processes (time domain). The other older alternative has been the frequency division duplexing (FDD) where the transmission is achieved in a different band to the reception, with a guard band, the duplex frequency spacing, separating the two bands (frequency domain). Each of the two methods has merits and demerits that are very well known in the literature. Later developments introduced hybrid systems of TDD and FDD and then the code division duplexing which is essentially a hybrid scheme of TDD and FDD through time–frequency coding of the sequence of bursts. Of recent, one of the aspects of the space–time premise [1] was the introduction of a new domain, the spatial domain which differentiates signals based on the spatial signatures, that is the channels' responses based on the physical coordinates of the space linking the receiving and transmitting elements. This defines the spatial division duplexing (SDD) which can be used to separate transmission and reception processes based on spatial geometry.

The last variant here, the SDD; could be considered as an IBFD communication technique [2], if the definition was extended to include the link exploiting different spatial routes. This however is said with conservative reservations about the detail that these channels are not identical in their response. Classifying SDD as an IBFD is sensible though, when considering in effect the overall information source/destination to the information destination/source duplex communication link. The performance delivered is that of a FD. Figure 3.1 depicts these topologies for these link techniques.

These concepts above, including the SDD; have been founded on relatively inaccurate assumptions such as channel reciprocity, wide sense stationarity, uncorrelated signals etc. as part of the WCT heritage. The problem with these lies in that, as the size of information (and consequently size of bursts in signalling domain) increases these concepts become impractical. Sought has been to find means to reduce the size of burst as compared to channel coherency in both time and frequency domains. For example many coding techniques aimed at increasing the per signal information content of a message, some exploited partial approaches, for example quasi-stationary channel quality indicator (CQI) in advanced long-term

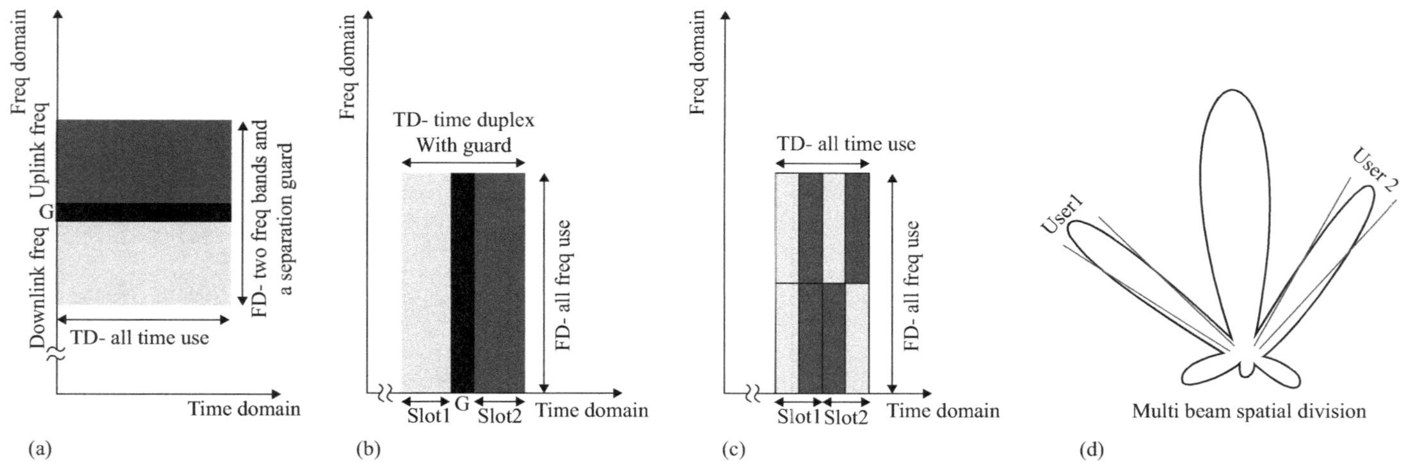

Figure 3.1 Conventional duplexing schemes: (a) frequency division duplexing (FDD), (b) time division duplexing (TDD), (c) frequency hop code division duplex (CCD), and (d) spatial division: users share all time all frequency

evolution (A-LTE). In this quasi-stationary CQI, the channel is digitized into a sequence of generalized response states that are a compromise delivering averages of the channel responses, but not the exact signatures of the channel.

The IBFD provides a practical solution to these and other challenges. The most obvious and immediate advantage of IBFD is that it provides immediate halving of the required channel bandwidth (in other words doubling the effective bandwidth and hence capacity). This in effect provides for fewer constraints on time coherency and bandwidth/frequency coherency of the channel, that is, it provides instantaneousness and mitigates the latency and delay problems. The IBFD has opened avenues for many flexible designs. The freeing of half the link time indeed has the potential to reshape the whole WCT. In the following, the feasible IBFD techniques and associated methodologies are closely examined.

3.2.1 The basic IBFD techniques

The most prominent IBFD technique was introduced in 2010/11 through researchers from University of Stanford [3,4]. The nomenclature used there was single-channel frequency duplexing. Full duplexing in the same bandwidth and through the same spatial channel (route) was achieved. The idea was based on a sequence of SI cancellation stages. The objective is to annihilate the self-signal as heard by the transceiver and reduce its strength compared to the received signal in the same band. This eluded the need for domain division in resources.

The original art's philosophy exploits antenna cancellation on the signalling domain to reduce the received power of the transmitted signal (SI) when seen on the transceiver front end. This is performed using phase difference through geometrical asymmetry (in the near field) of two antenna elements a placed at distances d and d plus half a wavelength from the receiving point, that is 180 degrees phase shift (see Figure 3.2).

In [4] a BALUN transformer was used instead to introduce the 180 degrees phase difference. In either case, the objective is taking advantage of the local destructive self-interference at the receiving element without influencing the far-field transmission, that is, the transceiver does not hear its own signals in its RF front-end. The indigenous art further exploited a combination of RF cancellation,

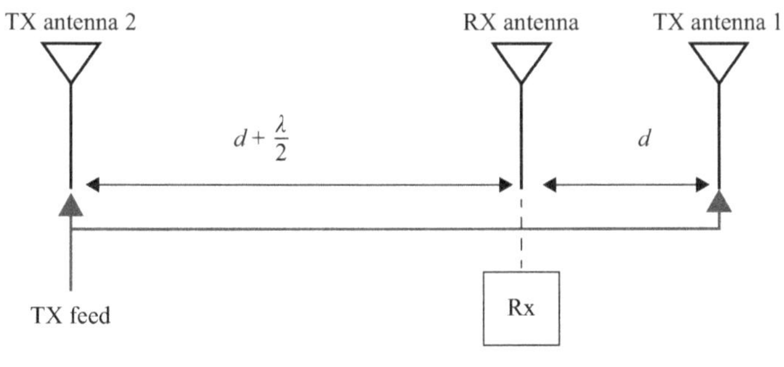

Figure 3.2 The antenna cancellation arrangement

analogue cancellation and digital cancellation techniques. These techniques have been advanced on and improved in later literature in combinations with RF suppression. The philosophy behind these techniques will be highlighted accordingly.

3.2.2 Antenna cancellation techniques (ACT)

These techniques are applied at the most front end of the transceiver unit, at the signalling domain in the EM emission stage. The concept depends on placing antenna elements in certain electro magnetic geometry so as to cause phase differences in the near-field zone. This phase difference when superimposed, effects a self-cancellation or nulling at the intended receiving point. The self-interference cancellation (SIC) happens in the near-field zone and exhibits minimal effect on the far-field pattern. Thus the transmission objective is not forfeited.

The antenna cancellation technique (ACT) are generally frequency sensitive since the superposition of phase differences critically depends on the wavelengths by default and these are based on the arrival time according to the geometrical positioning. In its indigenous art, only two transmitting antenna (Tx) and one receiving antenna (Rx) elements were used. The two transmitting elements are treated as main and auxiliary transmit points in the far field.

The following points summarize the features of the technique and the shortcomings in its indigenous form.

- It depends on semi-symmetrical geometries based on arrangement of antenna elements.
- The far-field effect is not influentially affected by such geometries.
- Although the original art assumed linear orientation of elements, using vector analysis the null can result as cancellation of components of two (or more) vectors summed at more than one null point in the vicinity; given the difference in wavelengths should always be a resultant of $\lambda/2$ (in the case of two vectors) and resolved to zero (in general) to cause destructive addition.
- In terms of size it requires at least half lambda plus the normal antenna separation.
- As this geometrical structure is based on the size of lambda, the behaviour is frequency dependent and therefore the application is not efficient when bandwidth is big.
- The errors in placement and phase delays are a mapped function of λ and frequency. Also the corresponding mismatches in amplitudes superimpose on the centre and reduce the effectiveness or accuracy of cancellation.

3.2.2.1 Extension of the indigenous art mathematical model to (N) antenna elements [5]

The multi-element antenna (MEA) theory has established itself a great deal in concurrent literature. Basically the use of more than one transmitting elements qualifies the technique to the MEA category of systems. The above model may then be extended to analyse the case of N transmit elements and possibilities of M null points for receiving. Below is an extended form of the model of [3] presented in [5].

The following variables can be defined:

Indexed number of antenna elements: $i = 1, 2, 3, \ldots, N$

Amplitude of signal from ith antenna is A_i

Attenuation of the A_i is α_i

The mismatch in amplitude of ith antenna compared to reference antenna 1 is ε^{A_i}

(the ratio $\boldsymbol{A_i/\alpha_i}$ is ideally supposed to be constant but as it is not then $\boldsymbol{A_1/\alpha_1} = A_{at}$ such that $\boldsymbol{A_i/\alpha_i} = \varepsilon^{A_i} + A_{at}$) where ε^{A_i} is a metric of the resultant amplitude mismatch added to the supposed constant ratio, that is the ratio of the amplitude to attenuation (how much loss in received signal) is equal to the constant ratio plus mismatch error.

The error in placement of elements causes error in phase; this phase error compared to phase of reference signal **1** is $\varepsilon^{\phi_{1i}}$. (In general form, $\varepsilon^{\phi_{1i}}$ can be borrowed to represent the phase shifts of antenna elements inclusive of error and inclusive of (π-difference) for the pair cancellation.)

The phase of the signal from $\boldsymbol{i}$th antenna: $\boldsymbol{\phi_i}$

And the phase constant $\boldsymbol{\phi_1}$ of the input reference signal **1** is stated as $\psi = \omega_c t + \phi_1$.

Analysis

Considering the target receiving point (the null); signals from $\boldsymbol{i}$th antennas are supposed to arrive in different phases

The phase difference with reference to signal 1 is $\phi_1 - \phi_2 = \varepsilon^{\phi_{12}}$ that is the phase error between signals 1 and 2.

Similarly $\phi_1 - \phi_3 = \varepsilon^{\phi_{13}}$

Rearranging we get

$\phi_1 = \phi_2 - \varepsilon^{\phi_{12}} = \phi_3 - \varepsilon^{\phi_{13}} = \phi_i - \varepsilon^{\phi_{1i}}$

$$\text{i.e. } \varepsilon^{\phi_{1i}} = \phi_1 - \phi_i \tag{3.1}$$

Thus expressing the received near-field signal seen at any point in the field; we get:

$$R(t) = \big[(A_{at} * X(t) * e^{j\psi}) + (\{A_{at} + \varepsilon^{A_2}\} * X(t) * e^{j\psi} * e^{j(\varepsilon^{\phi_{12}})}) + (\{A_{at} + \varepsilon^{A_3}\} * X(t) * e^{j\psi} * e^{j(\varepsilon^{\phi_{13}})}) + \cdots + (\{A_{at} + \varepsilon^{A_N}\} * X(t) * e^{j\psi} * e^{j(\varepsilon^{\phi_{1N}})})\big].$$

The first term is obtained by substituting $e^{j(\varepsilon^{\phi_{11}})} = e^{j(0)} = 1$ and $\varepsilon^{A_1} = 0$

$$R(t) = (X(t) * e^{j\psi}) * \begin{bmatrix} \{A_{at}\} + \{(A_{at} * e^{j(\varepsilon^{\phi_{12}})}) + (\varepsilon^{A_2} * e^{j(\varepsilon^{\phi_{12}})})\} + \{(A_{at} * e^{j(\varepsilon^{\phi_{13}})}) \\ + (\varepsilon^{A_3} * e^{j(\varepsilon^{\phi_{13}})})\} + \cdots + \{(A_{at} * e^{j(\varepsilon^{\phi_{1N}})}) + (\varepsilon^{A_N} * e^{j(\varepsilon^{\phi_{1N}})})\} \end{bmatrix}.$$

$$= (X(t) * e^{j\psi}) * \left(A_{at} + \sum_{i=2}^{N} A_{at} * e^{j(\varepsilon^{\phi_{1i}})}\right) + (X(t) * e^{j\psi}) * \left(\sum_{i=2}^{N} \varepsilon^{A_i} * e^{j(\varepsilon^{\phi_{1i}})}\right).$$

$$= (A_{at} * X(t) * e^{j\psi}) * \left(1 + \sum_{i=2}^{N} e^{j(\varepsilon^{\phi_{1i}})}\right) + (X(t) * e^{j\psi}) * \left(\sum_{i=2}^{N} \varepsilon^{A_i} * e^{j(\varepsilon^{\phi_{1i}})}\right) \tag{3.2}$$

Knowing that phase error $\varepsilon^{\phi_{1-1}} = 0$ and amplitude mismatch $\varepsilon^{A_1} = 0$ the above expression can be restated as

$$\boldsymbol{R}(\boldsymbol{t}) = \left(A_{at} * X(t) * e^{j\psi}\right) * \left(\sum_{i=1}^{N} e^{j(\varepsilon^{\phi_{1i}})}\right) + \left(X(t) * e^{j\psi}\right) * \left(\sum_{i=1}^{N} \varepsilon^{A_i} * e^{j(\varepsilon^{\phi_{1i}})}\right) \tag{3.3}$$

This reduces to

$$\boldsymbol{R}(\boldsymbol{t}) = \left(X(t) * e^{j\psi}\right) * \left\{\left(\sum_{i=1}^{N} \left(A_{at} + \varepsilon^{A_i}\right) * e^{j(\varepsilon^{\phi_{1i}})}\right)\right\} \tag{3.4}$$

And the complex conjugate of $R(t)$ is

$$\boldsymbol{R}^*(\boldsymbol{t}) = \left(X(t) * e^{-j\psi}\right) * \left\{\left(\sum_{k=1}^{N} \left(A_{at} + \varepsilon^{A_k}\right) * e^{-j(\varepsilon^{\phi_{1k}})}\right)\right\} \tag{3.5}$$

$$\begin{aligned}\boldsymbol{R}(\boldsymbol{t}) * \boldsymbol{R}^*(\boldsymbol{t}) = X(t)^2 * 1 * &\left\{\left(\sum_{i=1}^{N} \left(A_{at} + \varepsilon^{A_i}\right) * e^{j(\varepsilon^{\phi_{1i}})}\right)\right. \\ &\left. * \left(\sum_{k=1}^{N} \left(A_{at} + \varepsilon^{A_k}\right) * e^{-j(\varepsilon^{\phi_{1k}})}\right)\right\}\end{aligned} \tag{3.6}$$

The product in parenthesis {} can be reduced to

$$\begin{aligned}&\sum_{i=1}^{N}\sum_{k=1}^{N} \left(A_{at} + \varepsilon^{A_i}\right) * \left(A_{at} + \varepsilon^{A_k}\right) * e^{j((\varepsilon^{\phi_{1i}}) - (\varepsilon^{\phi_{1k}}))\ldots} \\ &= \sum_{i=1}^{N}\sum_{k=1}^{i-1} \left(A_{at} + \varepsilon^{A_k}\right) * \left(A_{at} + \varepsilon^{A_i}\right) * e^{j((\varepsilon^{\phi_{1i}}) - (\varepsilon^{\phi_{1k}}))} + \sum_{i=1}^{N}\left(\left(A_{at} + \varepsilon^{A_i}\right)^2\right) \\ &\quad + \sum_{i=1}^{N}\sum_{k=i+1}^{N} \left(A_{at} + \varepsilon^{A_k}\right) * \left(A_{at} + \varepsilon^{A_i}\right) * e^{j((\varepsilon^{\phi_{1i}}) - (\varepsilon^{\phi_{1k}}))}\end{aligned} \tag{3.7}$$

Using symmetry of conjugate sums

$$= \sum_{i=1}^{N}\left(\left(A_{at} + \varepsilon^{A_i}\right)^2\right) + 2 * \sum_{i=1}^{N-1}\sum_{k=i+1}^{N} \left(A_{at} + \varepsilon^{A_k}\right) * \left(A_{at} + \varepsilon^{A_i}\right) * \cos\left(\left(\varepsilon^{\phi_{1i}}\right) - \left(\varepsilon^{\phi_{1k}}\right)\right) \tag{3.8}$$

$$\begin{aligned}\boldsymbol{R}(\boldsymbol{t}) * \boldsymbol{R}^*(\boldsymbol{t}) = X(t)^2 * &\left\{\sum_{i=1}^{N}\left(\left(A_{at} + \varepsilon^{A_i}\right)^2\right) + 2 * \sum_{i=1}^{N-1}\sum_{k=i+1}^{N} \left(A_{at} + \varepsilon^{A_k}\right)\right. \\ &\left. * \left(A_{at} + \varepsilon^{A_i}\right) * \cos\left(\left(\varepsilon^{\phi_{1i}}\right) - \left(\varepsilon^{\phi_{1k}}\right)\right)\right\}\end{aligned} \tag{3.9}$$

where this expression can be equal to zero at locations where this *power metric function* (PMF) is equal to zero

$$\sum_{i=1}^{N}\left((A_{at}+\varepsilon^{A_i})^2\right)+2*\sum_{i=1}^{N-1}\sum_{k=i+1}^{N}(A_{at}+\varepsilon^{A_k})*(A_{at}+\varepsilon^{A_i}) * \cos\left((\varepsilon^{\Phi_{1i}})-(\varepsilon^{\Phi_{1k}})\right)=0 \tag{3.10}$$

$$\sum_{i=1}^{N}\left((\boldsymbol{A_{at}}+\varepsilon^{A_i})^2\right)=-\mathbf{2}*\sum_{i=1}^{N-1}\sum_{k=i+1}^{N}(\boldsymbol{A_{at}}+\varepsilon^{A_k})*(\boldsymbol{A_{at}}+\varepsilon^{A_i}) * \mathbf{cos}\left((\varepsilon^{\Phi_{1i}})-(\varepsilon^{\Phi_{1k}})\right) \tag{3.11}$$

Equation (3.11) defines the null function for an antenna array it is a PMF explaining the geometry conditions that produce nulls regardless of the time-varying amplitude of the transmitted signal. The number of possible nulls M is less than or equal to number of elements N. The importance and applications of this function would be revisited later.

Verification

When substituting $N=2$ and $\{\varepsilon^{\Phi_{1k}}=\pi+\overline{\varepsilon^{\Phi_{1-k}}}\}$ in (3.10) the result is obtaining the same power analysis for two elements in the appendix of [3]. This also means this generalized form can also incorporate pairs of π-phase differenced elements in symmetrical order.

Analysis of the power metric function (PMF) [6]

On the left hand side (LHS) of the nulling function (PMF) of (3.11), the minimum value of the coefficients sum $(\boldsymbol{A_{at}}+\varepsilon^{A_i})^2$ is realized when the amplitude mismatch is zero for elements, that is when $\varepsilon^{A_i}=\mathbf{0}$ for all i, and for which the value of lower boundary (UL1) of LHS is $UL1=\boldsymbol{N}*(\boldsymbol{A_{at}})^2$ and the maximum boundary (UL2) is obviously $UL2=N*(\boldsymbol{A_{at}}+\varepsilon^{A_{max}})^2$, where $\varepsilon^{A_{max}}$ is the maximum mismatch coefficient.

On the right hand side (RHS) of (3.11); for $\zeta_{ik}=(\varepsilon^{\Phi_{1i}})-(\varepsilon^{\Phi_{1k}})$, the cosine function fluctuates between -1 through zero to $+1$, that is the value fluctuates between the negative and positive of the maximum value which is $\boldsymbol{N}*(\boldsymbol{A_{at}}+\varepsilon^{A_{max}})^{\mathbf{2}}$. However, since the LHS is a positive function, all negative values of RHS will be discarded and lower boundary of RHS becomes same as that of LHS $[\boldsymbol{N}*(\boldsymbol{A_{at}})^2]$ for the equality to hold.

In essence, the solution is in finding the suitable phase perturbations of ζ_{ik}, for N degrees of freedom. Since cosine is an even function, the positive boundaries imply these phases' perturbations must be fluctuating in sign, that is, include a number of (π-plus) phase shift differences; and converges to a negative product such that the overall LHS is negated to a positive value. Also these phase perturbations are reasonably assumed small in magnitude. Finally, since ζ_{ik} depends on d, this objective can either be attained through proper allocation of antenna elements in the space (static null positioning) or alternatively by controlling the amplitudes and phases to impose the nulls in a predetermined manner (adaptive).

3.2.2.2 Grating nulls against pattern nulls

In this technique, the radiations of concern are all in the very near-field region. Thus the nulls obtained are rather grating nulls, not to be mistaken for far-field and Fresnel near-field pattern nulls. The grating nulls are those that result from phase additions of fields propagating in more than one direction [7]. Grating lobes are experienced when inter-elements spacing is very large thus increasing the aperture dimensions and thus increasing the span of near-field zone. Alternatively as in this case, these exist by exploiting radiation cancellation in the very near-field zone (condition $d + \lambda < 0.62\sqrt{(d + \lambda)^3/\lambda}$) [7, page 34]. In a way it can be simplified that grating nulls are created in the vicinity of the confinement zone of the array aperture. The pattern nulls however are attributed to the diffraction pattern.

It is established in literature that the relationship between Fourier transforms of radiating electric field and aperture distribution is exact [8,9]. This result is inferred from Parseval's theorem. This implies that the coefficients of the nulling function are the same for the radiating signal electric field strength and for the resultant pattern. This explains the ambiguity of the close similarity between the mathematics of the PMF and that of the far-field pattern.

3.2.2.3 Theoretical aspects: challenges and limitations

Like all the WCT aspects, it is always a compromise between merits and demerits and solutions to challenges. The most notable limitation of this technique is that it cannot stand on its own. The amount of annihilation in the self-signal power is quite small compared to usual weakness of reception through long-range communication links. In the very ideal situation of a very small antenna placement mismatch, the average reduction in self-signal strength is 30 dB [3,10]. However, the reduction in self-signal strength must be sufficient to avoid receiver front-end desensitization. Therefore, the technique is by default, complemented by other stages of SIC. These stages include variants comprising the RF suppression, RF cancellation, Analogue cancellation and the digital baseband cancellation. In terms of bandwidth, the technique by default is very sensitive to antenna placement and thus to frequency changes. In its original form, it cannot function in wide band mode. And the required antenna aperture of at least $(2d + \lambda/2)$ implies need for big surface area. Yet more, the covering range of the technique is poor. This is easy to infer, since more transmission power means restoring the ambiguity between the self-signal (stronger) and the received signal (weak and attenuated). A lot of research has been carried out to improve the performance parameters and to answer to the challenges as stated. The authors in [4,5,11] include different novel approaches to answer to challenges of the antenna cancellation challenges.

3.2.2.4 Technical implementations: challenges and limitations

On technical bases, there are yet physical challenges relating the physics of antenna theory. Strictly speaking, working in very near-field region has cost in issues such as mutual coupling and dielectric materials selection and the casting of elements. It is important to comprehend the nature of the antenna physics. When d is very small, and the aperture is small the wave nature assumed here becomes obsolete. The point is, magnetic and electric fields need space to be converted to waves. Before the wave emission process happens, these act as normal fields obeying EM induction principles, thus induces currents and magnetic fields to all points in the vicinity. In other words, a minimum space is required to enable all elements to resonate for this analysis to be valid.

The mutual coupling in the intended null points with implanted receiving nodes (elements) is an advantage, since it implies less attenuation and more exact signal images received (as these cancel in the null point *perfectly*). Mismatch in antenna placement and in amplitudes of signals is inevitable in the casting process. The more antenna elements are incorporated into the system, the heavier are these costs. Yet the progress of this technique is so quick that it has earned accommodation to protocols of the forthcoming 5G technologies. The boost in the research area is attributed to the tremendous potential and promise it entails; that is to re-shape the whole WCT a new.

3.2.3 Passive RF suppression techniques

RF suppression is a means to attain IBFD, through use of different channel routes or different polarizations that take place in the RF (EM radiation) domain. It is a subject of overlap with the alternative nomenclature; spatial division and polarization diversity division. Isolating the transmitted RF signals from the received RF signals at same frequency band using different routes or polarities makes the two terminal nodes communicate in actual effect, in a practically Full Duplex link. Separation here, does not mean directly annihilating the transmitted signal but eluding it in the receive channel. In concurrent literature, there have been four major lines relating RF suppression-IBFD that received more focus [10]. The philosophy behind these techniques is briefly delineated below. Further interest in detailed research works exploiting these approaches can be pursued in these References [12–15].

3.2.3.1 Directional and polarization isolation

The directional isolation in relation to IBFD in an early literature has been defined by Everett *et al.* in [2] as the positioning whereby: ‘the direction in which the base station (transceiver) transmits is (in general) different from the direction from which it receives’. This could be achieved dynamically through selective or directional antennas; or simply by physical placement in a directionally isolated manner.

On the other hand, polarity diversification is about transmitting information signals in different polarity to the information receiving interface, that is orienting the transmit elements and receive elements in different polarizations (orientations). Practically speaking, it has been reported that a maximum of six polarities can be accommodated in one single physical channel. It is also possible to combine the two techniques in a hybrid platform whereby each isolated direction entertains both polarization and isolation benefits. The two concepts are illustrated in Figure 3.3.

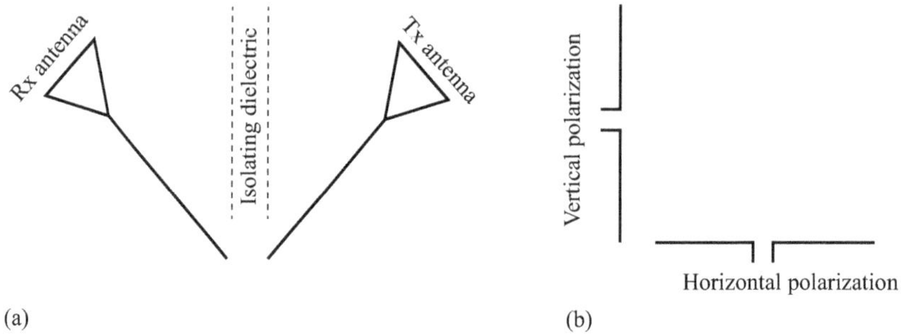

Figure 3.3 (a) Directional isolation by placing elements in orthogonal orientation and (b) orthogonality in polarity

Further readings

Ma *et al.* [10] reported another directional/polarization diversity art [12] which is adjoined to complementary stages to attain IBFD. The technique there is used in relaying context where the system consisted of a compact antenna relaying unit entertaining both directional isolation and polarization isolation. The relay re-routes repeated base station signals to a domestic unit (user equipment (UE)) and vice versa. The results showed up to 48-dB suppression of transmitted signals in both relaying links.

3.2.3.2 Adaptive isolation techniques

These are, as the name suggests, techniques that are adaptively controlled to effect isolation. The most commons include antenna and beam selections as dynamic ways (adaptive) to isolate the channel directions. Alternatively, the null space projection offers the opposite effect of the beam selection concept. A projected null space isolates or rejects the signal in the projection direction, whereas beam selection links the signal in the selected beam direction. Antenna selection is usually attained using switching techniques which have known demerits such as high insertion loss and the switching time latency where the angular spread is limited. The beam selection is usually attained using a network of adaptively controlled beamforming weights. With less insertion loss and better beam switching fluency, it outperforms the antenna selection. Both techniques select a direction over others and switch the RF receiving chain to incoming signals through switching to the corresponding selected antenna/beam. Null space projection is effected through use of filters (e.g. minimum mean square error (MMSE)) to prescribe nulls in the wanted or selected-to-reject direction. The effect is that receiving chains will null-out incoming signal is the prescribed direction and listen to sources in other directions [16,17].

Further readings

These techniques in complement to a natural suppression (directional isolation) stage have been the accomplishment of [13]. There the authors brought together this collection of techniques; in a practical implementation in a network relaying context. The natural isolation was attained through pre-design of spatial placement of elements. In a way, this is similar in approach to the art presented in [11]. This directional isolation in combination with the above techniques provided about 40-dB suppression. The dependence on the channel rank to make antenna and beam selection choices is a potential drawback for the technique. This is because it implies need for perfect channel state information (CSI) knowledge and this, as follows later, is supposedly an earning from IBFD not a cost as becomes the case here.

3.2.3.3 Beamforming using time domain waveforms

Herein, the beamforming is carried out using weights and waveforms developed in time domain. The indigenous terminology for this method is time-domain transmit beamforming introduced in the art of [14]. It is meant to differentiate this procedure from the conventional frequency domain transmit beamforming techniques (FDTB), another terminology used in the same art. FDTB is nothing but the usual beamforming techniques described in frequency domain mathematics. The difference is about the signalling system design and implications associated with engineering the system in the frequency domain as against time domain. The need for this arises from the fact that in conventional FDTB designs, the guard prefixes usually go un-cancelled in all antennas SIC schemes. These residuals leak as noise into the receiver chain and decrease the SINR. Thus, design of waveforms in time domain answers to this perspective and improves SIC performance.

Further readings

Hua *et al.* [14] treat in detail one such time-domain approach and provides results conforming better SIC performance. The technique is very similar to other suppression techniques except for the use of temporal waveform designs. The attained suppression goes up to 50 dB for a carrier.

3.2.3.4 Balanced feed networks in RF isolation context

Balanced feed networks are basically a cancellation method, not a suppression method. They are usually consisted of an organization of combinations of balanced phase shifters, attenuators, couplers, power dividers and/or circulators in a control loop feed network. The attenuators and phase shifters adjust the cancelling feedback whereas couplers, power dividers and circulators control the directions of RF signal flow. These are conventional in RF circuitry designs and are basically designed on principles of guided wave theory. They have demerits including associated insertion losses, leakages, and signal distortions. The circulators in general suffer more leakage than couplers. Despite the demerit of leakage, because of their lowest insertion loss, the circulators are practically the best of the possible alternatives, compared to the directional couplers or the power dividers.

Couplers, power dividers and circulators, all bring about isolation of paths (suppression) when properly organized. However, their organization is defined according to the related function. In the IBFD context, these are used to *isolate* (suppress) RF paths. This is complemented by the phase shifters and attenuators producing negating images that cancels out the transmitted signals in the receive chain simultaneously.

Further readings
The use of RF balanced feed network as isolation between RF chains lead to the introduction of a class of single-element antennas which performs both transmit/receive functions simultaneously that is it radiates and gets illuminated simultaneously. For example, the arrangement of [15] provides isolation between transmitting and receiving RF chains through use of a pair of three legs (directions) circulators. In addition, these are complemented with phase shifts effecting SIC in the receive RF chain. The antenna reflections of the transmitted wave and associated leakages from the receiving path are also self-cancelled in the arrangement.

The beauty of the technique is that it exploits only one-single antenna element, an important solution to compact sizes. The obvious deficiency of the technique is implementation cost as it embodies too many components. Knox [15] reported suppression values in 40–45-dB range. Related approaches are also found in [18–20].

3.2.4 *Active RF cancellation techniques*

In general, all cancellation techniques exploit images of the unwanted SI signals (and additive noise, also to be mitigated) and control their phase shifts to create destructive (phase reversed) images which when added to the interfering signals end up cancelling each other. The 'RF cancellations' are designated so, since they are executed in RF front end of the system that is, after up-conversion, in wave guiding and the signalling domain. In the following, four recent technical approaches in concurrency with developments in the IBFD theory are reviewed.

3.2.4.1 Echo image cancelling using baseband to RF up-conversion

This is an explicit logic of replica production. In this mechanism, a copy of the baseband signal is used to regenerate an image of the SI signal. This image is then reversed in phase by a 180-degree phase shifter (inverter), thus forming the cancelling signal which is passed through a different radio chain, up-converted and destructively added to the reception stream at the receiving terminal, for example [21] In theory, this is supposed to completely eliminate the SI. This however does not happen because of many channel factors, for example deformations in SI signal, non-linearities in the RF chain, path attenuations, mismatch of antenna elements, amplitudes mismatch, delay, temporal delays etc. The technique is observed to attain around 30-dB SI cancelation and involves extra cost for the parallel radio chain [21].

3.2.4.2 Feed-forward networks

Feed-forward networks are basically a class of control networks where the feed is *pre-calculated* and added externally in the forward direction of the source/sink flow stream. These are similar to the balanced feed networks except that the associated phase-shifter network is not a 'balanced' network. The phrase 'balanced' refers to the effect of using balanced phase shifters units which generate streams that have exact and balanced phase shifts, for example quadrature hybrids used in [15]. These are replaced here by attenuator and ordinary phase shifters, plus couplers instead of circulators.

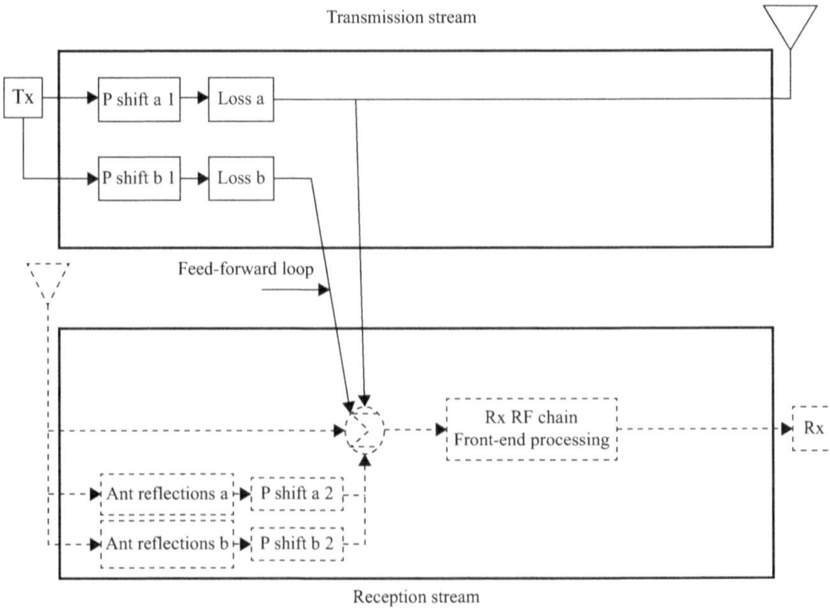

Figure 3.4 Feed-forward loop in the reception stream ('a' and 'b' are images circulated via different paths)

In feed-forward looping, the weights of parameters of concern (amplitudes and phases here) are generated on the basis of a control parameter or a comparative reference, for example a metric of the achieved SIC. The attenuator attenuates the amplitudes to a desired match. The phase shifters generate *adaptively* prescribed phase shifts on the cancelling image signal. The couplers *replicate* the self-signal, then route it forward – after passing through the attenuator/phase-shifters' network – to the received stream (*feeding forward*). This results in cancelling SI off the received signal. Figure 3.4 illustrates the general feed-forward control loop block diagram for the SIC where phase shifts here could be in quadrature as suggested in [15].

The adaptiveness of the attenuator and phase shifters units enables outside control to be imposed on the circuitry and excellent power handling capabilities. This marks a main difference between this feeding forward and the static balanced feed networks. It is also obvious this is a pure cancellation arrangement with no suppression involved. The technique is cited to offer up-to 75-dB cancellation, but this varies with bandwidth [22]. The main disadvantage is the cost of slotting in extra components.

3.2.4.3 RF analogue canceller variables' computations

Computing the RF analogue canceller variables is an essential processing algorithm. Feed networks, attenuators and phase shifters depend on these to adapt accordingly. An analogue canceller is about creating a perfect analogue negative image of the actual signal to be used to cancel it. The degrees of freedom of concern in the signals are their amplitudes and phases. Amplitudes get attenuated in the channel response, whereas phase shifts are function of a delay per channel and different paths. This delay is an implicit function of the path distance from emitting sources to the target illuminated sink points. Both of these could be modelled as tapped-inputs to channels. Since there are no constraints on the possible paths, phase perturbations and attenuations these formulate a non-convex problem presenting many possibilities of nulls and a computational challenge.

Multivariate analysis and advanced matrix algebra techniques come into call here; for example reformulation as a convex problem and using complex Wiener technique to solve for the global set of parameters as suggested in [23].

The computed weights are used to perfectly cancel the analogue signal; this implies the sensitivity of the algorithm choice and implementation. A complex algorithm may result in computational processing delays, whereas reduced complexity comes at a cost of less accurate estimates of cancelling variables.

McMichael and Kolodziej [23] reported a performance range of 49–65-dB SIC through the convex reformulation approach. The main drawback of this is the dependency on the CSI, which is necessary for a valid estimation of the attenuations and phase errors.

Similar algorithms flood the existing art. Alternative algorithms such as gradient descent algorithm which delivers sub optimal estimates and others are cited for example sake in [24–26].

3.2.4.4 Adaptive phase inversion cancellers

Like the feed networks, these exercise a control mechanism to optimize the cancellation. Although static phase shift circuits such as quadrature hybrids suffer from uneven bandwidth response, dynamic phase inversion, for example using BALUN transformers, is insensitive to bandwidth and power. The dynamic nature of the BALUN provides 180-degree phase shifts regardless of the frequency in use (and theoretically of power used too). For example, in [4], one of the indigenous arts, instead of exploiting spatial geometry to effect cancellation, a BALUN was used. Another example is the use of the electric balanced duplexers [27] providing relevant dynamicity in phase inversions.

The 180-degree (or related values of) phase shifts in essence generate streams of in-phase and anti-phase negating images which when superimposed at a target point or stream, cancel out perfectly. Usually, these require a control loop to add attenuations and time delay to the signal to compensate the transmitted air signal's naturally experienced attenuations and delays. The control loop is locked up using residual energy (received signal strength indicator) or the left over after cancellations, to adjust the amount of attenuations and estimate weights of phase shifters.

Advantages of these designs include eluding the frequency dependency and hence alleviating the bandwidth limitation. The number of required antenna elements is reduced, since no auxiliary elements or generators are needed to effect cancellation by placement. And theoretically these designs provide for better transmitted power cancellations, therefore a higher transmitted power and an improved range. The cancellations are more efficient since the phase shifts are also experienced by associated impairments and distortions, so bad images also cancel out. The technique is cited to offer at least 45-dB cancellation [4]. Arrangements of static quadrant hybrids can do the same inversions, although cost-wise they require more components, and they suffer higher insertion losses and less bandwidth efficiency.

3.2.5 Analogue cancellations

This method is carried out in the baseband before the analogue to digital converter (ADC) process. The analogue signal is cancelled by destructive addition to its reverse phase image. Relevant estimated attenuations and phase shifts are introduced to the cancellation image to improve the performance. The technique comes after many stages of *RF amplification and additional system noise* which undermines its efficiency. The brilliancy about antenna cancellations and RF cancellations over this is that it comes at the most front end of the system; thus, minimal noise is introduced and cancelation of *only* the unwanted Tx signals is more efficient. Brett *et al.* [28] present one implementation of this technique. The technique preceded the IBFD long ago to mitigate self-noise but could be readily incorporated as an additional annihilation technique in IBFD systems. The offered cancellation in [28] is about 10 dB only.

3.2.6 Digital baseband cancellations

These techniques have very well founded base in WCT. They have been used extensively as a baseband technique to eliminate Inter Symbol Interference. Here in IBFD context, it used for SIC where the cancellations take place at baseband after ADC process. It is important to note that these techniques are used as complementary techniques, not a stand-alone class of techniques. If the self-signal is not annihilated, it will surpass the dynamic range of the ADC and result in quantization noise much stronger than the weak far-field signal. The ambiguity of such far-field signal will not be resolved using digital band cancellations, since the cancellations do not suppress the resultant quantization noise. The received baseband signal can be expressed as a sum of original far-field signal, added far-field channel noise, added near-field self-interfering signal and channel noise added to the self-interfering signal. Of all these, the signal of interest is the far-field signal. For this to be decoded, the other signals must be removed. The power of the self-signal as said is much more than that of the strongly attenuated far-field signal, and to attain the IBFD means are focused on annihilating this self-signal to values efficiently lower than the wanted far-field signal. In digital terms, this is about nulling out or minimizing the power per message/frame, power per symbol, power per sample/bit; of those messages/symbols/bits decoded from the self-signal transmission and those erroneous noise bits convolved with them so as to obtain the pure far-field received stream of bits.

This objective, however, adds further constraints on the design. These include RF impairments, ADC resolution, power amplifier (PA) non-linearity, local oscillator (LO) phase noise, in-phase/quadratic-phase (I/Q) imbalance, jitter of ADC/DAC (digital to analogue converter), channel variations and channel delay profiles as main paradigms of radio chain impairments. The PA and I/Q impairments can be eluded by taking the reference signal feedback from the output of the PA, before up-conversion where these impairments take place. Alternatively, pre-distortion in the transmitter can be used to compensate these. Variations in near-field channel are considered of negligible effect if the frame length during processing is made small enough maintaining in concept the channel temporal coherency. However, LO phase noise, ADC resolution and ADC jitter are usually difficult to compensate and directly reduces the overall possible SI suppression. These three impairments in effect result in random phase distortions and ambiguities with respect to decoder reference phase. Their combined effect can be simplified as additive white Gaussian noise (AWGN). Remains the frequency offset caused by difference in oscillators timings; this is readily predictable in existing applied WCT [29,30].

3.2.6.1 Modelling the received digital signal in IBFD context

The expected receive signal is $\boldsymbol{r}(\boldsymbol{t}) = \left[r_{(0)}, r_{(1)}, r_{(2)}, \ldots, r_{(N-1)}\right]^T$ where the vector represents the received signal samples, and $\boldsymbol{N}$ is the number of samples contained in the frame. The self-signal (near-field signal) can be similarly represented as X where $\boldsymbol{X}(t) = \left[x_{(0)}, x_{(1)}, x_{(2)}, \ldots, x_{(N-1)}\right]^T$. If the associated channel response to the near-field signal is represented as $\boldsymbol{h}$, then the convolved self-signal seen in the RF chain would be $\boldsymbol{Xh}$. Similarly, $\boldsymbol{Y}(\boldsymbol{t}) = \left[y_{(0)}, y_{(1)}, y_{(2)}, \ldots, y_{(N-1)}\right]^T$ represents the intended received far-field signal, and $\boldsymbol{g}$ represents the associated channel response to the far-field signal both convolved as $\boldsymbol{Yg}$. Since offset in far-field signal frequency is translational to the symbol components, it has been modelled over many existing arts as a diagonal frequency offset multiplier that is $\boldsymbol{f} = \text{diag}\left[\left(1, e^{j2\pi\omega}, e^{2j2\pi\omega}, \ldots, e^{(N_f-1)j2\pi\omega}\right)\right]$. Thus, if AWGN noise Z is used to model the convolved noise and phase impairments due to the LO phase noise, ADC resolution and Jitter, and I stands for effect of all other impairments mentioned above, then the·received signal can therefore explicitly be modelled as

$$\boldsymbol{r}(\boldsymbol{t}) = \boldsymbol{Xh} + \boldsymbol{fYg} + \boldsymbol{z} + \boldsymbol{I} \tag{3.12}$$

where

$$\mathbf{I} = \mathbf{PA} + \frac{\mathbf{I}}{\mathbf{Q}} + \mathbf{ADC\ quantization\ errors} + \mathbf{channel\ variations} \\ + \mathbf{other\ system\ units\ and\ device\ impairments}.$$

If I is neglected then

$$\boldsymbol{r}(\boldsymbol{t}) \cong \boldsymbol{Xh} + \boldsymbol{fYg} + \boldsymbol{z} \tag{3.13}$$

The following headings are a review of readings in arts treating the variants of (3.12).

3.2.6.2 Recent techniques relating digital self-signal (Echo) cancellations

Echo cancellation is a terminology defining SIC in digital and analogue in-system zone. Basically it is similar in concept to the image up-conversion to RF in Section 3.2.4. In a similar manner, an exact image of the near-field (self-signal) baseband signal (without up-conversion) is negated and used to cancel out the positive image passing through the radio chain.

In (3.12), for any IBFD digital cancellation system; to recover $\boldsymbol{Y}$; then $\boldsymbol{Xh}, \boldsymbol{F}, \boldsymbol{g}, z$ and I should be removed from $\boldsymbol{r}(\boldsymbol{t})$. In practice, this has been a founded art, removing echo and surrounding noise. In IBFD, the receiver possesses prior knowledge of the X vector components and has ready access to accurate estimates of h. This prior knowledge can be appropriately exploited to bring about a good suppression of the self-signal. Traditional methods are adjusted to accommodate this objective. The following briefly highlights current research works that related existing art to the IBFD theory.

In [29], two stages of iterative echo canceller are proposed. The first estimates the near-field channel response $\boldsymbol{h}$ through use of least squares (LSs) algorithm and uses this to create and finite impulse response (FIR) filter of L steps to remove negative image of $\boldsymbol{Xh}$. The second stage exploits traditional system designs to decode Y out of '$\boldsymbol{fYg} + z$'. The art reported an increase in capacity performance (system efficiency) by 1.4–1.8 factor. This implies that impairments induced errors which caused the drop from the expected factor of 2.0. The art however omitted to report the exact range of self-signal suppression attained.

Li *et al.* [30] suggested the use of an adaptive least mean square (LMS) technique as a core digital cancellation technique. LMS is complemented with an adaptively controlled FIR filter that dynamically adjusts the negating image parameters to enhance the self-signal suppression. The technique is reported to provide a 20-dB suppression estimates.

Ahmed *et al.* [31] treated the problem of LO phase noise; basically it is an enhancement of existing literature which focuses on removing receiving RF chain LO phase noise using MMSE filters. The enhancement is in combining the LO phase noise mitigation in both transmission and reception RF chains through MMSE filters and cancelling out the local (transmitter) LO phase noise. This process is carried simultaneously with an LS digital self-signal image cancellation process. The technique is reported to provide 9-dB improvement over only self-signal (echo) cancellations.

In a similar dedicated impairment treatment, the authors in [32–34] focus on a different impairment each and integrate their solutions to these impairments into a digital cancellation scheme. Anttila *et al.* [32] tackle the non-linearities of PA(s) using parallel Hammerstein structure to suppress estimated non-linearities. The reported performance showed 10-dB higher transmit power. Ahmed *et al.* [33] on the other hand treated the non-linearities inclusive of those associated with PA and those associated with low noise amplifier, the phase noise and the ADC quantization noise in one model. These are eliminated using joint iterative channel estimates and successive iterative estimation of each of these non-linearities. The metric used to assess the technique was comparing it with the simulated ideal linear system; the reported results reflected a just 0.5-dB shortage of ideal performance. Another merit of this work is that it considered an OFDM signal. The art however is limited to the digital baseband performance only, without consideration of complementary prior to digital IBFD methods, and how they would relate to OFDM.

Korpi *et al.* [34] provided a unique pioneer modelling of computations for all associated impairments and parameters with stress on I/Q impairments. It proposed a set of wide-linear least-squares algorithms for complete impairments parameters' estimation and cancellations. The approach is featured by sharp decline in performance after median-high transmit power and is reported to offer about 15-dB transmit power increase. Reference [35] is closely associated with the art of [34]. It is additionally featured by impairment compensation methods, inclusion of thermal noise effects and a joint augmented cancellation algorithm that combines the widely linear scheme of [34] and a conjugate SIC algorithm. The reported performance cites a 15-dB higher transmit power but also featured a slow decline in performance after median-high transmit power, that is more stability in cancellation even when transmit power exceeds the design ranges.

These techniques and cited references above are among the frontiers of the known research works on the topic that have been published so far. This field however is versatile for many thoughts. For example, successive cancellation schemes that use more than one image in the SIC chain is an unvisited topic in IBFD context. Having worked-out to the bottom of the system, next will be the study of the possibilities of hybrid combinations of the above schemes and algorithms at different stages of the RF (passband) and baseband chains.

3.2.7 Hybrid combinations of techniques

As clearly pointed out, the incoming signals are much stronger than the received signals. For a Wi-Fi indoor link, this difference is on the range of 100–120 dB. As seen from the reviewed techniques above, at most the performance delivered does not exceed 75 dB for a singled out technique. The practical implementations of IBFD require careful matching of a sequence of annihilating techniques to maximize the resulting suppression/cancellation. It also requires a compromise between merits and demerits of these techniques. Most of the literature above exploited more than one technique at a time. Although some of these techniques are established in the prior WCT literature, for example analogue and digital SIC methods, for example successive interference cancellation, MMSE etc.; yet integrating these to attain IBFD is a new emerging field. In the following, features relating the integration of these techniques are considered. A further citation of two recently proposed integrations is presented with a brief highlight of these.

3.2.7.1 A platform for integration of IBFD techniques

Table 3.1 summarizes the features of the IBFD techniques. The antenna cancellations and the RF signalling cancellations share many similarities which in general make them replacement alternatives for the designer; but for a higher efficiency, integrating them would be a form of successive SIC in an iterative manner similar to the philosophy used popularly in direct sequence code division multiplexing baseband (digital) successive interference cancellation techniques [36]. The RF suppression is a mutually exclusive alternative to the ACT; more feasible in network applications and relatively longer distances whereas antenna cancellation is feasible both in systems and networks' designs. The digital baseband cancellations

Table 3.1 Comparison of IBFD techniques

Technique	System block	Description	Variants	Integrate-ability
Antenna cancellations	RF frontend; free space propagated EM waves	EM wave cancellation by destructive phase difference addition (free space)	1. Placement of element's spatial coordinate	Design-isolated from other system blocks i.e. integrate-able with other techniques except the RF suppression techniques
RF signalling suppression	RF frontend; free space propagated EM waves	Isolation of EM waves	1. Spatial direction 2. Polarization diversity	Design-isolated from other system blocks i.e. integrate-able with other techniques except the antenna cancellation
RF signalling cancellation	RF frontend; guided EM wave (before down conversion)	EM wave cancellation by destructive phase difference additions in RF domain	1. RF phase (inversion) shift circuits	Design-linked to system blocks, suffers insertion losses and reflected leakages, integrate-able with design complexity
Analogue baseband cancellation	Baseband; before ADC	Analogue electrical signals cancelled by phase difference destructively superimposed	1. Analogue phase shift/attenuators networks	Part of analogue base-band circuitry, integrate-able, design complexity
Digital baseband cancellation	Baseband; after ADC	Digitally coded signals subtracted (filtered) digitally using conventional techniques such as MMSE, LMS, CPE	1. MMSE 2. LMS 3. Common phase error (CPE) 4. Others	Part of digital base-band circuitry, integrate-able, less design complexity

are the commonly incorporated techniques in almost every art of IBFD currently presented. It is less costly in system design and more efficient than the analogue baseband cancellations.

3.2.7.2 A review of recently proposed IBFD hybridized methods

The presented art of [37] exploits hybrid techniques to attain FD. It hybridizes directional RF suppression with polarization RF suppression. Directional antennas are used and shielded with absorptive shielding (use of lossy materials) to attenuate the SI. The transmit and receive antennas operate in orthogonal polarization states, thus experiencing another degree of isolation. This combination substantially improved the SI suppression to a performance exceeding 70 dB, that without yet exploiting a digital or other cancellation stages.

Similarly the art of [38] presents a method hybridizing RF suppression by polarization with analogue cancellations. This presented a practical implementation of a small form-factor design for a mobile handset. The embodiment consisted of two dual polarization antennas (suppression by isolation) in addition to an analogue cancellation stage using an electrically balanced network composed of a hybrid transformer and the balanced network (a dummy load of resistors and capacitors). The transformer functions in a manner similar to the BALUN, however in the baseband domain. The transformer provides a 180-degree phase shifts to all the transmitted images and its nonlinearity products and noise generated in the transmitter. The balanced network reflects over the transformer to imitate the antenna impedance. High precision tuning of this balanced network (reactance plus resistance) will filter out the transmitted signal with a measured performance of more than 50 dB for this stage. A complementary digital cancellation stage improves the SI suppression to above 100 dB. One obvious drawback is the need for tuning, and which also narrows the practical bandwidth.

3.2.7.3 The nulling function [5,6]: a recursive seed to hybrid combinations

Rearranging the null function in (3.11) $\sum_{i=1}^{N}\left((A_{at}+\varepsilon^{A_i})^2\right)-2*\sum_{i=1}^{N-1}\sum_{k=i+1}^{N}(A_{at}+\varepsilon^{A_k})*(A_{at}+\varepsilon^{A_i})*\cos\left((\varepsilon^{\phi_{1i}})-(\varepsilon^{\phi_{1k}})\right)=0$. This function irrespective of the value of X(t) is the theoretical frame for the antenna cancellation method and also can be exploited in the RF cancellations as well (Figure 3.5).

This function can be restated as

$$f_m\left(\varepsilon^{A_m}(f),\varepsilon^{\phi_{1m}}(f)\right)=\sum_{i=1}^{N}f_i\left(\varepsilon^{A_i}(f),\varepsilon^{\phi_{1i}}(f)\right)=0\ldots\text{for } m=1,2,3,\ldots,M$$

where M=index number of nulls. That is, for a prescribed null with known spatial placements, the task would be to find the attenuation vector V_ε and the phase errors vector V_ϕ that will produce a null (static positioning), or to have known values of V_ε and V_ϕ and perturb these to satisfy a zero condition at a null whose spatial placement is reverse computed from V_ϕ (dynamic or adaptive nulling).

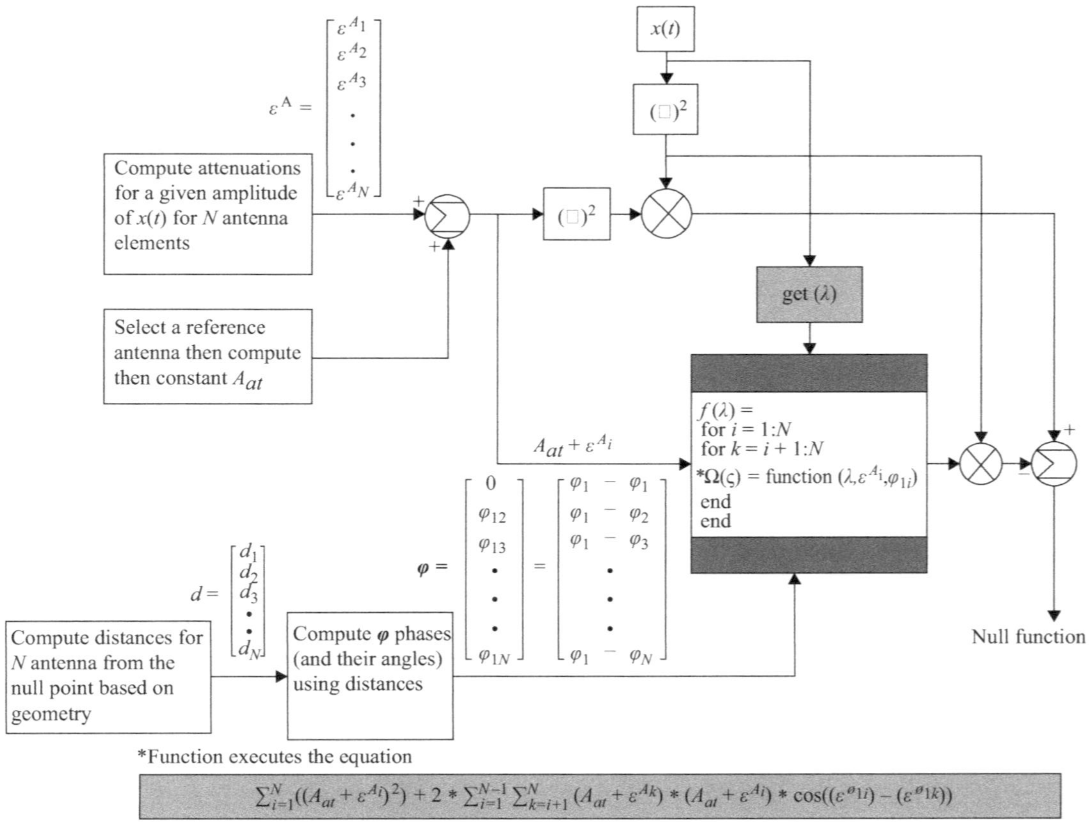

Figure 3.5 Pictorial model of the null function [6]

The function is a series of tapped channels (Figure 3.5) and can be easily implemented in a programmed unit. This can be directly implemented in an antenna cancellation system or in an RF cancellations system or in a hybrid combination of the two using same parameters and outputs. In static positioning, these vectors are computed once in the launching of a system, since the variations in the channels in confinement of the antenna elements experience negligible changes as long as the zone of concern is in the vicinity of the elements. Thus, this null function is a strong candidate for embodiment in hybrid IBFD systems.

3.3 The evolutionary impact of the IBFD techniques on WCT and associated developments

The obvious advantages of the IBFD have been highlighted in the beginning of this chapter. Henceforth, these are categorized according to their fields. Their impact is put to focus to provide an envision into the evolutionary process and model changes; these will add to the existing WCT fields. The IBFD is a front-end system technique; but its feasibility implies modifications in the physical layer protocols, networking protocols, architectures and topologies. This is true for point-to-point systems, intra-network, network-to-network and broadcast links. It is a technique that touches the heart of everything in WCT body.

3.3.1 The IBFD in the 5G networks

The *IBFD* has been incorporated as a fundamental 5G air technology. It is categorized in the advanced transmission technologies, the enabling technologies and included in the Radio Access Network (RAN) technologies. This considered a primary element of technologies intended for 5G. The IBFD is not exactly mature art but is vigorously developing. For example, a minimum value of 136-dB isolation between transmit and received RF chains is required for an outdoor application to function properly [39]. This has not yet been reported for IBFD techniques but the Single Antenna Element IBFD designs of University of Stanford have reported achieving 110 dB, so the gap is within reach. With respect to the field implementation of IBFD, of recent Kumu and the Deutsche Telekom executed realistic FD 5G field trials (September 2015) [40].

These results and the current research rigour are focused on certain areas. The most immediate exploitation is the Network Relaying where IBFD enhances the network communications a great deal as will follow. Protocols' design such as medium access layer (MAC or address layer) and upper hierarchical layers [41], to accommodate FD and the ability to opportunistically switch between FD and half duplex (HD) modes is another thriving area [42]. Less however has been written with respect to the CSI, even though it should have been the most immediate exploitation. On the system level, single-antenna techniques have gained favour, particularly for user equipment, although parallel work is carried in sought of MIMO FD compliant designs [5,11,19]. The IBFD in 5G is expected to function in almost every field, yet as an enabling technology it has an immediate impact on fields including network relaying, CSI, backhauling, cognitive networks, interlayer protocols, energy harvesting techniques etc.

Categorizing the IBFD from the perspective of its being an enabling technique, it entertains three link topologies which repeat themselves in different contexts, for example those relating the fields marked above. These link topologies are the relaying link, the bidirectional link and the multiple-access/broadcast link. In the relaying link, a source and a destination communicate and the relay enhances the communication according to the relaying protocol in use. In A-LTE (and the 5G), this topology is exploited in relaying, backhauling and CoMP. The second topology is the bidirectional topology where only two terminals exchange through a direct mutual link. This is exploited in ad-hoc scenarios and in multiple hub transparent relaying. The third topology, the multiple-access/broadcast link is a multi-point to point/point to multi-point link and is exploited in Base Stations, in the RAN architecture and global CSI frames in CoMP. In the following, the possibilities and potentials of the IBFD are explored and highlighted.

3.3.1.1 IBFD in the network relaying techniques

The IBFD techniques have direct influence on the TDD relaying protocols and the cooperative relaying (in-band, out-of-band, static, random, fixed and dynamic architectures). Since cooperative relaying is quite related to CoMPs (in form), this as well has direct influence on CoMP strategies and architectures.

Considering the three possible TDD relaying protocols relating the source/relay destination communication link trilogy, the possible slot uses are shown in Table 3.2.

In all the three TDD protocols, the second slot used in the main frame is called the relaying slot and is used in the strategy of the protocols. For example in protocol P1, the source transmits in only one slot, whereas the relay receives in that slot and transmits in the second slot. The destination receives in both slots. The merit here is the destination benefits from diversity to improve sensitivity as the combining of the two versions improves the SNR.

Focus has been subjected here on the TDD, since FDD by default forfeits the IBFD philosophy. Evaluating IBFD, it is explicit and obvious; the implementation of IBFD in relaying enhances the efficiency to almost a double since the function can be achieved in one slot. However, this is just a point in an overwhelming flood of possibilities. Enumerating some of these, the multi-hob relaying and isolation

Table 3.2 TDD in-band relaying protocols for type 1 relay

P1	**Slot 1**	**Slot 2**	**P2**	**Slot 1**	**Slot 2**	**P3**	**Slot 1**	**Slot 2**
Source	Tx		Source	Tx	Tx	Source	Tx	Tx
Relay	Rx	Tx	Relay	Rx	Tx	Relay	Rx	Tx
Destination	Rx	Rx	Destination		Rx	Destination	Rx	Rx

through different hobs' directions have been the subject of many recent researches, for example [43–46]. These do not stand among the hot topics in relaying but when examined in the perspective of feasibility of IBFD, their applications become vital. It is direct to visualize this: for example, the complications of multi-hob designs reduce into a single slot link design and minor latency concerns in the forwarding process. Another area of practical implementations relates the interlayer and cross layer optimization [41] in relaying as a group of nodes (network relaying) or an individual node to node communication. On protocol level, the singularity of channel simplifies the handshakes and related features a great deal. Alternatively, decode and forward (DF) scheme of cooperation, between evolved Node B (eNB) and Relay, may be exploited using the extra available slot for processing the coding. In general, this makes it possible to attain both the performance advantage of IBFD, whereas SNR and cut bound capacity are likely to improve.

Considering capacity for example, whether the relay is static or random, fixed or dynamic, the cut-set bound capacities are usually defined in terms of a time variable t. If the time-slot for a source to destination link is expressed as 100% then t is a variable between 0 and 100% whose value is determined by the HD slot allocations. All known capacity equations relating the broadcast scenario of relay link are optimized for optimized t. In IBFD this variable t is maximized to 100% and implies maximum cut-set capacities. t is optimized to 100% since the communication is now FD. And the achievable rates, also dependent on this parameter t, are both improved and reduced in complexity being a convex optimization problem for that sake.

Another relevant application of IBFD would be realized in considering hybridization of the protocols, for example in [47] it is mentioned that protocol P3 even though efficient with DF and code and forward strategies but is not efficient when *multiple relays* are used. If IBFD is used in the first slot to communicate between cooperative relays, the control signals, to attain joint transmission as a MIMO frame, whereas in the second slot's destination receives multiple transmissions from multiple relays but optimized to cancel inter-relay interference. Thus the benefits of P3 are exploited even in the case of multiple relays.

The applications of the IBFD technique however are not limited to the full realization of perfect IBFD; which is highly probable to come valid along the way. An immediate application of the technique can lend itself to the problem of self-interference in two-way relaying [48] where at least the minimum amount of SIC may improve the performance of the two-way relays.

Yet another example, IBFD provides an inquisitive perspective about the coding and precoding techniques. For example, in the up-link (UL) from UE to relay (in MAC), saving the IBFD time slot, in combination with a network coding technique (e.g. [48]) can help provide a transparent frame that makes all users dependent on a relay node as a one-point link (issue of hidden nodes).

The IBFD has opened so many avenues indeed. In essence, the ongoing rapid enhancements on the IBFD technology opens wide avenues to reshaping of existing TDD-related protocols and the relaying theory as a whole.

3.3.1.2 IBFD in the channel state information acquisition techniques

CSI is an essential backbone to cooperative and adaptive schemes such as relaying and CoMP, likewise to its being essential in the per node context in the conventional cells and adaptive coding and modulation techniques. The topic of CSI is of enormous potential to many parameters and network applications such as Capacity, CoMP, Relaying, Backhauling etc. Research has been carried in abundance on the topic, for example [49–52]. The CSI requires separation in domains either as FDD or TDD. FDD is usually associated with complexities in the nature of the feedback signal such as the low correlation between the UL and DL responses and asymmetry of the streams [53,54]. On the other hand, TDD CSI uses two time slots to send and receive information on the same frequency channel and assume that principle of channel reciprocity is valid within the transmission period. Efficient CSI is evaluated by virtue of how fast it returns feedback before the channel condition changes. TDD CSI is to a great extent of preference in CoMP and Cooperative relaying in 4G and A-LTE since it is asymmetrical and which allows different CSI rates communicated on the basis of the hierarchy of the link. For example the UE gives less data than would a relay than would an eNB. IBFD gives a promise of designing efficient CSI schemes that react to network instantaneously.

Considering the current status of IBFD as a rising technique, the operation of CSI functionality does not require as big bandwidth as would the control channels, regenerative relaying links, backhaul and data traffic. It will likely require a less complex design of dedicated IBFD receivers used only for CSI in cooperation with the network. The idea here is that these can be used in the forbidden periods, for example when relay is transmitting and the eNB is scheduled to receive. With IBFD an eNB for an example can communicate CSI using a single module for this purpose providing the relay its CSI during the transmission slot. The IBFD design requirements are less for such a strategy, since the technique is new and it will be more practical to consider using it with less constraints. It is worth remarking here that it leads to huge unrealistic bandwidth requirements if sought is to deploy IBFD in the direct networking communications considering the current state of the art.

The suggestion sought here is to design a complementary CSI architecture based on IBFD to enhance the different levels of performance such as efficient regenerative relaying, robust coding and precoding strategies, partial and full interference coordination, for example coordinated power control and/or coordinated beamforming etc. which are known constraints for the relaying and CoMP technologies [55]. The focus of this falls within the 5G frame of work and could be pursued with attention to the backward compatibility to existing CSI architectures (e.g. working on pre-matrix code indicator/CQI/rank indicator frames) that is the benefits of IBFD are to be utilized rather without necessarily changing the existing architectures except where most necessary.

Scenarios for such applications are many. For say, a situation where two messages of CSI are coordinated such that during the original architecture CSI time slots, a vector quantized [53] message is transmitted in the conventional transmission slots (forward stream), whereas during the feedback reception slots instantaneous full CSI is transmitted in the opposite direction (IBFD mode), and both CSI forwarded messages are correlated to give better estimates (something similar to Hybrid-Automatic Repeat Request – H-ARQ – design philosophy in use with forward error coding). Another scenario is in answering to the problem of sharing CSI between relay nodes in cooperative mode. Unlike the eNB link to UE, where the overhead is reduced by provisioning a 'sounding zone' [53], during which all UE(s) transmit full message CSI and the eNB assumes TDD reciprocity, this cannot be achieved in relaying context with multi-hops simply because TDD reciprocity is defeated. This technique can be implemented here using the downlink (eNB to Relays) as a 'sounding zone" implementing IBFD modules at relays and eNB.

Many avenues are fertile and valid here; another example yet is found in the implementation of multi-hop relaying of CSI using amplify and forward; for example in an ad-hoc like manner through different CSI links, when communicating global CSI in particular and the local CSI in general. And as for precoding, perfect CSI within channel TDD reciprocity at transmitter side gives roads to excellent precoding techniques and power allocation algorithms.

3.3.1.3 IBFD in the backhauling techniques

Backhauling is the parallel system that coordinates the exchange of network controls, operational information such as beamforming weights, CSI, CoMP coordination messages etc. The sensitivity of this system is measured on two factors; controls and information that need to be instantaneously coordinated and the size of network information communicated. Both factors have impact on influential parameters such as network latency, network capacity, mobility parameters, congestions, slot size etc. and which all influence the efficiency at the end user and over the whole network. The IBFD provides the two needed features – instantaneousness and doubling of bandwidth. Techniques like CoMP will be extremely enhanced when excellent backhauling system is provided. Schemes, such as joint decoding, joint coding, joint beamforming, etc., depend on the capacity of the backhaul and amount of shared information and time coherency of the information received. Excellent CSI that is instantaneous, detailed, full or almost full CSI, provides for enhanced time coherency of the global channel. The IBFD provides means to replace the expensive wired/optical backhaul structures with a more cost efficient, easier to deploy wireless structures. A good research relating the topic in 5G contexts is presented in [56].

3.3.1.4 IBFD in cognitive networks

The IBFD provides extra time slot per link since it enables transmission and reception at the same time slot. Thus during the UL and during the downlink the IBFD units communicate in dual mode that is transmit while receiving or receive while transmitting. Cognitive radio is an access technique enabling optimized sharing between users, and this is different from IBFD, being an enabling technique. Yet technology is about finding useful features to exploit. Cheng *et al.* [57] illustrated such an approach. The sharing protocols are improved if the extra time slots are used. The scheduling information such as availability of spectrum or request for use of spectrum can be communicated in these slots. This is more or less a concept of opportunistic IBFD. Once again the IBFD illustrates potentials for shaping all the WCT applications and the networking protocols and techniques.

3.3.1.5 IBFD and the energy harvesting techniques

The concept of energy harvesting is one of the thriving aspects of 5G, and which falls within the category of energy aware communications and green communications in general. The focus is on reuse of energy. RF Energy Harvesting Networks ('RF-EHN'-in the 5G nomenclature) are those networks which capture and store RF energy in the near field and those received through the far field. The objective is reuse of the energy captured as a way of preserving resources. The RF domain involves several energy conversion interfaces and processes (e.g. reception and transmission), and these processes are dispersive in nature. Dispersive link is the opposite of a deterministic link where in the deterministic link the energy flow has definite routes and definite targets and specific design constraints, that is the exact energy needed is consumed by the link entities. The reception process usually comprises energy reception and information reception simultaneously. There is intermittency in the information reception process during which the sought is to harvest the dispersed far-field RF energy by means such as inductive coupling, capacitive coupling, magneto dynamic coupling etc. Same means are utilized during the transmission process in the near field to harvest self-looped energy. However, the harvest can be also attained even when the communication process is active. Harvest can be affected in the in-band spectrum (i.e. in the same RF frequency) or the out of band spectrum (i.e. responsive to any EM frequency). The dispersive nature of the communication link however does not cease when using IBFD. Rather the energy crop is doubled by virtue of duplicity of the exchanged energy and increase in number of auxiliary elements (when using antenna cancellations); for example, Mohammadi *et al.* [58] examine a time switched harvest of energy in an IBFD relayed MIMO design. Also the SIC involves near-field loops which are excellent harvest resources, for example Maso *et al.* [59] add an energy harvester circuit between circulators and receiver chain during the suppression/cancellation algorithm. Once again IBFD illustrates leverage beyond just an enabling technique.

3.3.1.6 IBFD and the shaping of protocols

In view of these enormous possibilities, the IBFD is a physical layer technique whose impact influences all the protocol layers. In essence, time variable is a basic constituent of all existing protocols, and IBFD provides a whole free time-slot all around. A two-way hand shake principle becomes an immediate one way process in IBFD perspective. An ARQ control receives immediate response and feedback in the IBFD promise. Without IBFD only one link in each transmission in a neighbourhood is possible, because there is a need for time slot to listen and avoid collision. The IBFD allows many neighbours and links to coexist. It is a global revolution in WCT with many degrees of freedom that can be leveraged to manipulate existing protocols.

The core concept of most of the previous headings related modifications in the physical layer. The MAC protocols, however, are very sensitive and connected to any modifications in the physical layer. The brilliant recent survey made in [60], and which is an excellent starting point for a researcher, has reported a collection of proposals for MAC protocol modifications in IBFD. Table II (page 23) of [60] presents an excellent summary of these techniques and solutions and their characterizing features. Issues of concern included centrality of MAC protocols for network infrastructures and distributed MAC for ad-Hoc links and networks.

In network infrastructures, the backhaul network controls link the network entities in many joint activities; that is, the functional units are not isolated in the communication processes but exercise inter-dependency in functionalities. This relates more to the IBFD multiple-access/broadcast topology and the relaying topology where more than one user accesses an access point simultaneously. This simultaneousness results in the known problem of inter-user interference in the multiple access designs. The conventional HD resource allocation techniques therefore need a reshape; that calls for the centralism of MAC protocols. Reference [60] cited in References [4,61,62] to treat centralism issues in IBFD MAC protocols. For the asymmetric traffic and the hidden node problems, busy tone signalling is suggested in [4]. Continuous transmission in IBFD causes the phenomena of node starvation whereby nonstop communication of connected nodes, consume the resources which else could be leveraged during the cease of transmission/reception in HD links. Riihonen *et al.* [61] propose an opportunistic three-element IBFD scheme consisted of: shared random back-off, snooping to discover IBFD active transmissions and virtual contention resolution. Fukumoto and Bandai [62] enhanced a prior art optimized opportunistic IBFD scheme exploiting spatial resources. These research works treated ideal situations of either a fully hidden node or a fully conflicting node.

Kim *et al.* [63] related solutions to the practical situation of partial interference, through scheduling a hybrid (FD/HD) transmission protocol. It proposed the ‘Janus’ protocol which reduces collisions by a control algorithm which controls the packets’ transmission rate and timing accordingly. In addition, The Janus proposal also covers fairness issues and a policy to acknowledge received packets per cycle.

Scheduling issues and resource allocation in these IBFD link topologies have been researched in [64–67] and partially in [63]. Di *et al.* [64] was cited to relate the resource allocation as a joint optimization problem and a subcarrier matching problem; the latter is solved by using the ‘matching theory’. On another hand, Cheng *et al.* [65] researched optimum power allocations for a given quality of service delay constraint, whereas Liu *et al.* [66,67] developed an energy-efficient resource algorithm for orthogonal frequency division multiple access networks.

When considering ad-hoc networks, the link topology in use is the bidirectional topology; since there is no multiple access to the same point and eluding in this classification, the IBFD transparent ad-hoc inter-node communications which are basically a relaying topology. The MAC protocols in HD convention are distributive controls where each node is unaware of the transmission parameters and modes of the neighbours. This lack of means for coordination call for a new class of distributed

MAC protocols to accommodate IBFD. Challenges immediately rise in this sought, for example fairness with respect to the existing collision avoidance (CA) protocols such as the notorious carrier sensing multiple access/CA protocol. Kim *et al.* [60] cite a collection of useful research works and developed solutions relating distributed MAC protocols. To begin with, among others, the authors in [3,68,69] reported the advantages of IBFD from the perspective of the distributed MAC protocols such as solution to hidden node problem and eluding the handshaking process and associated delays. Radunovic *et al.* [68] introduced the Contraflow protocol which preceded the Janus proposal. It is an SIC-based solution that optimizes the spatial re-use. The major difference of it with Janus is that it is designed for a symmetric a-centric approach. Radunovic *et al.* [68] also explained the Contraflow-IBFD solution to the 'exposed node problem'. The 'exposed node problem' relates the performance of two nodes separated by long distance such that that incoherency and inconsistency of the channel parameters obliterate the conventional handshaking algorithms and break down the link.

Symmetric traffic environment is treated in [68,70,71]. A-symmetric traffic environment is treated in the previous references and more focused in [72]. Goyal *et al.* [73] treated the issue of inter-node interference when multiple nodes are communicating together.

These cited above and more yet illustrate the extensive research and rigorous developments in IBFD-compliant MAC layer protocols and solutions. Another protocol area which can leverage the technique is security area. IBFD provides excellent flexibility for security protocols; a one hot topic that will draw attention immediately and is expected to present many useful researches. Considering philosophies merging 5G with IoTs and the drift towards packet oriented communications, IBFD is a key technique towards a fully packet oriented networks as it eludes collusion, network latency, higher layer routing choices etc. There is an ongoing argument in the 5G as to whether IBFD should be transparent to upper protocol layers and confined within the physical and MAC layers so as to provide backward compatibility –OR– it should be extended to upper layers protocols to enhance resource managements in every level. It is, however, definite the technique impact is revolutionary and implies a reshape process to most protocols.

Further readings

In addition to the cited Reference [60], another parallel survey, though with more inclination towards the relaying aspect, is found in [74].

3.3.1.7 IBFD and the cloud/fog network computing

With developments in the concept of IoT and its convergence with 5G platforms, the decentralized programming techniques exploit the 5G architecture whether at the edge of the network as Fog or as centralized distributed network processing as in the cloud. Coordination of these with IBFD is direct to visualize since IBFD influences the MAC and routing protocols and backhauling techniques and the cloud access. Simeone *et al.* [75] related the Cloud Radio Access Networks (C-RAN), a class of cloud-based architectures proposed for the 5G and illustrated improved performance when merged with IBFD technique. The IBFD provides cloud with physical bandwidth, mitigated latency and more efficient and secure data handling protocols. The incorporation of IBFD protocols also comes in powerfully in the network virtualization protocols and provides an efficient security platform for that cause. This feature lends itself directly to cloud computations [76].

3.3.2 The potentials and deficiencies of the single antenna IBFD

IBFD can be attained using a single-antenna element interface using RF isolation techniques, whereby the transmission RF chain is isolated from the reception RF chain. This approach has actually resulted in the highest reported results of SIC techniques (up to 110 dB [15,19]). This approach has a collection of thrilling advantages. First and most important is efficiency and higher performance. Then the compactness of size, which readily gains favour in UE designs. The independence of the antenna element as a compact unit calls for questioning the feasibility of MIMO frame of work. This however is abstained by the cross-talk impairment and which requires conventional digital cancellation techniques. Yet as the number of antenna units increase the number of associated digital cancellation unit increases quadratically. The cascaded cancellation designs suggested an answer to this [19]. So in essence, both FD and MIMO benefits are feasible. The bandwidth of the antenna unit depends on the isolation technique and the antenna design. But even with an excellent antenna design and a relatively good range of bandwidth isolation techniques (e.g. the Electric Balancers [27]), the IBFD performance deteriorates for wideband applications. The sought therefore is for devising new ways to integrate the isolation into multi-element structures so as to obtain MIMO performance and IBFD in wide band and long-range design. The focus here on this technique is due to the fact that it is practically the lead technique for implementing IBFD with the best reported results.

3.4 Conclusion

The huge amounts of information transported over current and future wireless networks, call for efficient use of the limited spectrum. Any method, traditional or new, that provides a means of spectrum saving is certainly needed and will definitely help with this. Multi-level modulation, MIMO and IBFD techniques are examples of methods of efficient spectrum utilization. IBFD is a new method that waits to be applied in practical systems.

The merits of IBFD are immediate to observe. However, the attainment of IBFD is expensive in cost. All IBFD techniques require complementary stages; non (no stage) is a standalone technique that accomplishes the objective without further complementary stages. Even the techniques that relatively stand on their own (e.g. the single antenna element IBFD) face complications when trying to integrate their implementation with other useful techniques (the MIMO for example). Comparatively, the IBFD on its own is not sufficient to replace other techniques, it is more efficient to incorporate it in existing techniques rather than replace them. For example, if selection is to be made between MIMO and IBFD, MIMO would have the priority, since it delivers *N* times the normal bandwidth (*N* orthogonal channels), whereas IBFD at best doubles it. On the other hand, ability to integrate IBFD in MIMO frame enhances the capacity, mitigates the delay and delivers flexible design criterion. Hence, there is a compromise to make between the complexity associated with IBFD designs, and the cost of many extra components as against the delivered performance, cost effectiveness and cheaper and simpler alternatives.

IBFD has certain unique features that make it indispensable and irreplaceable in certain applications. The instantaneous duality and exclusion of delays provide the uniqueness when considering applications such as CSI and backhauling techniques. Continuity of the link in transparent relay nodes is another unique feature that is very promising in the relaying context. These unique features may justify the expensive cost and design complexity in the associated scenarios.

What is not disputable is that, if practical limitations of IBFD are overcome, the IBFD qualifies indisputably as the most influential technique on the contemporary WCT fundamentals. The technique is quickly developing and active research work has revealed so many routes and yet more is coming forth.

References

[1] El Sanousi, G. T. A. "A concise review of the space time premise and its impact on modern wireless communication theory." *Wireless and Optical Communications Networks, 2009. WOCN'09. IFIP International Conference on*. IEEE, 2009.

[2] Everett, E., *et al.* "Empowering full-duplex wireless communication by exploiting directional diversity." *Signals, Systems and Computers (ASILOMAR), 2011 Conference Record of the Forty Fifth Asilomar Conference on*. IEEE, 2011.

[3] Choi, J. I., Jain, M., Srinivasan, K., Levis, P., Katti, S. "Achieving single channel, full duplex wireless communication." *Proceedings of 16th Annual International Conference on Mobile Computing and Networking*. Mobicom, 2010.

[4] Jain, M., Choi, J. I., Kim, T., *et al.* "Practical, real-time, full duplex wireless." *Proceedings of 17th Annual International Conference on Mobile Computing and Networking*. Mobicom, 2011.

[5] Geili, T. A., Sanousi, E., Franz, H., Mohammed A. H. Abbas. "On The Antenna Cancellation Technique When Exploiting Array Geometries (1): The Basic Hexagonal Star Antenna Array Geometry," Unpublished.

[6] Geili, T. A., Sanousi, E., Franz, H., Mohammed A. H. Abbas. "On The Antenna Cancellation Technique When Exploiting Array Geometries (2): Bandwidth-Nested and Multiplicity-Nested Hexagonal Star Antenna Array Geometries," Unpublished.

[7] Balanis, C. A. *Antenna theory: analysis and design*. John Wiley & Sons, USA, 2005. Pages 34 and 351.

[8] Jull, E. V. "Aperture antennas and diffraction theory." No. 10. *IET* (1981). Page 19.

[9] Steyskal, H. "Synthesis of antenna patterns with prescribed nulls." *IEEE Transactions on Antennas and Propagation* 30.2 (1982): 273–279.

[10] Ma, Z., *et al.* "Key techniques for 5G wireless communications: network architecture, physical layer, and MAC layer perspectives." *Science China Information Sciences* 58.4 (2015): 1–20.

[11] Khandani, A. K. "Two-way (true full-duplex) wireless." *Information Theory (CWIT), 2013 13th Canadian Workshop on*. IEEE, 2013.

[12] Haneda, K., Kahra, E., Wyne, S., *et al.* "Measurement of loop-back interference channels for outdoor-to-indoor full-duplex radio relays." *Proceedings of the Fourth European Conference on Antennas and Propagation*. Barcelona, 2010. 1–5.

[13] Riihonen, T., Werner, S., Wichman, R. "Mitigation of loopback self-interference in full-duplex MIMO relays." *IEEE Transactions on Signal Processing* 59 (2011): 5983–5993.

[14] Hua, Y. B., Liang, P., Ma, Y. M., *et al.* "A method for broadband full-duplex MIMO radio." *IEEE Signal Processing Letters* 19 (2012): 793–796.

[15] Knox, M. E. "Single antenna full duplex communications using a common carrier." *Proceedings of IEEE 13th Annual Wireless and Microwave Technology Conference*. Cocoa Beach, 2012. 1–6.

[16] Steyskal, H. "Array pattern synthesis with near-field nulls." *Antennas and Propagation Society International Symposium, 1994. AP-S. Digest*. Vol. 1. IEEE, 1994.

[17] Steyskal, H., Shore, R., Haupt, R. L. "Methods for null control and their effects on the radiation pattern." *IEEE Transactions on Antennas and Propagation* 34.3 (1986): 404–409.

[18] Bharadia, D., McMilin, E., Katti, S. "Full duplex radios." *ACM SIGCOMM Computer Communication Review* 43.4 (2013): 375–386.

[19] Aryafar, E., *et al.* "MIDU: enabling MIMO full duplex." *Proceedings of the 18th Annual International Conference on Mobile Computing and Networking*. ACM, 2012.

[20] Etellisi, E. A., Elmansouri, M.A., Filipovic, D. S. "Wideband monostatic simultaneous transmit and receive (STAR) antenna." *IEEE Transactions on Antennas and Propagation* 64.1 (2016): 6–15.

[21] Duarte, M., Sabharwal, A. "Full-duplex wireless communications using off-the-shelf radios: feasibility and first results." *Proceedings of Forty Fourth Asilomar Conference on Signals, Systems and Computers*. Pacific Grove, 2010. 1558–1562.

[22] Phungamngern, N., Uthansakul, P., Uthansakul, M. "Digital and RF interference cancellation for single-channel full duplex transceiver using a single antenna." *Proceedings of 10th International Conference on Electrical Engineering/Electronics, Computer, Telecommunications and Information Technology*. Krabi, 2013. 1–5.

[23] McMichael, J. G., Kolodziej, K. E. "Optimal tuning of analog self-interference cancellers for full-duplex wireless communication." *Proceedings of 50th Annual Allerton Conference on Communication, Control, and Computing*. Allerton, 2012. 246–251.

[24] Gholian, A., Ma, Y., Hua, Y. "A numerical investigation of all-analog radio self-interference cancellation." *Signal Processing Advances in Wireless Communications (SPAWC), 2014 IEEE 15th International Workshop on*. IEEE, 2014.

[25] Wang, J., Zhao, H., Tang, Y. "A RF adaptive least mean square algorithm for self-interference cancellation in co-frequency co-time full duplex systems." *Communications (ICC), 2014 IEEE International Conference on*. IEEE, 2014.

[26] Zhang, Z., *et al*. "Differential evolution algorithm based pilot aided weighted data fusion scheme for uplink base-station cooperation system." *Wireless Communications and Signal Processing (WCSP), 2014 Sixth International Conference on*. IEEE, 2014.

[27] Laughlin, L., *et al*. "Optimum single antenna full duplex using hybrid junctions." *IEEE Journal on Selected Areas in Communications* 32.9 (2014): 1653–1661.

[28] Brett, K., Jorma, L., Behnaam, A. "An analog baseband approach for designing full-duplex radios." *Proceedings of 2013 Asilomar Conference on Signals, Systems and Computers*. Pacific Grove, 2013. 987–991.

[29] Li, S. H., Murch, R. D. "Full-duplex wireless communication using transmitter output based echo cancellation." *Proceedings of IEEE Global Telecommunications Conference*. Houston, 2011. 1–5.

[30] Li, N., Zhu, W. H., Han, H. H. "Digital interference cancellation in single channel, full duplex wireless communication." *Proceedings of 8th International Conference on Wireless Communication, Networking and Mobile Computing*. Shanghai, 2012. 1–4.

[31] Ahmed, E., Eltawil, A. M., Sabharwal, A. "Self-interference cancellation with phase noise induced ICI suppression for full-duplex systems." *Proceedings of IEEE Global Telecommunications Conference*. Atlanta, 2013. 3384–3388.

[32] Anttila, L., Korpi, D., Syrjala, V., *et al*. "Cancellation of power amplifier induced nonlinear self-interference in full-duplex transceivers." *Proceedings of Asilomar Conference on Signals, Systems and Computers*. Pacific Grove, 2013. 1193–1198.

[33] Ahmed, E., Eltawil, A. M., Sabharwal, A. "Self-interference cancellation with nonlinear distortion suppression for full-duplex systems." *Proceedings of Asilomar Conference on Signals, Systems and Computers*. Pacific Grove, 2013. 1199–1203.

[34] Korpi, D., Anttila, L., Syrjala, V., *et al*. "Widely-linear digital self-interference cancellation in direct-conversion full-duplex transceiver." *IEEE Journal on Selected Areas in Communications* 32 (2014): 1674–1687.

[35] Korpi, D., Anttila, L., Valkama, M. "Feasibility of in-band full-duplex radio transceivers with imperfect RF components: analysis and enhanced cancellation algorithms." *Proceedings of Ninth International Conference on*

Cognitive Radio Oriented Wireless Networks and Communications. Oulu, 2014. 532–538.

[36] Holtzman, J. M. "DS/CDMA successive interference cancellation." *Spread Spectrum Techniques and Applications, 1994. IEEE ISSSTA' 94, IEEE Third International Symposium on*. IEEE, 1994.

[37] Everett, E., Sahai, A., Sabharwal, A. "Passive self-interference suppression for full-duplex infrastructure nodes." *IEEE Transactions on Wireless Communications* 13 (2014): 680–694.

[38] van Liempd, B., Debaillie, B., Craninckx, J., *et al.* "RF self-interference cancellation for full-duplex." *Proceedings of Ninth International Conference on Cognitive Radio Oriented Wireless Networks and Communications*. Oulu, 2014. 526–531.

[39] "Understanding 5G" page 34; a white paper by Anristu.com published at http://www.goanritsu.com/5G, visited in 8 January 2017.

[40] Available at http://www.sktelecom.com/en/press/detail.do?idx=1119, visited on 8 January 2017.

[41] Sahai, A., Patel, G., Sabharwal, A. "Pushing the limits of full-duplex: design and real-time implementation." *arXiv preprint arXiv:1107.0607* (2011).

[42] Karakus, C., Diggavi, S. "Opportunistic scheduling for full-duplex uplink–downlink networks." *arXiv preprint arXiv:1504.05898* (2015).

[43] Fang, X., Yang, D., Xue, G. "Distributed algorithms for multipath routing in full-duplex wireless networks." *Eighth IEEE International Conference on Mobile Ad-Hoc and Sensor Systems*, 2011.

[44] Ju, H., *et al.* "Full duplexity in beamforming-based multi-hop relay networks." *IEEE Journal on Selected Areas in Communications 30.8* (2012): 1554–1565.

[45] Sekhar, K. K. Simultaneous Signaling and Channel Estimation for In-Band Full-Duplex Communications Employing Adaptive Spatial Protection. *Dissertation, Arizona State University*, 2014.

[46] Song, Y., *et al.* "Lattice coding for the two-way line network." *IEEE Journal on Selected Areas in Communications* 32.9 (2014): 1707–1720.

[47] Simoens, S. Cooperative MIMO Communications: Information Theoretical Limits and Practical Coding Strategies. *PhD thesis, Technical University of Catalonia* (UPC), Jan 2008. Ch. 5.

[48] Li, J., Hu, Z., Wang, Y. "DF/AF cooperative relay in LTE-A", *IEEE 71st Vehicular Technology Conference*. VTC 2010-Spring, 2010.

[49] George, E. M., *et al.* "Performance analysis of multi-antenna multi-hop relay networks using antenna selection with delayed channel state information at the transmitter." *Communications and Signal Processing (ICCSP), 2015 International Conference on*. IEEE, 2015.

[50] Jiang, Q., *et al.* "Overlaid-pilot based channel state information feedback for multicell cooperative networks." *Personal, Indoor, and Mobile Radio Communications (PIMRC), 2015 IEEE 26th Annual International Symposium on*. IEEE, 2015.

[51] Love, D. J., *et al.* "An overview of limited feedback in wireless communication systems." *IEEE Journal on IEEE Journal on Selected Areas in Communications 26.8* (2008): 1341–1365.

[52] Komulainen, P., *et al.* "CSI acquisition concepts for advanced antenna schemes in the WINNER+ project." *Future Network and Mobile Summit, 2010*. IEEE, 2010.

[53] Wang, Y., Frattasi, S., Sorensen, T.B., Mogensen, P.E. "Network time-synchronization in TDD based LTE-advanced systems." *IEEE 69th Vehicular Technology Conference*, 2009. VTC Spring, 2009.

[54] Frank, P., Müller, A., Speidel, J. "Performance of CSI-based multi-user MIMO for the LTE downlink." *Proceedings of the 6th International Wireless Communications and Mobile Computing Conference*. ACM, 2010.

[55] Tran, T.-T., Shin, Y., Shin, O.-S. "Overview of enabling technologies for 3GPP LTE-advanced." *EURASIP Journal on Wireless Communications and Networking* 2012.1 (2012): 1–12.

[56] Pitaval, R.-A., *et al.* "Full-duplex self-backhauling for small-cell 5G networks." *Wireless Communications, IEEE* 22.5 (2015): 83–89.

[57] Cheng, W., Zhang, X., Zhang, H. "Full duplex wireless communications for cognitive radio networks." *arXiv preprint arXiv:1105.0034* (2011).

[58] Mohammadi, M., *et al.* "Full-duplex MIMO relaying powered by wireless energy transfer." *arXiv preprint arXiv:1505.00080* (2015).

[59] Maso, M., *et al.* "Energy-recycling full-duplex radios for next-generation networks." *IEEE Journal on Selected Areas in Communications* 33.12 (2015): 2948–2962.

[60] Kim, D., Lee, H., Hong, D.-K. "A survey of in-band full-duplex transmission: from the perspective of PHY and MAC layers." *IEEE Communications Surveys & Tutorials* 17.4 (2015): 2017–2046.

[61] Riihonen, T., Werner, S., Wichman, R. "Hybrid full-duplex/half-duplex relaying with transmit power adaptation." *IEEE Transactions on Wireless Communications 10.9* (2011): 3074–3085.

[62] Fukumoto, M., Bandai, M. "MIMO full-duplex wireless: node architecture and medium access control protocol." *Proc. Seventh International Conference on Mobile Computing and Ubiquitous Networking*. 2014. 76–77.

[63] Kim, J. Y., Mashayekhi, O., Qu, H., Kazadiieva, M., Levis, P. "Janus: a novel MAC protocol for full duplex radio." *CSTR*, 2.7 (2013): 23.

[64] Di, B., Bayat, S., Song, L., Li, Y. "Radio resource allocation for full-duplex OFDMA networks using matching theory." *Proc. IEEE Conference on Computer Communications Workshops*. 2014. 197–198.

[65] Cheng, W., Zhang, X., Zhang, H. "QoS driven power allocation over full-duplex wireless links." *Proc. IEEE International Conference on Communications*. 2012. 5286–5290.

[66] Liu, G., Ji, H., Yu, F. R., Li, Y., Xie, R. "Energy-efficient resource allocation in full-duplex relaying networks." *Proc. IEEE International Conference on Communications (ICC)*. 2014. 2400–2405.

[67] Liu, G., Yu, F., Ji, H., Leung, V. "Energy-efficient resource allocation in cellular networks with shared full-duplex relaying." *IEEE Transactions on Vehicular Technology* 64.8 (2015): 3711–3724.
[68] Radunovic, B., Gunawardena, D., Proutiere, A., Singh, N., Balan, V., Key, P. "Efficiency and fairness in distributed wireless networks through self-interference cancellation and scheduling." Technical Report MSR-TR-2009-27. Microsoft Research, 2009.
[69] Miura, K., Bandai, M. "Node architecture and MAC protocol for full duplex wireless and directional antennas." *Proc. IEEE 23rd International Symposium on Personal Indoor and Mobile Radio Communications (PIMRC).* 2012. 369–374.
[70] Singh, N., Gunawardena, D., Proutiere, A., Radunovic, B., Balan, H., Key, P. "Efficient and fair MAC for wireless networks with self-interference cancellation." *Proc. International Symposium on Modeling and Optimization in Mobile, Ad Hoc and Wireless Networks (WiOpt).* 2011. 94–101.
[71] Kim, S., Stark, W. E. "On the performance of full duplex wireless networks." *Proc. 47th Annual Conference on Information Sciences and Systems (CISS).* 2013. 1–6.
[72] Cheng, W., Zhang, X., Zhang, H. "RTS/FCTS mechanism based full duplex MAC protocol for wireless networks." *Proc. IEEE Globecom Workshops (GC Wkshps)*. 2013. 5017–5022.
[73] Goyal, S., Liu, P., Gurbuz, O., Erkip, E., Panwar, S. "A distributed MAC protocol for full duplex radio." *Proc. 47th Asilomar Conference on Signals, Systems and Computers*. 2013. 788–792.
[74] Liu, G.-P., *et al.* "In-band full-duplex relaying: a survey, research issues and challenges." *IEEE Communications Surveys & Tutorials* 17.2 (2015): 500–524.
[75] Simeone, O., Erkip, E., Shamai, S. "Full-duplex cloud radio access networks: an information-theoretic viewpoint." *Wireless Communications Letters, IEEE* 3.4 (2014). 413–416.
[76] Liu, G., *et al.* "In-band full-duplex relaying for 5G cellular networks with wireless virtualization." *Network, IEEE* 29.6 (2015): 54–61.

Chapter 4

Latency delay evaluation for cloudlet-based architectures in mobile cloud computing environments

Hayat Routaib, Essaid Sabir, Elarbi Badidi and Mohammed Elkoutbi

The vision of ubiquitous computing in interactive mobile cloud applications and Internet of Things (IoT) based systems is still difficult to achieve. The difficulty lies in the use of cloud services in mobile devices, which impacts the issues of performance, scalability, availability, and lack of resources in mobile computing environments. Despite the astonishing advancement achieved in IoT technology, there is still much to do. Some IoT-based systems, which rely on a variety of mobile devices, need to work even when the connection is temporarily unavailable or under-degraded. Besides, mobile cloud service providers can reduce network latency by moving some of their services close to the user. To cope with this challenge, we propose in this chapter the usage of small clouds known as cloudlets, and we describe two cloudlet-based architectures, which allow leveraging the geographical proximity of cloud services to mobile users. We model the network latency of the different components of the two architectures using a continuous-time Markov chain (CTMC). These components are essentially the user nodes, the cloudlets, and the principal cloud. For each architecture, we simulate queries submitted by mobile users to a search engine, and we estimate the incurred delay by using the CTMC state models.

4.1 Introduction

Accessing information at any moment and place was a dream for many years since the emergence of computer science. With the current proliferation of wireless broadband networks and the impressive progress in mobile computing and cloud computing, mobile cloud computing (MCC) is being considered as the most promising technology for achieving this goal. Nowadays, users worldwide access their e-mail, the web, and many other services while they are on the move using their laptops, smartphones, tablets, and other mobile devices. Nevertheless, mobile devices are facing many challenges as they lack required capabilities regarding storage, battery life, and bandwidth [1]. Limited resources hamper the quality of services (QoS) significantly.

The cloud computing paradigm is extensively known as the next generation computing infrastructure. It allows users to employ computing resources (servers, networks, and storage) as a utility, platforms that include middleware services and operating systems, and software applications offered by cloud providers at low cost. It enables the delivery of virtualized services that scale up and down dynamically.

Given the benefits of cloud computing and the widespread utilization of Internet-enabled smartphones and tablets, the MCC paradigm is introduced as the integration of various cloud computing services into the mobile environment. This concept allows mobile users to access computationally intensive data processing and storage services via wireless and cellular networks [2]. MCC leads to improved battery life due to workload offloading, infinite storage, and high-speed data processing capability on the cloud.

The increasing availability of Internet access on mobile devices is enabling consumers to access a growing number of cloud applications while they are on the move. The estimated amount of mobile data traffic that tablets will generate by 2017 is 1.3 EB/month, which is 1.5 times higher than the entire amount of mobile data traffic in 2012 (885 PB/month) [3]. Mobile devices are rapidly becoming the main computing platform. As a consequence, optimizing these devices to better access cloud services is critical as the majority of MCC applications are still created based on the standard web with extensions for mobility support. The access to a cloud service requires from the mobile user to establish a connection to a cellular network such as 3G, which results in high latency, high cost, and significant energy consumption. Radio and battery technologies are continually improving. However, it is expected that they will remain the bottleneck in future mobile systems [4]. To cope with this challenge, we consider small clouds known as cloudlets [5] to which mobile users might connect using a 5G cellular network. This new technology has been proposed to enhance the communication latency, offer high-speed access to services, use the Internet of Things (IoT) technologies, and provide high frequencies to machine to machine connections used by devices in smart homes [6].

A cloudlet is a small scale cloud datacenter at the edge of the Internet that allows caching data and program codes to permit mobile users to access powerful computing resources with lower latency. A cloudlet has the capabilities of self-management and faster access control [7].

In this chapter, we propose a hierarchical and a ring cloudlet-based architectures that can be configured to respond the needs of mobile users, and we compare their performance concerning latency delay using a single and multiple requests scenarios. The goal of this comparison is to identify the most efficient and flexible architecture for data access and data synchronization between the cloudlets and the mobile users.

The remainder of the chapter is organized as follows. Section 4.2 describes related work on the issues of MCC and different cloudlet-based architectures. Section 4.3 presents hierarchical and ring cloudlet-based architectures and describes our proposed mathematical model of latency delay for single and multiple requests.

Section 4.4 presents numerical results. Finally, Section 4.5 concludes the paper and highlights future work.

4.2 Related work

Over the last few years, several researchers investigated the adoption of MCC. Also, many research works proposed cloudlet-based architectures [8–10]. As discussed by Sakr *et al.* [11], a mobile cloud (MC) needs to scale the resource requirements of different mobile devices with the demands of cloud-based mobile applications dynamically and guarantee a minimum level of availability and QoS. To take advantage of the cloud, mobile users need to define and specify their acceptable levels of QoS. However, these requirements are not enough to satisfy mobile cloud needs for additional aspects such as mobility, low connectivity, and limited sources of power [12].

A cloudlet-based architecture can address and alleviate these issues. Soyata *et al.* [12] implemented the Mobile Cloud Hybrid Architecture (MOCHA) cloudlet-based architecture, which aims to improve the response time for face recognition applications. However, the MOCHA architecture does not take into account the possible failure of one or more cloudlets, which can hamper the execution of applications. Verbelen *et al.* [13] proposed a more dynamic cloudlet-based scenario where mobile devices in the cloudlet network could cooperate. They also presented a new cloudlet-based architecture, which manages applications at the component level by distributing the application components among the cloudlets of the architecture. The drawback of this work is that it lacks communication between cloudlets. Yang *et al.* [14] proposed a new network architecture that integrates distributed and local cloudlets to bring cloud resources much closer to end users. The proposed system benefits from the advantages of wireless mesh networks regarding cost, efficiency, rapid deployment, self-organization, and low-latency access to cloud services.

Fesehaye *et al.* [15] investigated the impact of cloudlets on interactive mobile cloud applications by using services such as file editing, video streaming, and collaborative chatting. Their simulation results show the data transfer delay and system throughout through two cloudlet wireless hops for a single request of video streaming, file editing, and collaborative chatting. Moreover, they used 99 cloudlets forming peer-to-peer networks on 670 m $\times$ 670 m mobility region, which is good for a small number of cloudlets. But for professional systems with sensitive data, this type of architecture is not advisable as it cannot determine the whole accessibility setting of the entire network.

Sarkar *et al.* [16] assessed the applicability of the newly proposed paradigm of fog computing to IoT latency-sensitive applications. They specified a mathematical model to represent the fog computing network regarding power consumption, service latency, and cost. Then, they evaluated its performance by considering a high number of Internet-connected mobile devices demanding real-time service.

Corsaro *et al.* [17] introduced cloud, fog, and mist computing architectures for the IoT. They explained their applicability with real-world use cases, assessed their technological maturity, and highlighted the areas that should alleviate the connectivity, bandwidth, and latency challenges faced by industrials demanding consumer IoT applications.

As far as we know, the existing approaches don't rely on the concept of cloudlets and local clouds to leverage the geographical nearness of resources to mobile users, enhance the mobile user experience, and improve data synchronization among the cloudlets. In this work, we propose a hierarchical and a ring cloudlet-based architectures that a cloud provider can configure according to the geographical location of resources using new routing algorithms for mobile search applications. We focus on the routing algorithm for multiple requests to assess the performance and the efficiency of the two architectures.

4.3 Cloudlet architectures

In this section, we describe our proposed cloudlet-based architecture, which aims to reduce the latency and facilitate access to data stored in the cloud by mobile users as opposed to the classical architecture. The cloudlets play the role of intermediaries between mobile users and the cloud. They facilitate communication and offloading some tasks such as synchronization, on to the cloud in a transparent way for the users. Mobile users do not need to know where their requests and tasks are executed. Some tasks might be executed on the main cloud while others are partially executed on the cloudlets. This partitioning depends on the availability of data and applications on the cloudlets. The mobile user might communicate with the cloudlet via 5G connection. It is expected that the new 5G air interface and spectrum will be combined with WiFi the long-term evolution to provide universal high-rate coverage and a seamless user experience, provide about 1,000 times higher wireless area capacity, and save up the entire of energy consumption per service.

To demonstrate and illustrate the importance and the advantages of cloudlets with regards to latency and access to cloud services, we propose a hierarchical cloudlet based architecture, which connects mobile users to their closest cloudlet(s) through WiFi connections. Similarly, each cloudlet connects to other cloudlets through WiFi connections.

In this work, we use continuous-time Markov chains (CTMC) to represent and model the different states of mobile users and cloudlets as well as their interactions. Nodes represent mobile users and cloudlets. Mobile users' nodes might change over time according to the information they receive. Cloudlet nodes might change depending on their current state that can be "on" or "off." Several factors such as fast-fading, interferences, mobility pattern, and collisions might hamper wireless communications. We model these factors with a parameter that simulates their influence rate λ on cloudlet communication. In addition, the model takes into consideration sudden failures of cloudlet nodes.

Throughout the paper the following assumptions will hold for both architectures:

- Let $N+1$ be the total number of nodes ($N-1$ cloudlets nodes, one cloud node and one mobile node) in a communication scenario of a mobile user.

- The Markov chain is in state $i \in [0, 3]$ for hierarchical architecture where i is the level of transmission request and $i \in [1, 2, \ldots, N - 1]$ for ring architecture.
- The processes $\{N_t(i), t \geq 0\}$ are mutually independent homogeneous Poisson processes with rate $\lambda \geq 0$ which counts the number of arrival requests and the time that these requests occur in a given time interval. $N_t(i)$ is a node at level i at instant t for a given request, with $N(i+1) = N(i) + 1$ and $N(0) = 1$.

Indeed, when the cloudlet is in the operational status and goes off instantly; we can describe this situation by a CTMC with two states as shown in Figure 4.1. The first state is during the cloudlet's operational status (value 1) and the second one is when the cloudlet becomes inoperative (value 0). Therefore, the probability for the cloudlet to be in the operational status is $P_F(t) = (1/2)(1 + e^{-2t})$ and the probability to be in the inoperative status is $P_P(t) = (1/2)(1 - e^{-2t})$, where t is the time of the periodical update.

4.3.1 Hierarchical architecture

Figure 4.2 depicts our proposed integrated architecture, which is organized into a hierarchical multi-tiered tree topology. Figure 4.2 illustrates a four-tier hierarchical topology that we have used in our simulation experiments.

In Figure 4.3, cloudlet nodes are organized into a tree structure. The root or top level is reserved for a designated cloudlet called the super-cloudlet. This cloudlet is connected to several regional cloudlets. The regional cloudlets are the child nodes of the super-cloudlet. Depending on the availability of resources and the size of the covered geographical area, regional cloudlets can be organized into multiple levels on the tree. In such cases, each region is managed by a designated regional cloudlet, which in turn manages other cloudlets. Each regional cloudlet, at the lower level of the tree structure, manages multiple mobile users. When mobile users connect to the cloud their connection requests are automatically routed to an appropriate regional cloudlet based on their geographical location. Cloud providers specify and configure the different geographic partitions.

Consider a Markov chain is in state $i = 0, 1, 2, 3$ where i is the level of transmission request, and $j \in [1, \ldots, N - 3]$ is the occurrence number of the parallel cloudlets at the third level. The central cloudlet is a major cloudlet that monitors all its sub-cloudlets and is the only one that is connected to the general cloud. We assume that the central cloudlet does not go off and the general cloud is at the same

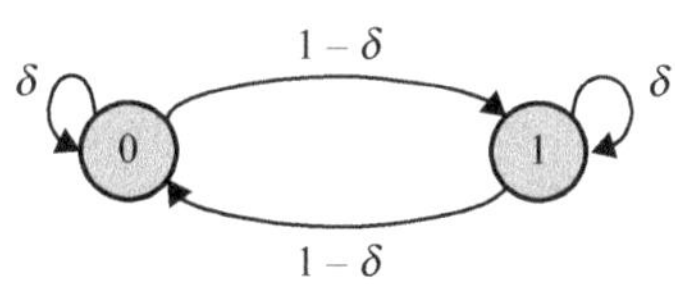

Figure 4.1 Markov chain of cloudlet states

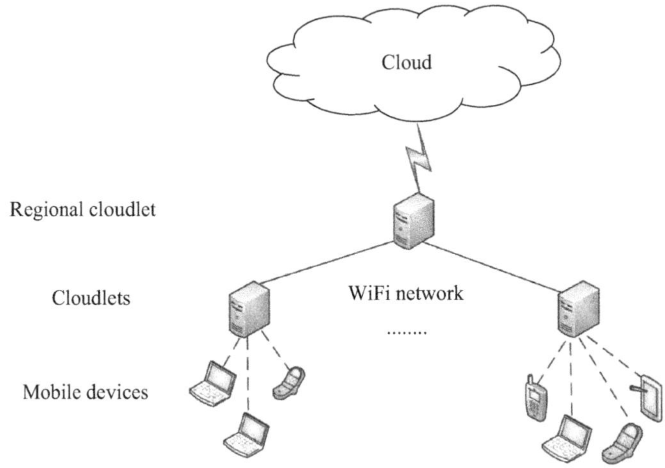

Figure 4.2 Hierarchical topology for MC communication

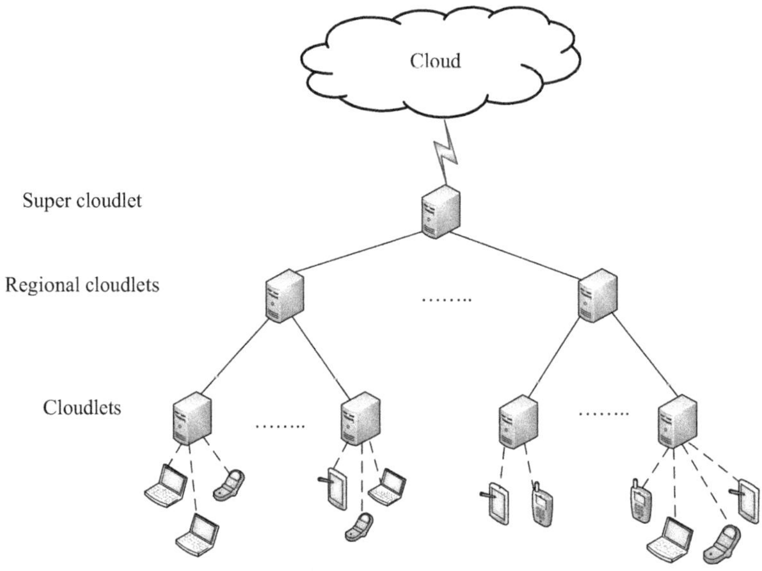

Figure 4.3 Tree topology for MC communication

level as parallel cloudlets when $j = C$ (i.e., the occurrence number j is pointed on the general cloud). Each other cloudlet may be in the operational state (value 1) or the inoperative state (value 0). For that reason, we use a two-dimensional Markov chain as shown in Figure 4.4.

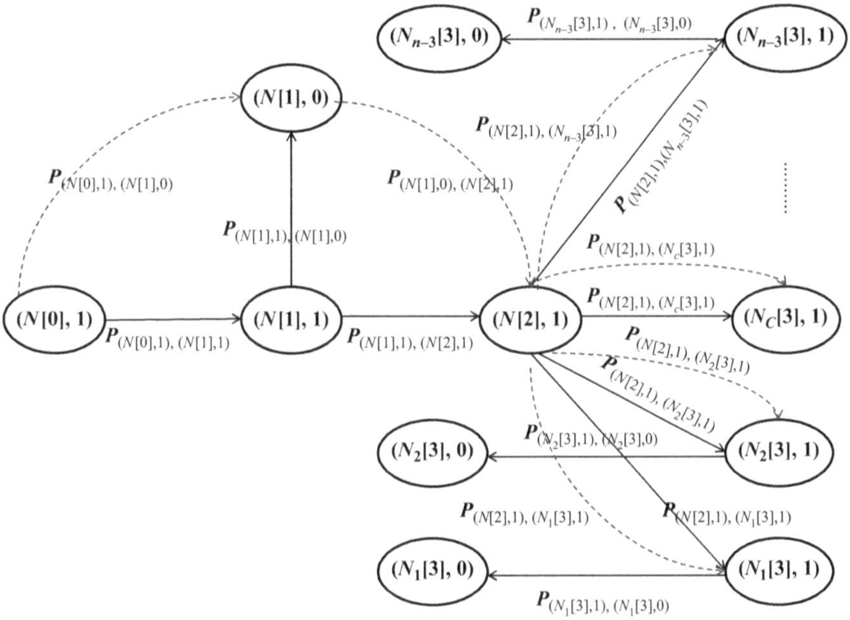

Figure 4.4 Markov chain model for the hierarchical architecture

4.3.1.1 Formulation of latency delay for one request

To simplify this Markov chain model, we introduce a system of equations. These equations describe several transmission rates between the mobile user, the general cloud, and every cloudlet which is in operative state.

$$R_{(N(i),a),\left(N_j(i+1),b\right)} = \begin{cases} \lambda P_F(t) & i = j = 0 \\ & a = b = 1 \\ \lambda^2 & i = 1; j = 0 \\ & a = b = 1 \\ \lambda^3 P_F(t) & i = 2; j \in [1, \ldots, N-3] \\ & a = b = 1 \\ \lambda^3 & i = 2; j = C \\ & a = b = 1 \end{cases} \tag{4.1}$$

When the cloudlet selected by the mobile user goes off, the transmission rate between nodes changes immediately and becomes

$$Q_{(N(i),a),\left(N_j(i+1),b\right)} = \begin{cases} 0 & i = j = 0 \\ & a = 1; b = 0 \\ \lambda & i = 1; j = 0 \\ & a = 0; b = 1 \\ \lambda^2 P_F(t) & i = 2; j \in [1, \ldots, N-3] \\ & a = b = 1 \\ \lambda^2 & i = 2; j = C \\ & a = b = 1 \end{cases} \tag{4.2}$$

Let λ_i be the rate of the ith request transition from mobile node $N[0]$ to destination node $N[i]$. The total rate of requests scattered with the only non-null one-step transition probabilities is

$$Pr_{(N(i),a),\left(N_j(i+1),b\right)} = \begin{cases} \lambda^{(1-N)} \dfrac{P_F(t)}{N-2}, & i = j = 0 \\ & a = b = 1 \\ \lambda^{(2-N)} \left(1 - \dfrac{P_F(t)}{N-2}\right), & i = 1; j = 0 \\ & a = b = 1 \\ \lambda^{(3-N)} \tilde{\boldsymbol{\Phi}}, & i = 2; a = b = 1 \\ & j \in [1, \ldots, N-3] \\ \lambda^{(3-N)} \breve{\boldsymbol{\Phi}}, & i = 2; j = C \\ & a = b = 1 \end{cases} \tag{4.3}$$

We define Br as the transition probability when the first cloudlet selected by the mobile user goes off.

$$Br_{(N(i),a),\left(N_j(i+1),b\right)} = \begin{cases} 0 & i = j = 0 \\ & a = 1; b = 0 \\ \lambda^{(1-N)}, & i = 1; j = 0 \\ & a = 0; b = 1 \\ \lambda^{(2-N)} \boldsymbol{\Phi}, & i = 2; a = b = 1 \\ & j \in [1, \ldots, N-3] \\ \lambda^{(2-N)} \breve{\boldsymbol{\Phi}}, & i = 2; j = C \\ & a = b = 1 \end{cases} \tag{4.4}$$

where $\Phi = \left(1 - \prod_{k=1}^{(N-3)-h}(((k-1)P_F(t))/(N-2))\right)$, $\tilde{\Phi} = \left(1 - \prod_{k=1}^{(N-3)-h}(kP_F(t)/(N-2))\right)$, $\mathring{\Phi} = \left(1 - \prod_{k=1}^{(N-1)-h}(kP_F(t)/(N-2))\right)$, and $\breve{\Phi} = (1 - \prod_{k=1}^{(N-2)-h}((k-1)P_F(t)/(N-2)))$.
We know that

$$Pr\left[X_c = N_{j,F}(3)\right] = Pr\left[X_c = N_{k,F}(3)\right], \forall j \neq k, j, k \in [1, \ldots, N-3]$$

Finally, we define the probability that $N[i]$ is a destination node as follows:

$$\begin{aligned} Pr[X_c = N(i)] = &\sum_{j=1}^{(N-3)-h}\left(\frac{\lambda^2}{\lambda^N}\Phi + \frac{\lambda^3}{\lambda^N}\tilde{\Phi}\right) + \frac{\lambda}{\lambda^N} \\ &+ \frac{\lambda^3}{\lambda^N}\mathring{\Phi} + \frac{\lambda^2}{\lambda^N}\left(1 - \frac{P_F(t)}{(N-2)}\right) \\ &+ \frac{\lambda^2}{\lambda^N}\breve{\Phi} + \frac{\lambda}{\lambda^N}\left(\frac{P_F(t)}{(N-2)}\right). \end{aligned} \tag{4.5}$$

The node $N[i]$ receives request at time t_i, define $\tau_i = t_{i+1} - t_i$ where τ_i is exponentially distributed with intensity λ^N and $T_{m1,c} = \sum_{i=1}^{k}\tau_i$. $T_{m1,c}$ is the delay message of the hierarchical architecture. It represents the time needed to send the request from the source node to $N[k]$ node. By assuming that the node $N[i]$ is the destination node of the request, and using the Laplace–Setieljes transform (LST) of $T_{m1,c}$, the latency delay is (for $\theta \geq 0$)

$$\begin{aligned} T^*_{m1,c} &= E[e^{-\theta T_{m1,c}}] \\ &= \sum_{k=1}^{N} E\left[e^{-\theta T_{m1,c}} \mid X_c = k\right] Pr[X_c = N(i)] \\ &= \sum_{k=1}^{N} E\left[e^{-\theta\sum_{i=1}^{k}\tau_i} \mid X_c = k\right] Pr[X_c = N(i)] \\ &= \sum_{k=1}^{N}\left(\frac{\lambda^N}{\lambda^N + \theta}\right)^k Pr[X_c = N(i)] \end{aligned} \tag{4.6}$$

Moreover, the results of the expected destination node and the expected latency delay can be computed as follows:

$$\begin{aligned} \frac{\partial T^*_{m1,c}}{\partial\theta}|_{\theta=0} &= \frac{\partial E[e^{-\theta T_{m1,c}}]}{\partial\theta}|_{\theta=0} = E\left[\frac{\partial e^{-\theta T_{m1,c}}}{\partial\theta}|_{0=0}\right] \\ &= E\left[-T_{m1,c}\right] = -E\left[T_{m1,c}\right] \end{aligned}$$

Therefore, the expected latency can be expressed as follows:

$$E\left[T_{m1,c}\right] = -\frac{\partial T^{*}_{m1,c}}{\partial \theta}\Big|_{\theta=0} = \frac{N}{\lambda^{N}} Pr[X_c = N(i)] \tag{4.7}$$

The expected destination node is given by

$$\begin{aligned} E[X_c] &= \sum_{i=1}^{N} i Pr[X_c = N(i)] = N Pr[X_c = N(i)] \\ &= \lambda^{N} E\left[T_{m1,c}\right] \end{aligned} \tag{4.8}$$

4.3.1.2 Formulation of the latency delay for multiple requests submission

- First scenario: sending multiple requests

This scenario happens when the cloudlet selected by the mobile user sends multiple requests to its central cloudlet. This latter attempts to handle these requests and find suitable responses. As shown in Figure 4.5, we consider the case with $R = 1{,}000$ requests, and we model the latency delay as follows:

$$\begin{aligned} P_r^R[X_c = N(i)] &= \sum_{l=1}^{R} l Pr[X_c = N(i)] \\ &= \sum_{l=1}^{R} l \left(\sum_{j=1}^{(N-3)-h} \left(\frac{\lambda^2}{\lambda^N} \Phi + \frac{\lambda^3}{\lambda^N} \tilde{\Phi} \right) + \frac{\lambda}{\lambda^N} \right) \\ &\quad + \sum_{l=1}^{R} l \left(\frac{\lambda^3}{\lambda^N} \mathring{\Phi} + \frac{\lambda^2}{\lambda^N} \left(1 - \frac{P_F(t)}{(N-2)} \right) \right) \\ &\quad + \sum_{l=1}^{R} l \left(\frac{\lambda^2}{\lambda^N} \breve{\Phi} + \frac{\lambda}{\lambda^N} \left(\frac{P_F(t)}{(N-2)} \right) \right). \end{aligned} \tag{4.9}$$

$$T^{R*}_{m1,c} = \sum_{k=1}^{N} \left(\frac{\lambda^N}{\lambda^N + \theta} \right)^{k} P_r^R[X_c = N(i)] \tag{4.10}$$

- Second scenario: sending a single request

This scenario occurs when each cloudlet sends a single request to its central cloudlet. If the sending cloudlet is in the operational state, then the central cloudlet will receive r requests where $r \in [1, \ldots, N-2]$ as depicted in Figure 4.6. Otherwise,

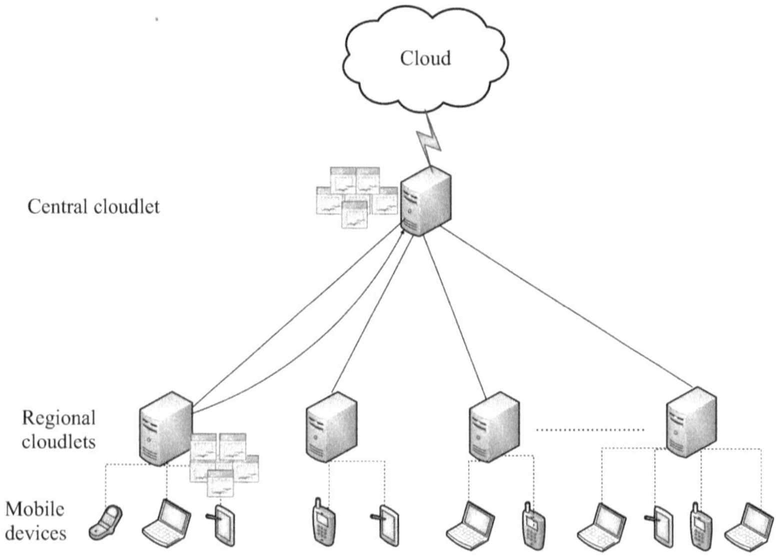

Figure 4.5 Multiple requests submission in the hierarchical architecture

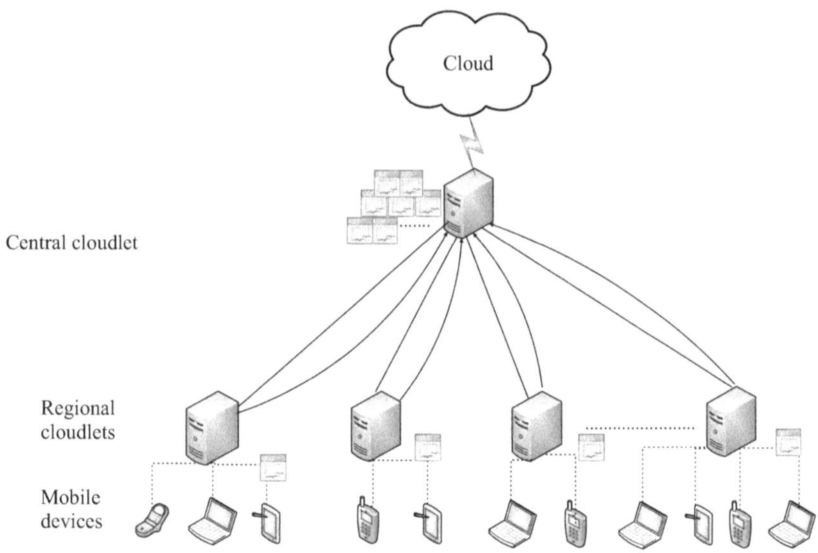

Figure 4.6 Single request submission in the hierarchical architecture

the central cloudlet will receive r requests where $r \in [1, \ldots, N-3]$. Therefore, the latency delay can be expressed as follows:

$$Pr_{(N(i),a),(N_j(i+1),b)} = \begin{cases} \lambda^{(1-N)} \dfrac{P_F(t)}{N-2}, & i=j=0 \\ & a=b=1 \\ \lambda^{(2-N)} \left(1 - \dfrac{P_F(t)}{N-2}\right), & i=1; j=0 \\ & a=b=1 \\ r\lambda^{(3-N)} \tilde{\boldsymbol{\Phi}}, & i=2; a=b=1 \\ & j \in [1, \ldots, N-3] \\ (r+1)\lambda^{(3-N)} \mathring{\boldsymbol{\Phi}}, & i=2; j=C \\ & a=b=1 \end{cases} \tag{4.11}$$

$$Br_{(N(i),a),(N_j(i+1),b)} = \begin{cases} 0 & i=j=0 \\ & a=1; b=0 \\ \lambda^{(1-N)}, & i=1; j=0 \\ & a=0; b=1 \\ (r+1)\lambda^{(2-N)} \boldsymbol{\Phi}, & i=2; a=b=1 \\ & j \in [1, \ldots, N-3] \\ (r+1)\lambda^{(2-N)} \breve{\boldsymbol{\Phi}}, & i=2; j=C \\ & a=b=1 \end{cases} \tag{4.12}$$

Finally, we describe the probability that $N[i]$ is a destination node for r requests as follows:

$$\begin{aligned} P_r^r[X_c = N(i)] = & \sum_{r=1}^{(N-3)-h} \left(\frac{(r+1)\lambda^2}{\lambda^N} \boldsymbol{\Phi} + \frac{r\lambda^3}{\lambda^N} \tilde{\boldsymbol{\Phi}} \right) \\ & + \sum_{r=1}^{(N-3)-h} \left(\frac{(r+1)\lambda^2}{\lambda^N} \breve{\boldsymbol{\Phi}} \right) \\ & + \sum_{r=1}^{(N-2)-h} \left(\frac{(r+1)\lambda^3}{\lambda^N} \mathring{\boldsymbol{\Phi}} \right) \\ & + \frac{\lambda^2}{\lambda^N} \left(1 - \frac{P_F(t)}{(N-2)} \right) \\ & + \frac{\lambda}{\lambda^N} \left(1 + \frac{P_F(t)}{(N-2)} \right). \end{aligned} \tag{4.13}$$

$$T^{r*}_{m1,c} = \sum_{k=1}^{N} \left(\frac{\lambda^N}{\lambda^N + \theta} \right)^k P^r_r[X_c = N(i)] \tag{4.14}$$

4.3.2 Ring architecture

In the cloudlet-based ring architecture, each cloudlet is connected to two other cloudlets, to a set of mobile nodes, and to the main cloud as shown in Figures 4.7 and 4.8. Data in the ring moves from one cloudlet to another, and each cloudlet treats every data along the way. Furthermore, cloud provider specifies at the level of mobile node a list of available cloudlets holding the primary, the left secondary, and the right secondary cloudlet which are the neighbors of primary cloudlet. The nearest operating cloudlet, that is connected to mobile node, is called the primary cloudlet. If the first communication with the primary cannot be established due to its damage, then after some time, mobile node tries to connect to other nearest cloudlets whether the left secondary or the right one which will play the role of primary. In this case, to avoid and limit the damages of network failures each cloudlet transfers requests to all its sibling nodes. We suggest that the ring architecture will have a dual link.

We consider a Markov chain in state $i = 0, 1, 2, \ldots, N-1$ where i is the level of transmission of the request, knowing that the general cloud is at the same level as the second cloudlet selected by the mobile user. If the primary cloudlet of mobile nodes goes off, then there will be any assignments of transmission levels to the cloud and the source will send directly to its left and right cloudlets. Let N be the total number of $(N-1)$ cloudlets adding the general cloud in the system. When mobile node which is at 0 level of transmission, it sends a request to its primary cloudlet, we assume that this primary cloudlet will send the request to their left and right neighbors with rate 50%, but this request will be sending just to k cloudlet in one direction and to $(N-k)$ cloudlets in other directions as depicted by Figure 4.7. Each cloudlet may be in the operational state or a inoperative state. For that reason, we consider two-dimensional Markov chain. We define λ^{ith} as the rate of the request transition from mobile node $N[0]$ to destination node $N[i]$, and λ^N as the total rate of request changes scattered in the entire architecture. The only non-null one-step transition probabilities are

$$Pr = \begin{cases} \lambda^{(1-(N+1))} \dfrac{P_F(t)}{N-1}, & i = j = 0 \\ & a = b = 1 \\ \dfrac{1}{3}\lambda^{((i+1)-(N+1))}\Psi, & i \geq 1; j = 0 \\ & a = b = 1 \\ \dfrac{1}{3}\lambda^{(2-(N+1))}, & i = 1; j = C \\ & a = b = 1 \\ \dfrac{1}{3}\lambda^{(((N-1)-k+1)-(N+1))}\mathring{\Psi}, & i = (N-1) - k + 1 \\ & j = 0; i+1 = k; a = b = 1 \end{cases} \tag{4.15}$$

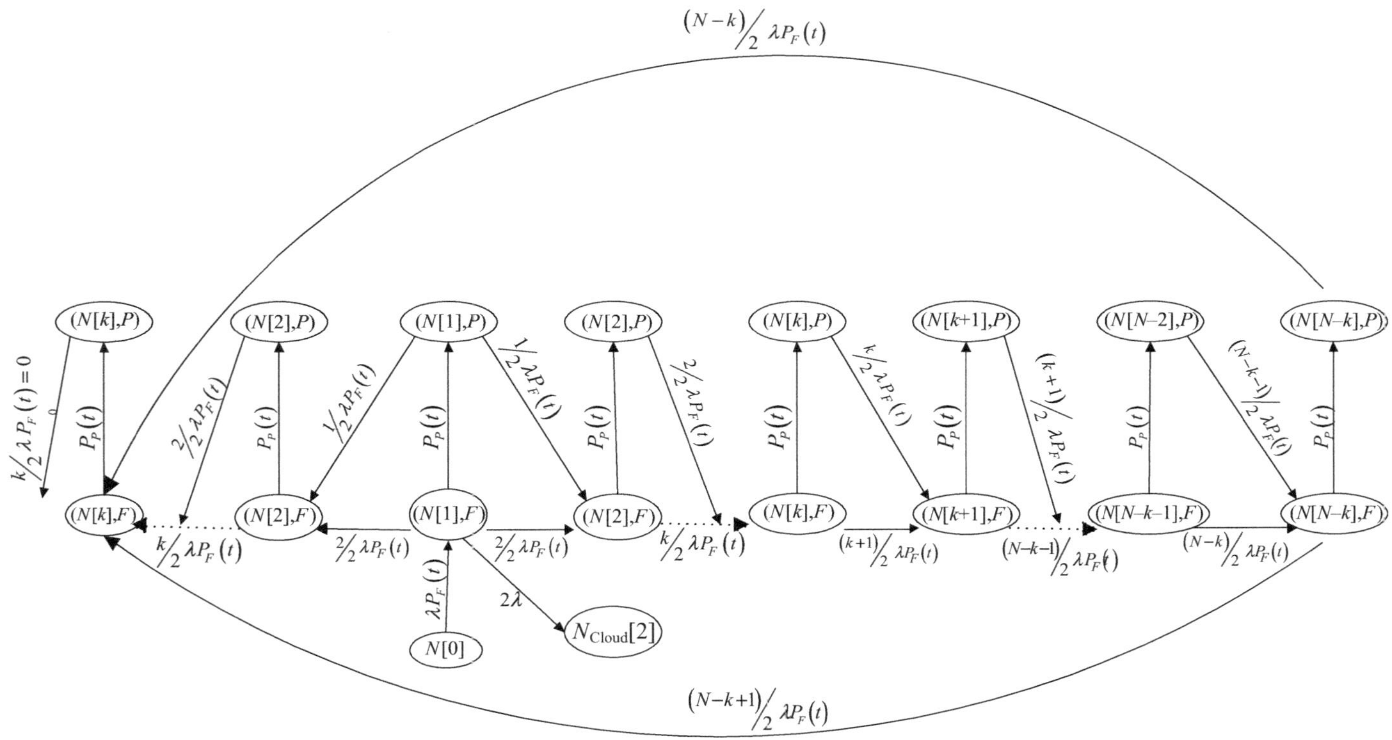

Figure 4.7 Markov chain model for the ring architecture

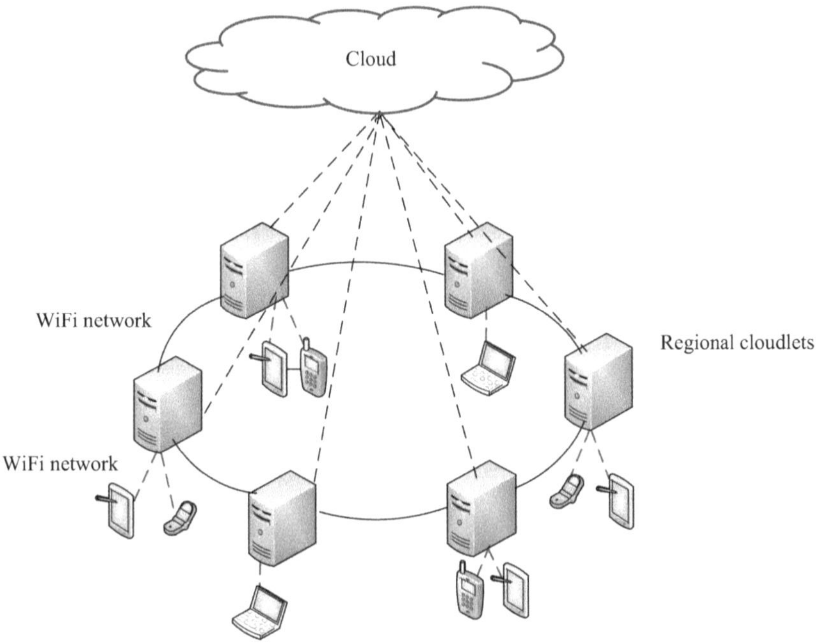

Figure 4.8 Ring topology for MC communication

$$Br = \begin{cases} 0 & i = j = 0 \\ & a = 1; b = 0 \\ \frac{1}{3}\lambda^{(i-N)}\tilde{\Psi}, & i \geq 2; j = 0 \\ & a = 0; b = 1 \\ \frac{1}{3}\lambda^{(2-N)}, & i = 2; j = C \\ & a = 0; b = 1 \\ \frac{1}{3}\lambda^{((N-1)-k)-N}\breve{\Psi}, & i = (N-1) - k; i + 1 = k \\ & a = b = 1; j = 0 \end{cases} \tag{4.16}$$

where $\Psi = \left(1 - \prod_{h=1}^{i} (hP_F(t)/(N-1))\right)$, $\tilde{\Psi} = \left(1 - \prod_{h=2}^{i} (((h-1)P_F(t))/(N-1))\right)$, $\mathring{\Psi} = \left(1 - \prod_{h=1}^{(N-1)-k+1} (hP_F(t)/(N-1))\right)$, and $\breve{\Psi} = \left(1 - \prod_{h=2}^{(N-1)-k+1} ((h-1)P_F(t)/(N-1))\right)$.

Finally, we describe the probability that $N[i]$ is a destination node as follows:

$$\begin{aligned} Pr[X_r = N(i)] &= \sum_{i=1}^{k-1} \frac{\lambda^{(i+1)}}{3\lambda^{(N+1)}}\Psi + \sum_{i=2}^{k-1} \frac{\lambda^{i}}{3\lambda^{N}}\tilde{\Psi} \\ &\quad + \sum_{i=1}^{(N-1)-k} \frac{\lambda^{(i+1)}}{3\lambda^{(N+1)}}\Psi + \frac{\lambda}{\lambda^{(N+1)}}\left(\frac{P_F(t)}{(N-1)}\right) \\ &\quad + \sum_{i=2}^{(N-1)-k} \frac{\lambda^{i}}{3\lambda^{N}}\tilde{\Psi} + \frac{\lambda^{(N-1)-k}}{3\lambda^{N}}\breve{\Psi} \\ &\quad + \frac{\lambda^{(N-1)-k+1}}{3\lambda^{(N+1)}}\mathring{\Psi} + \frac{\lambda^{2}}{3\lambda^{N}}\left(\frac{\lambda+1}{\lambda}\right) \end{aligned} \tag{4.17}$$

The node $N[i]$ receives request at time t_i, we define $\tau_i = t_{i+1} - t_i$ where τ_i is exponentially distributed with intensity λ^N and $T_{m1,r} = \sum_{i=1}^{k} \tau_i$. $T_{m1,r}$ is the delay of cloud-based ring architecture. It is defined as the time needed to send the request from the source node to the $N[k]$ node. By assuming that the node $N[i]$ is the destination node of the request, and using the LST of $T_{m1,r}$, the latency delay can be expressed for $\theta \geq 0$ as follows:

$$\begin{aligned} T^*_{m1,r} &= E[e^{-\theta T_{m1,r}}] \\ &= \sum_{k=1}^{N} E\left[e^{-\theta T_{m1,r}} | X_r = k\right] Pr[X_r = N(i)] \\ &= \sum_{k-1}^{N} E\left[e^{-\theta\sum_{i=1}^{k}\tau_i} | X_r = k\right] Pr[X_r = N(i)] \\ &= \sum_{k=1}^{N} \left(\frac{\lambda^N}{\lambda^N + \theta}\right)^k Pr[X_r = N(i)] \end{aligned} \tag{4.18}$$

Moreover, the expected latency delay can be computed as follows:

$$\begin{aligned} \frac{\partial T^*_{m1,r}}{\partial\theta}|_{\theta=0} &= \frac{\partial E[e^{-\theta T_{m1,r}}]}{\partial\theta}|_{\theta=0} = E\left[\frac{\partial e^{-\theta T_{m1,r}}}{\partial\theta}|_{\theta=0}\right] \\ &= E\left[-T_{m1,r}\right] = -E\left[T_{m1,r}\right] \end{aligned} \tag{4.19}$$

Therefore, the expected latency delay has the following expression:

$$E\left[T_{m1,r}\right] = -\frac{\partial T^*_{m1,r}}{\partial\theta}|_{\theta=0} = \frac{N}{\lambda^N} Pr[X_r = N(i)] \tag{4.20}$$

The expected destination node is given by

$$
\begin{aligned}
E[X_r] &= \sum_{i=1}^{N} iPr[X_r = N(i)] = NPr[X_r = N(i)] \\
&= \lambda^N E\left[T_{m1,r}\right]
\end{aligned}
\tag{4.21}
$$

4.3.2.1 Formulation of the latency delay for multiple requests submission

- First scenario: sending multiple requests

This scenario occurs when a cloudlet sends multiple requests $R = 1{,}000$ at the same time to its left and right neighbors cloudlets. Theses cloudlets attempt to send received requests and their own requests to their respective neighbors as shown in Figure 4.9. We model the latency delay in this case as follows:

$$
\begin{aligned}
P_r^R[X_r = N(i)] &= \sum_{l=1}^{R} lP_r[X_r = N(i)] \\
&= \sum_{l=1}^{R} l\left(\sum_{i=1}^{k-1} \frac{\lambda^{(i+1)}}{3\lambda^{(N+1)}}\Psi + \sum_{i=2}^{k-1} \frac{\lambda^i}{3\lambda^N}\tilde{\Psi}\right) \\
&\quad + \sum_{l=1}^{R} l\left(\sum_{i=1}^{(N-1)-k} \frac{\lambda^{(i+1)}}{3\lambda^{(N+1)}}\Psi\right) \\
&\quad + \sum_{l=1}^{R} l\left(\frac{\lambda^{(N-1)-k}}{3\lambda^N}\breve{\Psi}\right) \\
&\quad + \sum_{l=1}^{R} l\left(\sum_{i=2}^{(N-1)-k} \frac{\lambda^i}{3\lambda^N}\tilde{\Psi}\right) \\
&\quad + \sum_{l=1}^{R} l\left(\frac{\lambda}{\lambda^{(N+1)}}\left(\frac{P_F(t)}{(N-1)}\right)\right) \\
&\quad + \sum_{l=1}^{R} l\left(\frac{\lambda^{(N-1)-k+1}}{3\lambda^{(N+1)}}\mathring{\Psi} + \frac{\lambda^2}{3\lambda^N}\left(\frac{\lambda+1}{\lambda}\right)\right)
\end{aligned}
\tag{4.22}
$$

$$
T_{m1,r}^{R*} = \sum_{k=1}^{N} \left(\frac{\lambda^N}{\lambda^N + \theta}\right)^k P_r^R[X_r = N(i)]
\tag{4.23}
$$

- Second scenario: sending a single request

This scenario happens when the primary cloudlet sends a single request to its left and right neighbors cloudlets, which send the received request and their own requests to their respective left and right neighbors sequentially. The process continues this way

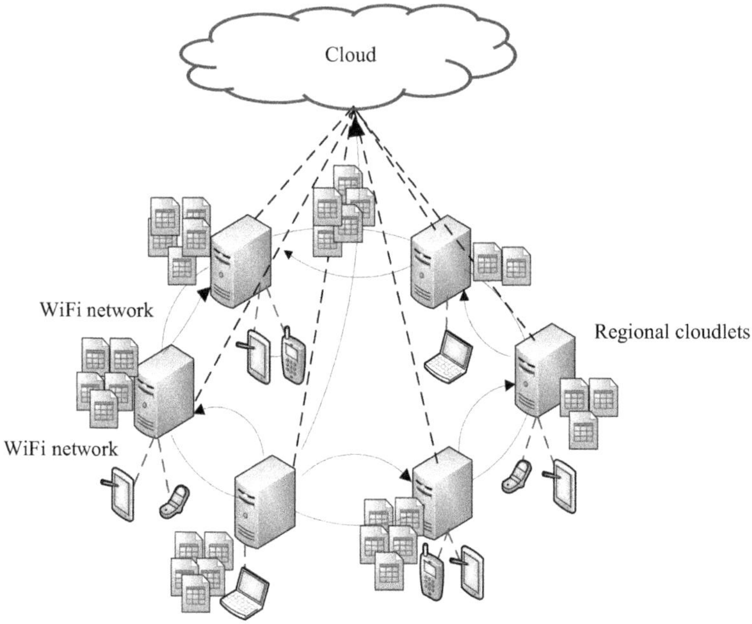

Figure 4.9 Multiple requests submission in the ring architecture

until the $((N-1)-k)$th cloudlet receives r requests, where $r \in [1,\ldots,(N-1)-k]$. Similarly, the kth cloudlet receives r requests, where $r \in [1,\ldots,k]$ as depicted in Figure 4.10. The latency delay can be expressed in this case as follows:

$$Pr = \begin{cases} \lambda^{(1-(N+1))}\dfrac{P_F(t)}{N-1}, & i=j=0 \\ & a=b=1 \\ \dfrac{i}{3}\lambda^{((i+1)-(N+1))}\Psi, & i \geq 1; j=0 \\ & a=b=1 \\ \dfrac{1}{3}\lambda^{(2-(N+1))}, & i=1; j=C \\ & a=b=1 \\ \dfrac{((N-1)-k)}{3\lambda^{(N-1)}}\lambda^{((N-1)-k+1)}\mathring{\Psi}, & i=(N-1)-k+1 \\ & j=0; i+1=k; a=b=1 \end{cases} \quad (4.24)$$

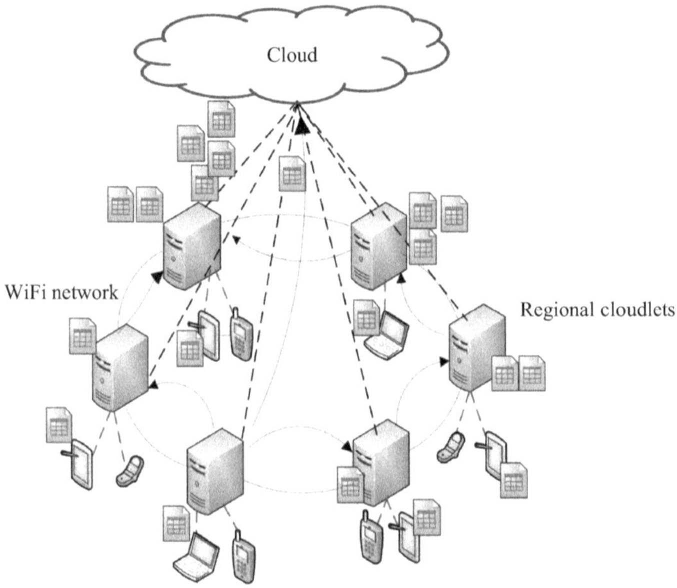

Figure 4.10 Single request submission in the ring architecture

$$Br = \begin{cases} 0 & i = j = 0 \\ & a = 1; b = 0 \\ \frac{i}{3}\lambda^{(i-N)}\tilde{\Psi}, & i \geq 2; j = 0 \\ & a = 0; b = 1 \\ \frac{1}{3}\lambda^{(2-N)}, & i = 2; j = C \\ & a = 0; b = 1 \\ \frac{((N-1)-k)}{3}\lambda^{((N-1)-k)-N}\breve{\Psi}, & i = (N-1)-k; i+1 = k \\ & a = b = 1; j = 0 \end{cases} \tag{4.25}$$

Finally, we describe the probability that $N[i]$ is a destination node for r requests as follows:

$$
\begin{aligned}
P_r^r[X_r = N(i)] &= \sum_{i=1}^{k-1} \frac{i\lambda^{(i+1)}}{3\lambda^{(N+1)}}\Psi + \sum_{i=1}^{(N-1)-k} \frac{i\lambda^{(i+1)}}{3\lambda^{(N+1)}}\Psi \\
&+ \sum_{i=2}^{k-1} \frac{i\lambda^{i}}{3\lambda^{N}}\tilde{\Psi} + \frac{((N-1)-k)\lambda^{(N-1)-k}}{3\lambda^{N}}\breve{\Psi} \\
&+ \sum_{i=2}^{(N-1)-k} \frac{i\lambda^{i}}{3\lambda^{N}}\tilde{\Psi} + \frac{\lambda}{\lambda^{(N+1)}}\left(\frac{P_F(t)}{(N-1)}\right) \\
&+ \frac{((N-1)-k)\lambda^{(N-1)-k+1}}{3\lambda^{(N+1)}}\mathring{\Psi} \\
&+ \frac{\lambda^2}{3\lambda^N}\left(\frac{\lambda+1}{\lambda}\right) \\
T_{m1,r}^{r*} &= \sum_{k=1}^{N}\left(\frac{\lambda^N}{\lambda^N+\theta}\right)^k P_r^r[X_r = N(i)]
\end{aligned}
\tag{4.26}
$$

4.4 Numerical results

To simulate request transfer for each the above cloudlet-based architectures, we consider scenarios where a mobile user uses a search application and sends requests to different cloudlets, which may be in the operational or inoperative states. We evaluate the latency delay of the user request using LST and a model of the probability of sending the request to the appropriate destination node. We assume that the requests are sent according to a Poisson distribution with rate λ. We use the following values for λ: 13%, 28%, 50%, and 75%.

As shown in Figure 4.11, we notice that the latency delay of one request for the hierarchical architecture begins when the number of cloudlets $N = 3$, as the major elements of sending are: the selected cloudlet, the central cloudlet and the general cloud. When $\theta = 0.3$, we observe that the latency delay progressively increases with λ. In fact, when λ gradually increases the latency delay also increases at the same time, which means that the latency delay becomes more larger when the rate λ of fast fading and collisions grows gradually. For example: when $\lambda = 75\%$ the latency delay increases from 400 to 600 ms for $N = 3$. For $N \geq 5$, it becomes stationary with values between 1,000 and 1,200 ms due to the structure of the hierarchical architecture. Concerning the curves of $\theta = 0.5$ and $\theta = 0.75$, they have the same shape as the curve mentioned above. However, when $\theta = 0.75$ the latency delay begins approximatively with 250 ms for $\lambda = 75\%$ and grows gradually until it becomes stable with a value between 400 and 450 ms. When $\theta = 0.5$, the latency delay begins with a value between 300 and 400 ms for $\lambda = 75\%$ and increases gradually until it becomes stable with a value between 400 and 450 ms. These results show that the latency delay decreases gradually when θ grows progressively, which means that the latency delay becomes more smaller when the probability θ of no interference interruptions raises

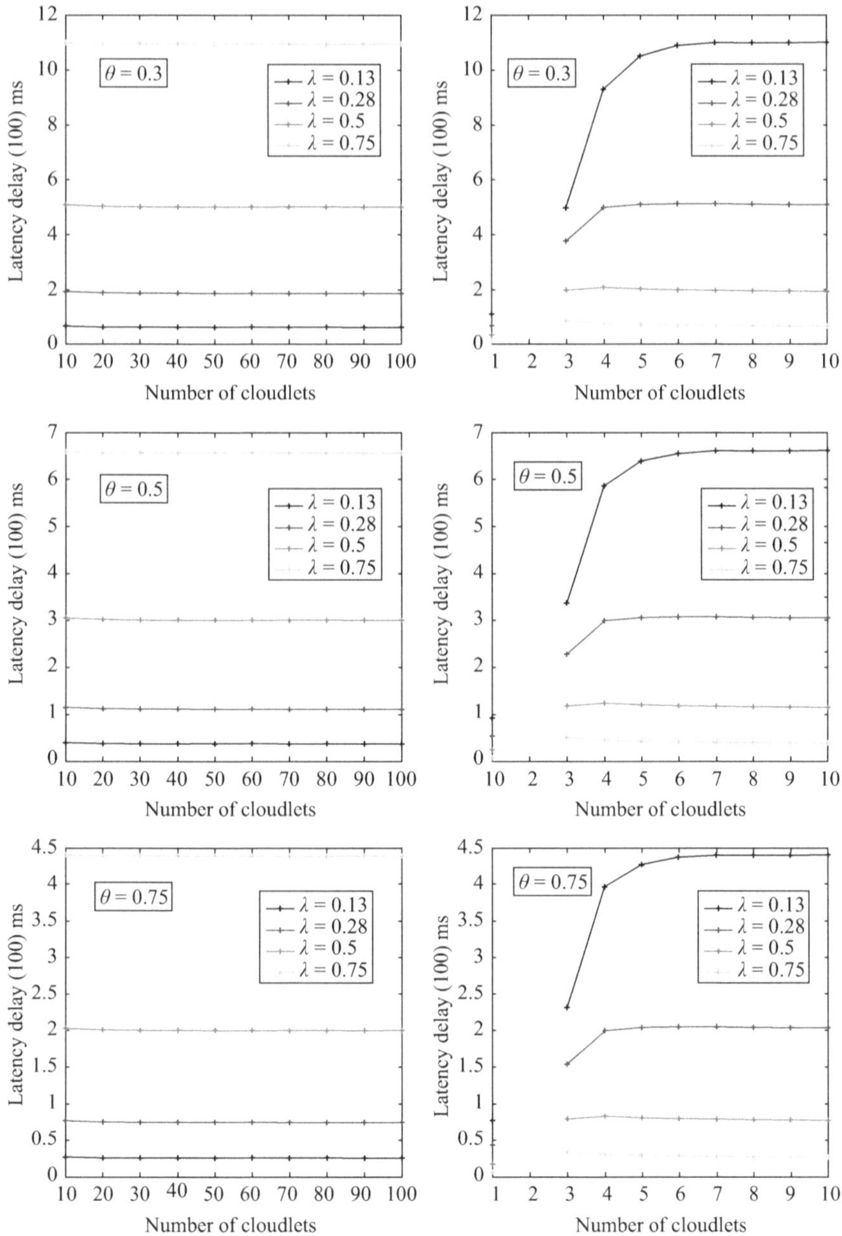

Figure 4.11 Latency delay for the hierarchical architecture

slowly. Although the latency delay of one request for the ring architecture is higher than the hierarchical architecture (as shown in Table 4.1 especially when $\theta = 0.3$), it increases gradually for $\lambda = 75\%$ until it becomes stationary with 1,600 ms as a value. For other values of λ, the latency delay of the ring architecture is smaller than the hierarchical one. Likewise, for $N = 2$, the latency delay of the ring architecture starts

Table 4.1 Simulation results of the hierarchical and ring cloudlet-based architectures

Results (%)	Hierarchical architecture		Ring architecture	
	$\theta = 30\%$	$\theta = 75\%$	$\theta = 30\%$	$\theta = 75\%$
$\lambda = 75$	$L \approx 1{,}100$ ms	$L \approx 450$ ms	$L \approx 1{,}600$ ms	$L \approx 640$ ms
$\lambda = 13$	$L \approx 80$ ms	$L \approx 25$ ms	$L \approx 90$ ms	$L \approx 35$ ms

L, the parameter of latency delay.

Table 4.2 Simulation results of the ring and hierarchical architecture in the first scenario

Results (%)	Hierarchical architecture		Ring architecture	
	$\theta = 30\%$	$\theta = 75\%$	$\theta = 30\%$	$\theta = 75\%$
$\lambda = 75$	$L \approx 10{,}600$ ms	$L \approx 4{,}350$ ms	$L \approx 12{,}800$ ms	$L \approx 5{,}100$ ms
$\lambda = 13$	$L \approx 6{,}000$ ms	$L \approx 300$ ms	$L \approx 800$ ms	$L \approx 30$ ms

L, the parameter of latency delay.

with a value less than the corresponding value of the hierarchical architecture as depicted in Figure 4.12. Therefore, the ring architecture is initially more efficient than the hierarchical architecture because of the elements involved in sending the request (i.e., the selected cloudlet by the mobile user and the general cloud). Moreover, we notice that for each value of θ, the latency delay increases progressively according to the development of λ values. Also, the latency delay becomes more smaller when the probability θ of no interference interruptions grows gradually. So, the comparison of the latency delays in both architectures shows that the hierarchical architecture is efficient when the number of cloudlets is significant, and the rate λ of fast fading and collisions is 75%. In contrast, the ring architecture is more responsive and more efficient when the number of cloudlets is initially less than 9 and the rate λ of fast fading and collisions is between 13% and 50%.

The first scenario, mentioned above, occurs when the selected cloudlet by the mobile user sends 1,000 requests to its central cloudlet, in the case of the hierarchical architecture, or to its left and right secondary cloudlets in the case of the ring architecture. As it is depicted in Figures 4.13 and 4.14, for the hierarchical architecture, when θ decreases progressively the latency delay grows rapidly. Also, when λ increases gradually, the latency delay raises at the same time. Therefore, the latency delay becomes more higher when the rate λ of fast fading and collisions is high. It decreases rapidly when the rate θ of no interference interruptions grows gradually. For the ring architecture, the curves of the latency delay have the same shape as the hierarchical one. However, when $\theta = 75\%$ and λ varies between 13% and 50%, the latency delay in the ring architecture is smaller than in the hierarchical one as it is depicted in Table 4.2. It means that when the rate λ of fast fading and collisions varies between 13% and 50% in the first scenario, the ring architecture is more efficient than the hierarchical architecture.

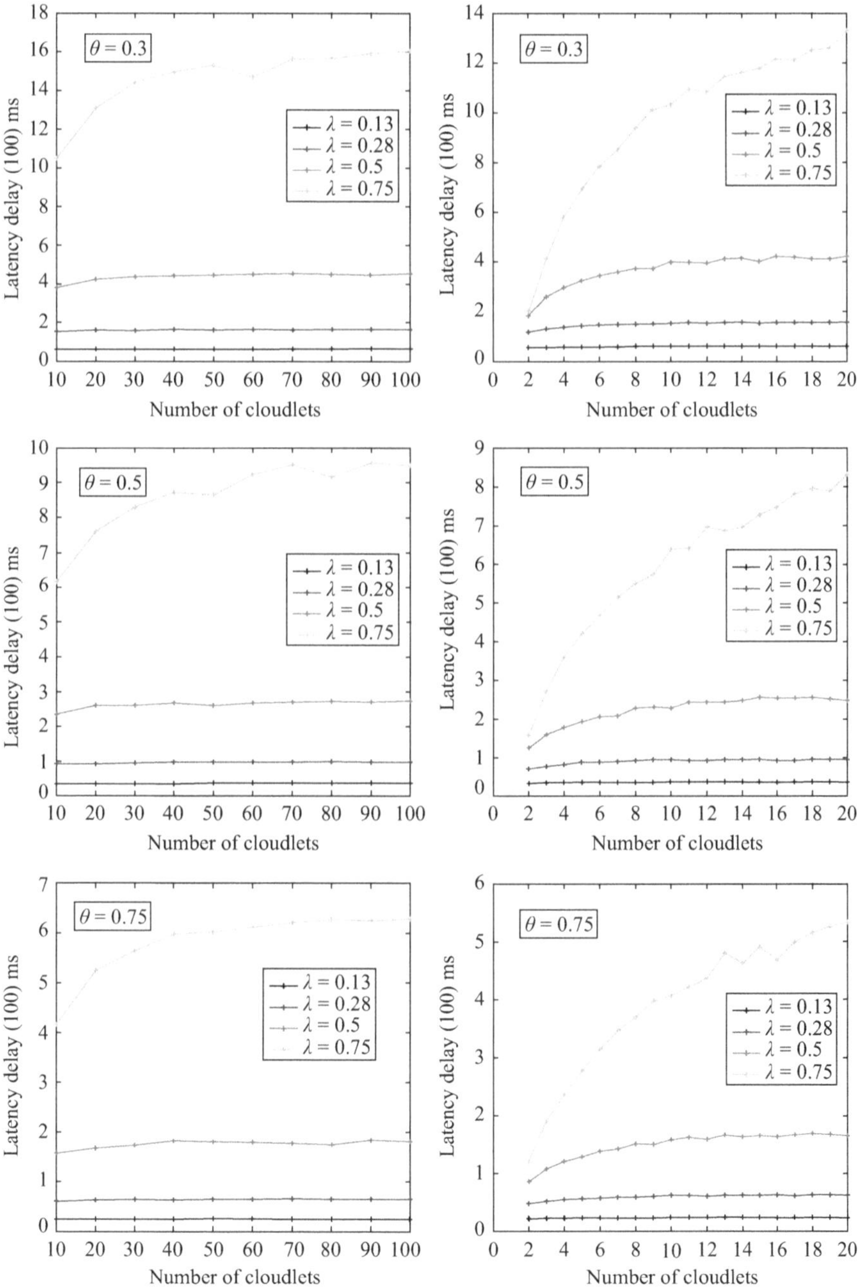

Figure 4.12 Latency delay for the ring architecture

In the second scenario, each cloudlet sends a single request to its central cloudlet (for the hierarchical architecture) or its left and right secondary cloudlets (respectively for the ring architecture). As shown in Figures 4.15 and 4.16, we notice that the latency delay of the hierarchical architecture grows rapidly and is higher than

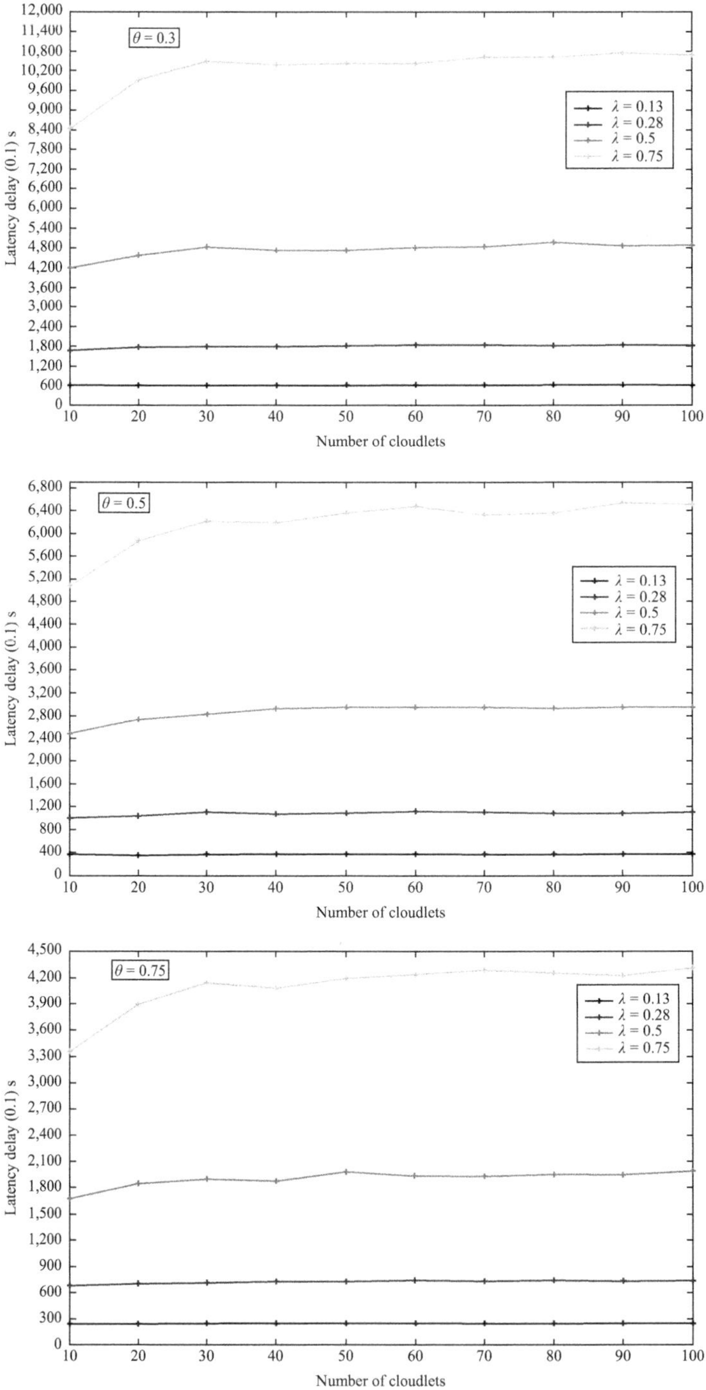

Figure 4.13 Latency delay for multiple requests submission in the hierarchical architecture

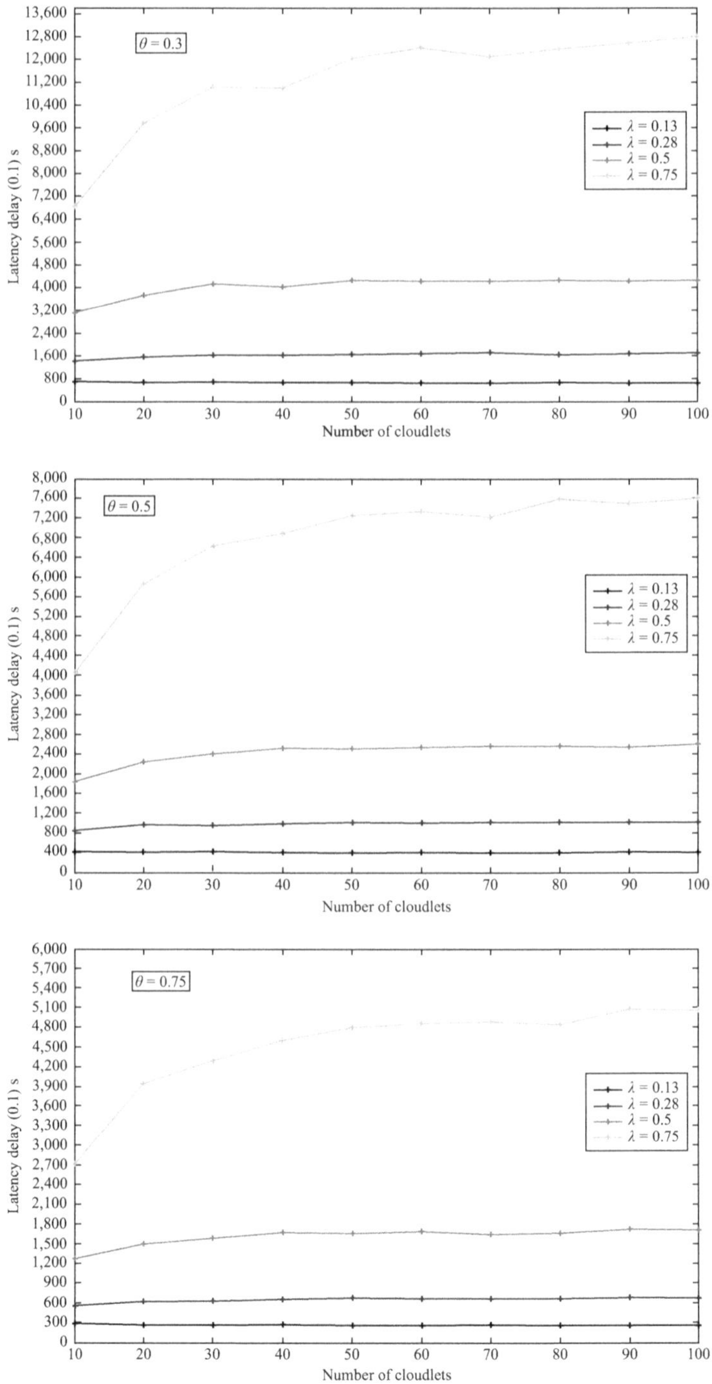

Figure 4.14 Latency delay for multiple requests submission in the ring architecture

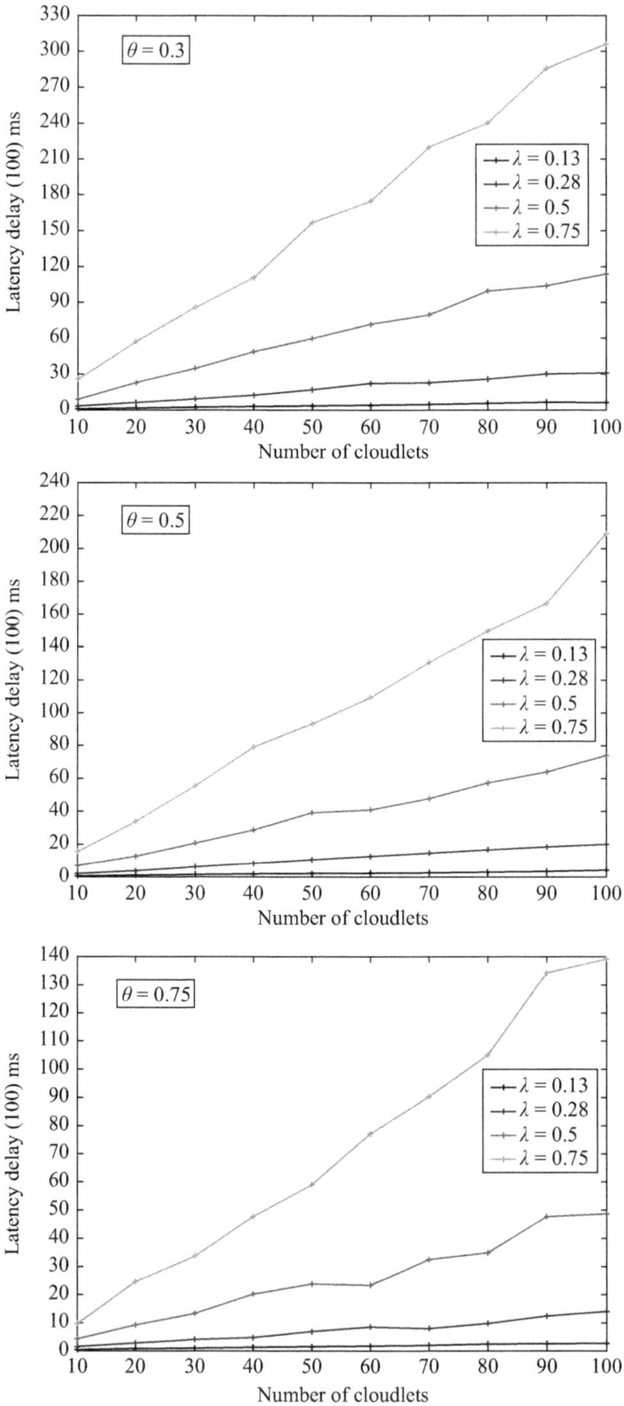

Figure 4.15 Latency delay for single request submission in the hierarchical architecture

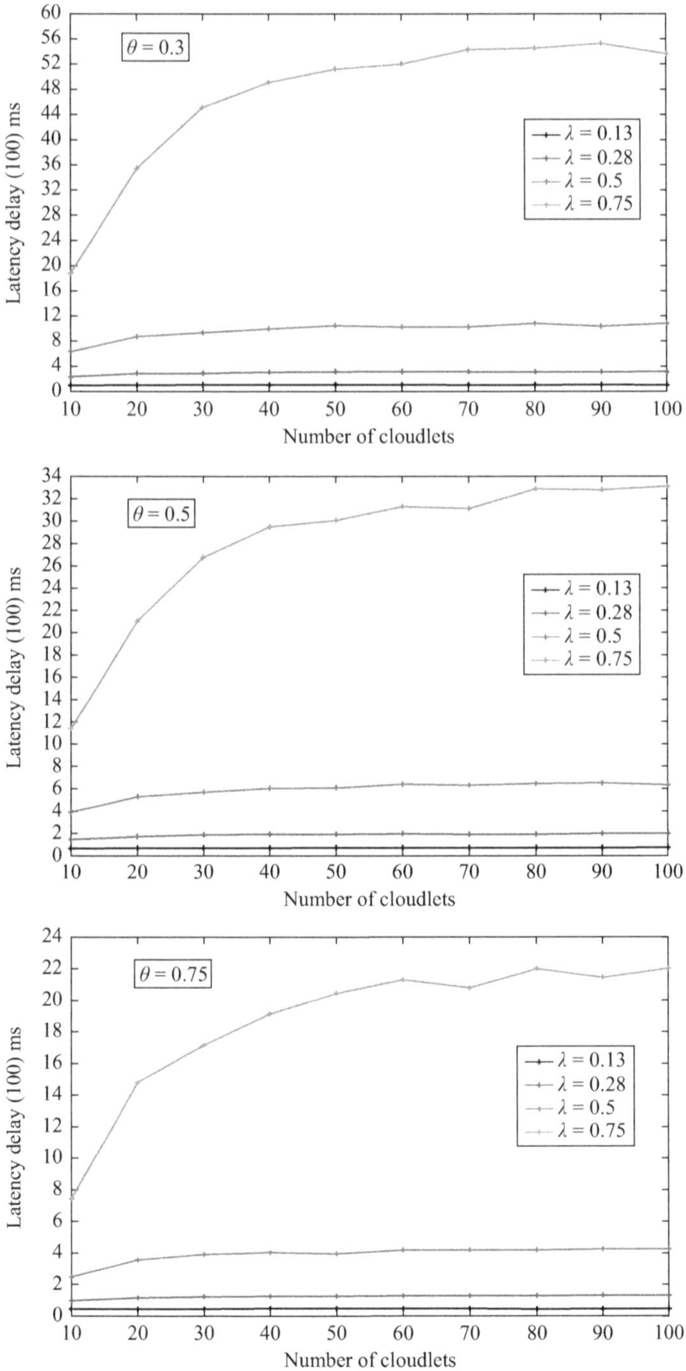

Figure 4.16 Latency delay for single request submission in the ring architecture

Table 4.3 Simulation results of the ring and hierarchical architecture in the second scenario

Results (%)	Hierarchical architecture		Ring architecture	
	$\theta = 30\%$	$\theta = 75\%$	$\theta = 30\%$	$\theta = 75\%$
$\lambda = 75$	$L \approx 306$ ms	$L \approx 140$ ms	$L \approx 54$ ms	$L \approx 22$ ms
$\lambda = 13$	$L \approx 10$ ms	$L \approx 5$ ms	$L \approx 2$ ms	$L \approx 1$ ms

L, the parameter of latency delay.

the latency delay of the ring architecture, that can be shown in Table 4.3. Therefore, the ring architecture is more flexible and more efficient while each cloudlet sends request at the same time, whereas the hierarchical architecture is not efficient because of the bottleneck of communicating requests to the central cloudlet.

4.5 Conclusion

In this paper, we have proposed two cloudlet-based architectures, hierarchical, and ring, which exploit users' proximity to improve the mobile user cloud experience. We model the latency delay of the two cloudlet-based architectures using the bidimensional Markov chain, and we implement two different scenarios of submitting user requests. This work represents a proof of concept for the use of cloudlets and compares the performance of the two architectures. In the first scenario of submitting multiple requests, the performance of the two architectures changes according to the fast fading variation. In the second scenario in which a single request is sent, the ring architecture is more responsive and more efficient than the hierarchical architecture. As a future work, we intend to study the use of hybrid architectures with multitiered topologies consisting of multiple groups of cloudlets organized into rings. We also plan to test the deployment of cloudlet architectures using the virtual machine technology with different mobile applications.

References

[1] Mahadev, S.: 'Mobile computing: The next decade'. The First ACM Workshop on Mobile Cloud Computing and Services (MCS 10), New York, NY, USA, 2010, pp. 5–6.

[2] Dinh, H.T., Lee, C., Niyato, D., and Wang, P.: 'A survey of mobile cloud computing: architecture, applications, and approaches'. *Wireless Communications and Mobile Computing*, 2011, vol. 13, pp. 1587–1611, 2011, DOI:10.1002/wcm.1203.

[3] Cisco VNI Forecast and Methodology, 2015–2020, http://www.cisco.com/en/US/solutions/collateral/ns341/ns525/ns537/ns705/ns827/white-paper-c11-520862.html, accessed April 2015.

[4] Satyanarayanan, M.: 'Fundamental challenges in mobile computing'. The Fifth Annual ACM Symposium on Principles of Distributed Computing (PODC 96), New York, NY, USA, 1996, pp. 1–7.
[5] Koukoumidis, E., Lymberopoulos, D., Strauss, K., Liu, J., and Burger, D.: 'Pocket cloudlets'. SIGARCH Computer Architecture News, 2011, pp. 171–184.
[6] Dewar, C., and Warren, D.: 'Understanding 5G: Perspectives on future technological advancements in mobile'. GSMA Intelligence, December 2014.
[7] Satyanarayanan, M., Bahl, P., Caceres, R., and N. Davies: 'The case for VM-based cloudlets in mobile computing'. *IEEE Pervasive Computing Transactions*, vol. 8, no. 4, pp. 14–23, Oct.–Dec. 2009, doi: 10.1109/MPRV.2009.82.
[8] Niroshinie, F., Seng, W. L., and Wenny, R.: 'Mobile cloud computing: A survey'. Future Generation Computer Systems, 29(1):84–106, 2013. ISSN 0167-739X.10.1016/j.future.2012.05.023.
[9] Khan, K., Wang, Q., and Grecos, C.: 'Experimental framework of integrated cloudlets and wireless mesh networks'. Telecommunications Forum (TELFOR), Serbia, 2012, pp. 190–193.
[10] Mehendale, H., Paranjpe, A. and Vempala, S. 'Lifenet: A flexible ad hoc networking solution for transient environments'. ACM SIGCOMM Computer Communication Review, pp. 446–447, 2011.
[11] Sakr, S., Liu, A., Batista, D. M., and Alomari, M.: 'A survey of large scale data management approaches in cloud environments'. Communications Surveys and Tutorials, IEEE, 2011, pp. 311–336.
[12] Soyata, T., Muraleedharan, R., Funai, C., Minseok, K., and Heinzelman, W.: 'Cloud-vision: Real-time face recognition using a mobile cloudlet-cloud acceleration architecture'. Computers and Communications (ISCC), 2012 IEEE Symposium on. 1–4 July 2012, pp. 59–66.
[13] Verbelen, T., Simoens, P., De Turck, F., and Dhoedt, B.: 'Cloudlets: Bringing the cloud to the mobile user'. Third ACM Workshop on Mobile Cloud Computing and Services Proceedings, 2012, pp. 29–35.
[14] Yang, Z., Zhao, B.Y., Xing, Y., Ding, S., Xiao, F., and Yafei, D.: 'Amazingstore: Available, low-cost online storage service using cloudlets'. The Ninth International Conference on Peer-to-Peer Systems (IPTPS 10). USENIX Association. Berkeley, CA, USA, 2010.
[15] Fesehaye, D., Gao, Y., Nahrstedt, K., and Wang, G.: 'Impact of cloudlets on interactive mobile cloud applications'. The 16th International Enterprise Distributed Object Computing Conference, IEEE, 2012, pp. 123–132.
[16] Sarkar, S., Chatterjee, S., and Misra, S.: 'Assessment of the suitability of fog computing in the context of Internet of Things'. IEEE Transactions on Cloud Computing, vol. PP, no. 99, pp. 1–1, doi: 10.1109/TCC.2015.2485206.
[17] Corsaro, A.: 'Cloudy, foggy and misty Internet of Things'. Proceedings of the Seventh ACM/SPEC on International Conference on Performance Engineering (ICPE'16). ACM, 2016.

Chapter 5

Survey on software-defined networking and network functions virtualisation in 5G emerging mobile computing

Eugen Borcoci

5.1 Introduction

This chapter is a *survey* of significant and recent proposals, studies and trials, concerning application of emergent software-defined networking (SDN) and network functions virtualisation (NFV) technologies in the domain of 5G networking.

The 5G architectures and technologies are identified today as main wireless network candidates [1] able to respond to novel and challenging requirements: high-capacity, low-latency, good vertical and horizontal scalability, universal support for data and media applications and services (in fixed and mobile environment). The domains include vehicular, that is vehicle-to-vehicle (V2V), vehicle-to-infrastructure and vehicle–to-'whatever', and more general machine-to-machine (M2M) and also Internet of Things (IoT) communications. The 5G networks should be flexible and open, able to accommodate heterogeneous access technologies and should provide open communication systems including cellular networks, clouds and data centres, home networks and gateways (GWs), satellite systems and others. The 5G networks should be adaptable to users' and services' changing needs to handle application-driven networks. The management and control (M&C) in a flexible way is a major requirement. The novel systems should also satisfy security, privacy, resiliency, robustness and data integrity requirements.

SDN and NFV are currently seen as important complementary technologies to implement the 5G architectures.

SDN separates the control plane (CPl) and data (forwarding) plane (DPl), thus enabling external control of data flows through logical software entities, that is remote vendor-neutral controllers. SDN abstracts the network components, their functions and the protocol to manage the forwarding plane. The SDN-centralised up-to-date view upon the network makes it suitable to perform network management, while allowing flexible modification of the network behaviour through the CPl. SDN-type control in 5G wireless networks is attractive, given its ability to support network virtualisation, automating and creation of new services on top of the virtualised resources. Routing and data processing functions of wireless infrastructure can be provided by software packages running in general-purpose computers or even in the cloud.

SDN is useful in wireless-distributed networks, such as mobile ad-hoc, sensor, device-to-device and vehicular networks. SDN will enable the management of heterogeneous network (HetNet) nodes (i.e. macrocell, picocell, etc.) and heterogeneous backhaul connectivity such as fibre, wireless and others.

NFV is a complementary technology to SDN; it can enhance and make the 5G networking more flexible by virtualising many network functions (NFs) and deploying them into software packages. Such functions can be dynamically assembled and chained to implement legacy services or novel ones. This approach will allow for higher flexibility and also resilience in operation and management of the mobile networks. NFV allows transparent migration between either virtual machines or real machines. Implementing mobile NFs in data centres provides more flexibility in terms of resource management, assignment and scaling. NFV can be a candidate for virtualising the core network as well as centralising some processing within radio access networks (RANs). Cloud-based radio access network (CRAN) can use virtualised software modules, running in different virtual machines. Combining NFV with SDN may offload the centralised location within networks nodes which require high-performance connections between radio access (RA) point and data centres.

This chapter will contain in the first part a summary of 5G requirements and challenges, with emphasis on those supposed to be (partially) solved through SDN/NFV control. Some relevant 5G use cases and services will be summarised. A short presentation of SDN and NFV concepts and architectures is done related to layering, CPl and DPl issues, network operation systems (NOS) and software technologies, virtualisation, north-bound and south-bound interfaces, function chaining, scalability and real-time issues and so on. The second part of the chapter will explore various architectures and implementations based on SDN/NFV in 5G environment; distribution versus centralisation; unified CPl concepts in 5G/SDN, heterogeneous CRANs, cellular 5G with SDN control, SDN approach for mobile cloud computing in 5G, backward compatibility and deployment issues. Some future open directions of research will be presented in the conclusions.

Given the limited space of this chapter, several aspects related to 5G technologies and services have not been discussed: security and privacy, details on scalability, reliability, mobility, IoT services, M2M and device-to-device (D2D) communications and cognitive radio network (CRN) aspects.

5.2 Summary of 5G technology

5.2.1 Requirements and challenges

The 5G evolution of mobile broadband networks will bring new unique network and service capabilities [1–4]. It will ensure user experience continuity in various situations like high mobility (e.g. in trains), dense or sparsely populated areas, or heterogeneous technologies. The application range is broad, targeting manufacturing, automotive, energy, food and agriculture, education, city management, government, healthcare, public transportation and so forth.

5G will support IoT, being capable to connect a massive number of sensors and rendering devices and actuators with stringent energy and transmission constraints.

Mission critical services can be served by 5G, due to its high reliability, global coverage and/or very low latency (today these are still handled by specific networks) and public safety. 5G will integrate networking, computing and storage resources into one programmable and unified infrastructure allowing optimised and dynamic usage of all distributed resources, convergence of fixed, mobile and broadcast services, support for multi-tenancy models, enabling players collaboration and leveraging on the characteristic of current cloud computing, thus leading to a single digital market. Additional 5G requirements are related to sustainability and scalability, energy consumption reduction and energy harvesting.

The 5G technology exposes some disruptive capabilities such as an order of magnitude improvement in performance (more capacity, mobility and accuracy of terminal location, lower latency, increased reliability and availability); connection of many devices simultaneously; help citizens to manage their personal data, tune their exposure over the internet and protect their privacy; enhanced spectral efficiency (SE) and high energy efficiency with respect to 4G; reduce service creation time and facilitate the integration of various players delivering parts of a service; to be built on more efficient hardware and inter-working in heterogeneous environments. The 5G key technological components include heterogeneous set of integrated air interfaces, cellular and satellite solutions, seamless handover between heterogeneous wireless access technologies, simultaneous use of different radio access technologies (RAT) and ultra-dense networks with numerous small cells (this require new interference mitigation, backhauling and installation techniques).

5G will be fully driven by software: a unified operating system is needed, in a number of points of presence (PoPs), especially at the network edge. To achieve the required performance, scalability and agility, the 5G can rely on technologies like SDN, NFV, mobile edge computing and fog computing. The 5G networking concepts will ease and optimise the network management based on cognitive features, advanced automation capabilities of operation through algorithms that optimise complex business objectives (e.g. end-to-end (E2E) energy consumption), data analytics and big data techniques (monitor the users quality of experience (QoE) through new metrics, combining network data and behavioural data while guaranteeing privacy).

According to summary figures of 5G, very ambitious goals/challenges are 1,000$\times$ in mobile data volume per geographical area reaching a target $\geq$10 Tb/s/km^2; 1,000$\times$ in number of connected devices reaching a density $\geq$1 M terminals/km^2; 100$\times$ in user data rate reaching a peak terminal data rate $\geq$10 Gb/s; 1/10$\times$ in energy consumption compared to year 2010; 1/5$\times$ in end-to-end latency reaching 5 ms, for example tactile Internet and radio link latency reaching a target $\leq$1 ms for special use cases like V2V communication; 1/5$\times$ in network management OPEX; 1/1,000$\times$ in service deployment time reaching a complete deployment in $\leq$90 min.

5.2.2 Key enablers and general design principles for a 5G network architecture

To solve the challenges presented in the previous sub-section, some key enablers have been identified [2–4] as efficient spectrum utilisation, usage of massive/3D multiple input multiple output (MIMO) and new air interfaces (e.g. new waveform, advanced multiple access), use of optical network technologies, simple access points (APs), small cells and local offload, advanced traffic management, caching and pre-fetching for content, control/data plane split, use of SDN/NFV/cloud technologies, third parties and/or user deployment models, enable new business models in a programmable manner, service-oriented network capabilities, application programmers interfaces (APIs) should be available at different levels (resources, connectivity and service enablers), energy-efficient hardware and management techniques and big data-driven network intelligence (NI).

Several 5G design principles have been defined [2] to deliver the solutions: design new air interface and new multiple access scheme and L1/L2 techniques optimised for high frequencies, latency and massive connectivity; use high frequencies and other spectrum options (e.g. pooling and aggregation); utilisation of optical transmission and switching where possible (in fronthaul and backhaul); bring communicating endpoints closer together in order to reduce E2E latencies; apply virtualisation principles; address coverage and capacity issues separately; minimise the number of network layers and pool resources; minimise functionalities performed by APs in order to make them more simple; maximise energy efficiency across all network entities; use intelligent agents to manage QoE, routing, mobility and resource allocation; efficient design of the non-access stratum (NAS) protocols (to reduce E2E latency), services and service complexity.

There are several types of architectures for 5G networks [4]: multi-tier, cloud-based architectures [5], CRN-based and D2D communication based. They can be combined in operation. *In this chapter, only the first two are discussed.* Commonly, two-tier architectures are proposed in many studies, models and implementation, where a macrocell base station (MBS) is in the top-tier and small-cell base stations in the lower tier work under MBS supervision. A macrocell covers all the small cells of different types, for example femtocell and picocell, while microcell and both tiers share an identical frequency band. The small cell enhances the coverage and services of a macrocell. In addition, D2D communication and CRN-based communication may enhance a two-tier architecture to a multi-tier architecture.

Cloud computing principles, suppose an infrastructure providing on-demand, easy and scalable access to a shared pool of configurable resources, while users do not worry about the management of resources. The cloud-based architectures and technologies are also attractive for 5G. The first approach has been to build CRANs for 5G networks [4,5,29,30]. The basic CRAN idea is to execute most of the MBS functions in the cloud, and hence divide the functionality of an MBS into a control layer and a data layer. Here, the SDN principles will naturally apply. The functions of the control and the data layers are executed in a cloud and in an MBS, respectively. A CRAN can provide a dynamic service allocation scheme for scaling the network without installing costly network devices.

5.3 Software-defined networking (SDN)

SDN [6–9] architecture is mainly focused on enhancing the network programmability in data centres and also in different types of networks. SDN clearly separates the CPl from DPl, by moving the CPl functions outside the traditional network nodes (routers, switches – called also *forwarders*) to some external logical software entity called *controller* (executed on general purpose computing hardware). Thus, SDN decouples the control software from specific networking hardware produced by different vendors and makes the DPl programmable.

The controller communicates (via a secure channel) with several forwarders and instructs them what actions to perform on the data plane flows (by filling in the forwarders the so-called flow tables). The actions to be performed on a given flow in a forwarder are determined by the match between the packet fields (one or more fields match search can be enforced) and flow tables records. Inside a forwarder, one or several flow tables may exist, assembled in a processing pipeline followed by a data packet. The networking components and their functions are represented to higher layers as abstractions, able to capture the requirements of the common software switches and routers. Therefore, a controller creates a centralised unified (and systematically updated) abstract view upon the network status and allows a coherent and flexible modification of the DPl behaviour by modifying on-fly the flow tables. Special communication protocols between the controller and forwarders have been designed, a typical examples being the *OpenFlow* [6,7], developed by open networking foundation (ONF) [10]. Note that also, other protocols have been proposed for the same purpose [8].

The SDN technology offers several important advantages [6] like high-performance, granular traffic control across multiple vendors' network devices; centralised M&C, common APIs abstracting the underlying networking details; network programmability opportunities offered to operators, enterprises, independent software vendors and users; network evolution into an extensible vendor-independent service delivery platform; increased network reliability and security; and better end-user experience. However, SDN technology has still problems under study related to centralisation (the controller is actually a single point of failure) – inducing issues about reliability, horizontal and vertical scalability, real-time capability of network control, backward compatibility and security [6,7].

5.3.1 SDN architecture

Figure 5.1 depicts the overall SDN architecture. The SDN CPl has two major interfaces: *northbound* and *southbound*. The northbound interfaces provide higher level abstractions to the applications. Several application examples are shown in Figure 5.1 for network management/control: routing, traffic engineering and quality of services (QoS) control. However, other additional applications can be linked at the northbound interface. The CPl southbound interface supports the communication with network devices (forwarding elements (FE), i.e. switches or routers).

A typical implementation of the southbound interface is the OpenFlow standard, which has been defined by ONF, in several versions 1.0–1.5 [10]. In large networks, where several controllers are necessary, additional east–west interfaces are defined to support inter-controller communications.

The CPl could be split into two sub-layers: abstraction/virtualisation sub-layer and NOS sub-layer [6,8,9]. The NOS is a distributed system that creates a

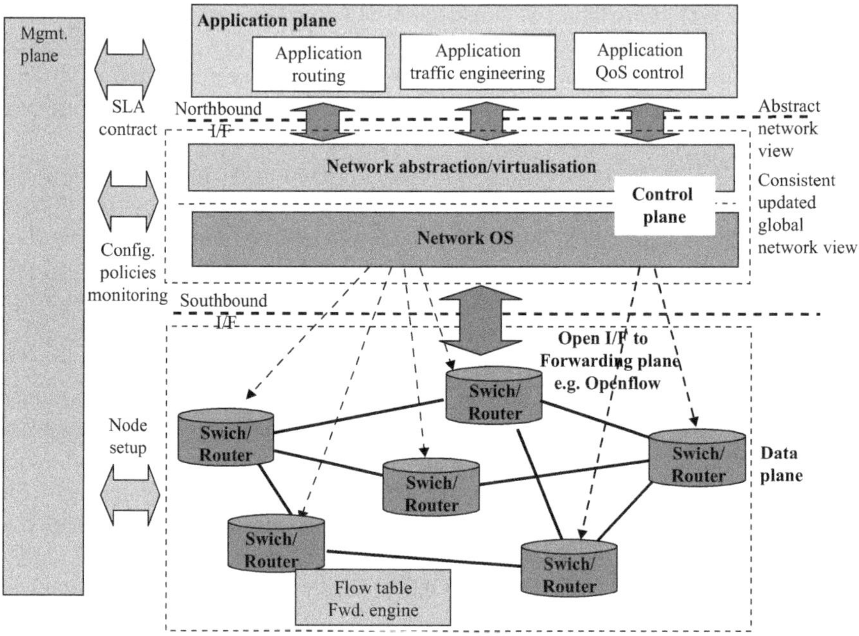

Figure 5.1 Generic architecture of software-defined networking

consistent, updated network view. It can be executed on servers (controllers) in the network. There are many examples of NOS implementations: NOX, Onix, Hyper-Flow, Floodlight, Trema, Kandoo, Beacon and Maestro [6]. The NOS uses forwarding abstraction in order to collect state information from FEs and generates commands to FEs. The virtualisation sub-layer creates an abstract network view for the application plane.

The OpenFlow is the first SDN standard implementing the CPl–DPl interface. It allows direct access to the DPl of network devices, that is FE, both physical and virtual (hypervisor-based) and allows one to move CPl out of the FEs to logically centralised control software. The OpenFlow specifies the basic primitives to be used by an external software application to programme the DPl (similar to the instruction set of a CPU). The *flow concept* identifies the network traffic based on pre-defined match rules that can be statically or dynamically programmed by the SDN control software. The DPl can be programmed on a per-flow basis (to provide – if wanted – extremely granular control), or in aggregating mode, thus enabling the network to respond to real-time changes at the application, user and session levels. The IT administrator can define how the data traffic should flow through FEs, based on parameters such as usage patterns, applications and cloud resources. The FEs no longer need to understand and process the intelligent protocol standards but simply accept instructions from their SDN controller. The management plane (MPl) in Figure 5.1 serves to manage the service level agreements associated with applications to configure and monitor the functionalities and performances of the

CPl – under guidelines of the network provider policies. The MPl generates low-level commands for network elements configurations.

SDN power consists in creating the possibility to manage the entire network through orchestration and provisioning systems. More flexibility is achieved for virtualised networking, self-service provisioning, dynamic (on-demand) various resource allocation and secure services. SDN can be equally applied in cloud (data centres) computing, in carrier wide area networks and more recently in wireless domain.

5.3.2 Benefits of SDN architecture for 5G

In 5G wireless networks environment, the SDN separation of the control logic from vendor-specific hardware is valuable, allowing to build open and vendor-neutral software controllers. SDN provides virtualisation capabilities, enabling automation and creation of new services on top of the virtualised resources in secure and trusted networks.

Historically, the CPl/DPl separation partially existed in the wireless networks. The Internet Engineering Task Force (IETF) standardised the control and provisioning of wireless access points (CAPWAP – RFC 5415) protocol which centralises the control in wireless networks. CAPWAP is technology-agnostic and requires specific bindings for each considered access standard; however, for the time being, only the binding for 802.11 has been defined. Radio configuration is expressed in terms of management information base elements, for example operating channel or the transmission power, beacon interval or medium access control (MAC) contention parameters. The control frames are delivered to a central controller having MAC functions, in a way similar to OpenFlow interface.

However, SDN offers a more complete framework than CAPWAP, by possibility of moving routing and data processing functions of wireless infrastructure into software packages in general servers or in clouds. The CPl consists of network management and optimisation tools implemented on the network servers. The DPl can be composed by base stations (BSs) which are software-defined (SD-BSs) [2,9], RAN controlled in SDN style and SDN forwarders (software switches) in the cellular core network part. Their L1–L3 control functions can be implemented in software on general-purpose computers and/or remote data centres. Several important 5G objectives can be well supported by SDN control: convergence of HetNets, evolution capability, adaptiveness, infrastructure-as-a-service (in cloud computing – like environment) and energy savings.

5.4 Network functions virtualisation (NFV)

NFV [12–15,18] is a recent architectural development, aiming to reduce the time to market of new services and improve network service (NS) provisioning flexibility. NFV decouples the software implementation of NFs from the underlying hardware by using virtualisation technologies and commercial off-the-shelf (COTS) programmable hardware (general-purpose servers, storage and switches). Therefore, various network-related functions, for example traffic load balancing, network address translation, firewalling, intrusion detection, domain name service, caching and others can be delivered in software and deployed on general-purpose servers. Currently, NFV exposes also several challenges, such as the network performance guarantees for virtual appliances, dynamic instantiation and migration and efficient placement of *virtual network functions (VNFs).*

The NFV general objectives are to improve capital efficiencies versus dedicated hardware implementation solutions by using COTS hardware to provide VNFs, through software virtualisation; sharing of hardware and reducing the number of different hardware (HW) architectures; improving the flexibility in assigning VNFs to hardware – thus realising a better vertical and horizontal system scalability, decoupling the networking functionalities from location, enabling time of day reuse and enhancing resilience through virtualisation and facilitating resource sharing; to support rapid service innovations through software-based service deployment; automation and operating procedures to increase the operational efficiency; reducing the power consumption by migrating workloads and powering down unused hardware; defining standardised and open interfaces between VNFs infrastructure and management entities.

The main actor involved in development of the NFV specifications is the European Telecommunications Standards Institute (ETSI) NFV group global (operators-initiated) *industry specification group (ISG)*, under the auspices of ETSI, having about 200 members (2014) and including 28 Tier-1 carriers (and mobile operators), service providers (SPs) and cable industry entities.

ETSI [13] has defined the NFV framework as totality of all entities, reference points, information models and other constructs defined by the specifications published by the ETSI ISG NFV. In NFV, the NSs are provisioned differently with respect to current networks practice. The software and hardware are decoupled; therefore, a network element is no longer a collection of integrated hardware and software modules, so they may evolve independently.

ETSI has defined several *functional blocks (FBs)* [15]. A *network function (NF)* is defined as an FB within a network infrastructure having well-defined external interfaces and well-defined functional behaviour (today an NF is often a network node or physical appliance). NFV applies the principle of separating NFs from the hardware they run on by using virtual hardware abstraction. Flexible NF deployment is possible.

The software/hardware detachment allows to re-assign and share the infrastructure resources. The hardware and software can perform different functions at various times. The HW resources pool is already in place and installed at some *network function virtualisation infrastructure (NFVI) – PoPs;* the immediate consequence is that the actual NF software instantiation can be automated. A network point of presence is a location where an NF is implemented as either a physical network function (PNF) or a VNF. The NFVI is the totality of the hardware and software components which build up the environment in which VNFs are deployed. NFV can leverage the different cloud and network technologies currently available. All these help network operators (NOs) to faster deploy new NSs over the same physical platform.

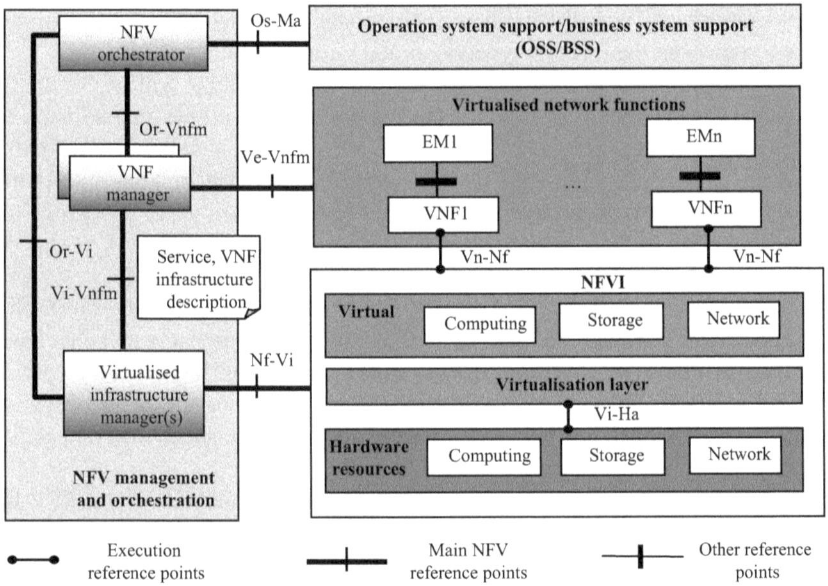

Figure 5.2 NFV reference architectural framework [15]

The NFV can support dynamic operations: the NFs are performed by instantiable SW components, which provide greater flexibility (in comparison to traditional networking procedures) to scale the actual VNF performance in a more dynamic way. A finer granularity can be obtained, for instance, according to the actual traffic.

Figure 5.2 shows the NFV reference architecture defined by ETSI [15], where the main FBs and reference points (interfaces) can be seen. Several working domains can be defined:

Operations and business support systems – VNF module contains the software implementations of NFs and runs over the NFVI. This module contains different element management entities and VNF.

NFV infrastructure (NFVI) includes all hardware and software components, building up the environment in which VNFs are deployed, and it can span across several locations, for example places where data centres reside. The network providing connectivity between these locations is regarded to be part of the NFVI. An NFVI component is an NFVI hardware resource that is not field replaceable, but is distinguishable as a COTS component at manufacturing time. The *virtualisation layer (VL)* is an important component of the NFV. It abstracts the hardware HW resources and decouples the VNF software from the underlying hardware, thus ensuring a HW-independent lifecycle for the VNFs. The VL is responsible for abstracting and logically partitioning physical (PHY) resources, commonly as a HW abstraction layer, enabling the software that implements the VNF to use the underlying virtualised infrastructure, providing virtualised resources to the VNF, so that the latter can be executed. The VL allows the software of the VNFs to be deployed on different physical hardware resources. Typically, this type of functionality is provided for computing and storage resources in the form of hypervisors and virtual machines (VMs).

NFV Management and orchestration (NFV-MANO) deals with orchestration and lifecycle management of physical and/or software resources that support the infrastructure virtualisation and the VNFs lifecycle management. NFV-MANO focuses on all virtualisation-specific management tasks and includes the partial managers for the data plane layers: *virtualised infrastructure manager (VIM), virtualised network function manager (VNFM)* and *NFV orchestrator (NFVO).* The NFVO optimises the resource allocation, that is manages the NS lifecycle, VNF lifecycle (supported by the VNFM) and NFVI resources (supported by the VIM).

An *NS* is a composition of NFs defined by its functional and behavioural specification. The NSs contribute to the behaviour of the higher layer service, which is characterised by at least performance, dependability and security specifications. The individual NF behaviour plus a network infrastructure composition mechanism determines the end-to-end (E2E) NS behaviour.

Many NFs existent in a legacy environment can be virtualised in the NFV framework: 3GPP evolved packet core (EPC) network elements, like mobility management (MM) entity (MME), serving gateway (SGW), packet data network gateway (PGW); residential gateway in home networks and conventional NFs (dynamic host configuration protocol servers, firewalls, etc.). The functional behaviour and the external operational interfaces of a PNF and a VNF are expected to be the same. A VNF may have one or several internal components, for example one VNF can be deployed over multiple VMs (each VM hosts a single VNF component).

The NFV technology is expected to provide strong support for several use cases [13,14]. The NFVI supports several use cases and fields of application already identified by the NFV ISG while providing a stable platform for the VNF evolution. It also provides a multi-tenant infrastructure, leveraging standard IT virtualisation technology that may support multiple use cases and fields of application simultaneously. The cloud-related use cases are *NFVI as a service, VNF as a service and service chains (VNF forwarding graphs).* The mobile use cases could be virtualisation of the *mobile core network* and/or of the mobile BSs. *Virtualisation of content delivery networks* is also supported by NFV. For access and residential environment, one can mention *virtualisation of the home environment* and *virtualisation of the fixed access NFs.*

5.5 SDN–NFV cooperation

While SDN separates the control and forwarding planes thus offering a centralised network view, NFV is primarily focused on optimising and making more flexible the NSs. Although NFV is intended to optimise the deployment of NFs (such as firewalls, DNS, load balancers, BSs, etc.), SDN is focused on optimising the underlying networks.

The SDN/NFV cooperation is of high interest in order to obtain flexible and programmable systems, taking benefits from both technologies [16–21].

Both architectures are optimised for the dynamic cloud environment at carrier scale. Several major standardisation organisations, forums and groups are active in both NFV/SDN areas, and cooperation between them currently becomes stronger: ETSI NFV ISG, the IETF and the ONF, as well as major industry-led open-source projects as *OpenStack* [11] and *OpenDaylight* [17].

Note that NFV is complementary to SDN, but not dependent on it (or vice versa). NFV can be implemented without an SDN, although the two concepts and solutions can be combined with potentially greater value. NFV goals can be achieved using non-SDN mechanisms, relying on the techniques currently in use in many data centres, but SDN separation CPl/DPl can enhance the performance, simplify compatibility with existing deployments and facilitate operation and maintenance (O&M). NFV is able to support SDN by providing the infrastructure upon which the SDN software can be run, while NFV aligns closely with the SDN objectives to use commodity servers and switches.

Figure 5.3 shows a high-level view of the NFV/SDN map in the ONF vision [10]. Deployment of NFV requires large-scale dynamic network connectivity both in the physical and virtual layers to inter-connect VNF endpoints. As Figure 5.3 shows, there are many complementary industry efforts focused on establishing an open NFV/SDN ecosystem.

The *OpenDaylight* project is a collaborative open-source project hosted by The Linux Foundation having as objective to accelerate the adoption of SDN and to create a foundation for NFV. It supports open standards, such as the OpenFlow, and delivers a common open-source framework and platform for SDN across the industry for customers, partners and developers. The first code from the Open-Daylight Project, named Hydrogen, was released in February 2014 [8,9]. Expected modules set include an open controller, a virtual overlay network, protocol plug-ins and switch device enhancements.

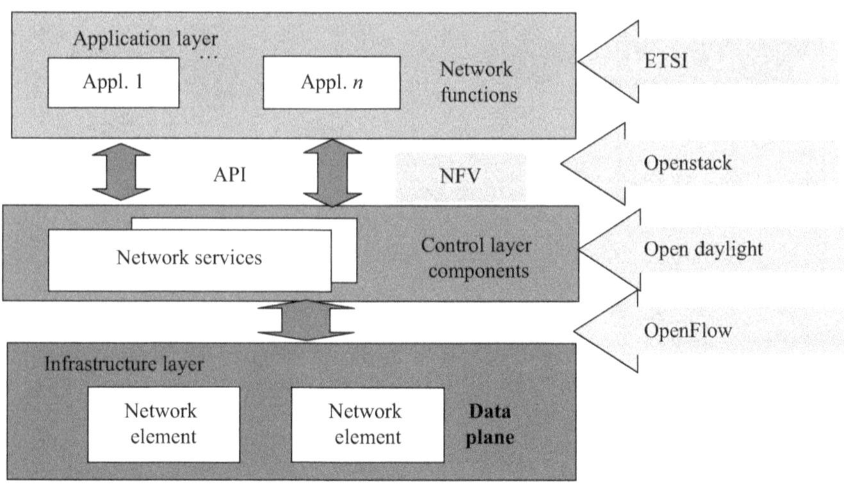

Figure 5.3 NFV and SDN industry map [10]

OpenStack is an open-source software supporting the deployment and management of a cloud (both private and public) infrastructure as a service (IaaS) platform. It fulfils two main requirements: massive scalability and simplicity of implementation. The software platform consists of inter-related components that control hardware pools of processing, storage and networking resources throughout a data centre. OpenStack is configurable – the user can choose whether or not to implement several services offered by the software.

The components are user-configurable that can be made through the API. Users either manage it through a web-based dashboard, through command-line tools or through a restful API. The tool is flexible, able to cooperate with other software. It supports different hypervisors (Xen, VMware or kernel-based VM [KVM]) for instance and several virtualisation technologies. The OpenStack is modular, offering a set of services (components). Some examples are the following (a full list is in [11]): *compute (code name: Nova)* – a cloud computing fabric controller; *object storage (code name: Swift)* – a scalable redundant storage system; *network management (code name: Neutron)* enabling connectivity between VMs, through virtual nodes. OpenStack is currently managed by the OpenStack Foundation [11]. Figure 5.4 (adapted after [17]) shows an example of embedding SDN modules inside NFV framework.

Note that both SDN controller and OpenFlow VNF switch can be realised at VNF layer, while an OpenFlow pSwitch is placed at hardware resource layer. At VL, special components are controlling the virtual network (VNet) and OpenFlow vSwitch. The NFV M&O coordinates the configuration of the SDN-related

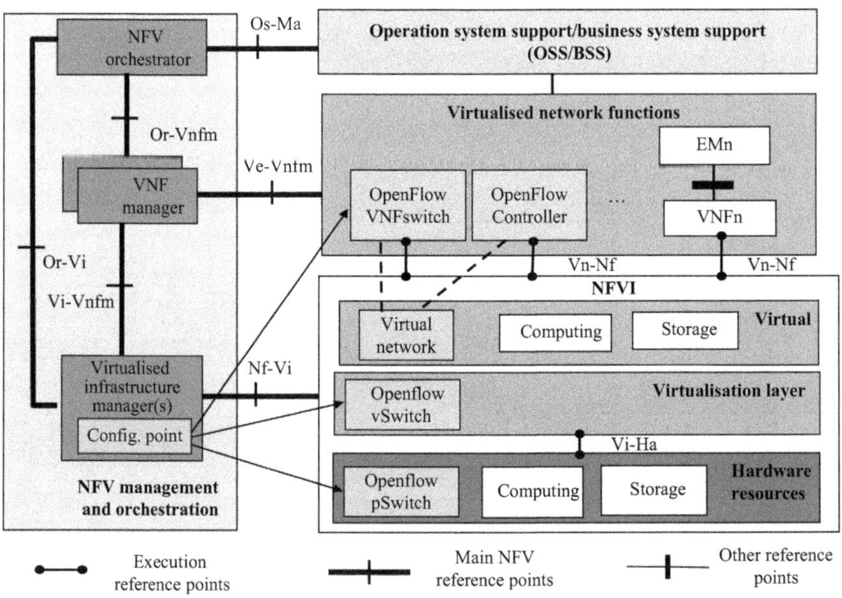

Figure 5.4 NFV–SDN cooperation example

components. A VNF utilises the virtualised resources and may use different VMs connected via virtual network.

5.6 SDN- and NFV-based architectures in 5G

Several studies have been recently dedicated to SDN- [22–26] and NFV [27,29,30]-based 5G architectures. However, part of them are high level (generic), lacking of focus; key details on the novel enabling technologies are not provided and still many open research issues are not yet covered. This sub-section attempts to present some relevant approaches.

5.6.1 General requirements and framework

While SDN and NFV technologies are very promising with respect of 5G development, several challenges, different from wire-line environment exist in the context of wireless environment, mobile and cellular networks [26]. Specific issues exist, such as management of the radio resources, interference problems, user mobility, radio resources scarcity, real-time response of SDN-like control, scalability related to the management of traffic coming from high number of users and so on. The network M&C must keep a lot of states required for MM; it should monitor the flows and detect if the user traffic exceeds its pre-assigned quota, assure different level of QoS, enforce congestion control, optimise resource utilisation and to perform the billing.

5.6.1.1 End-to-end SDN in a wired-wireless scenario

Different virtualised domains, spanning the same geographical area, should be inter-connected [27] and integrated to form a cloud infrastructure. The wireless access virtualised domains will be integrated with the wire-line virtualised domains, extending the cloud computing concept – *network as a service* (NaaS). The physical infrastructure can contain three major components: data centres, core networks and access networks. Virtualisation methods of hosts, core network and access network are respectively applied, resulting in slices (virtual infrastructures) that can offer to the users (fixed or mobile), virtualised servers, core network and access wireless networks. The applications or services are no longer bounded to a given domain or layer. However, open problems still exist, mainly in management area. Orchestration of different protocols and standards used in different controllers should be realised in order to achieve E2E manageability promised by the virtualisation of technologies.

The SDN-like control and virtualisation of the underlying infrastructure (in wireless domain, different RATs might exist) – allow multiple SPs to simultaneously control and configure the underlying infrastructure (each SP has its slice). In this way, the service evolutions could be achieved by gradually applying changes in each network slice, thus lowering the backward compatibility issues.

An E2E integration scenario for 5G mobile systems needs to have radio high data rates combined with advanced wireless/wired management functionalities.

The backhaul/fronthaul segments should be integrated, flexible and programmable, able to adapt to service requirements and traffic conditions. SDN separates the bearer from control functions and allows centralised management and automatic configuration of several types of gateways: cell site gateway (CSG) and small cell site gateway (SCSG) on the aggregation site gateway (ASG) [27]. Other scenarios integrate WLANs – seen as a primary access method. Enhancements are expected having as objective to provide the same responsiveness and SLA of wired connections.

A number of requirements and also expectations are related to SDN and NFV approach applied in 5G integrated networks, as presented below.

Flexibility and network programmability due to SDN/NFV: wired/wireless integration will be facilitated by SDN orchestration; one can realise more effective adaptation strategies and dynamic capability to react in a coordinated way to business and application needs, while offering the NaaS.

Wired and wireless network management: a unified common view and control of the wired and wireless network can be performed by a single SDN orchestrator, having different specific southbound interfaces. Unified control of both mobile/backhaul/fronthaul access segments will be realised, expected to simplify network operations, lowering the operational costs and increase the degree of management actions automation.

Unified policy enforcement: due to unified management, policies can be defined and enforced only once and applied across the whole network. Group-based policy model will become a standard approach integrated within SDN solutions.

Independency of the operator from vendors: SDN southbound universal interfaces will significantly facilitate inter-operability among different vendor devices for an operator's network. However, the problem of southbound interfaces universality of the SDN forwarders is still open.

Performance improvement: the network throughput can be improved for users located in overlapped service areas by enabling advanced programmability of migration and handoff strategies. Download rates can be increased by activating multiple parallel streams. The SDN control and NFV can implement power-saving solutions (e.g. traffic migrations and sleep configurations) during high traffic peaks.

Flexible and customised applications: via its standard northbound and open APIs, SDN may offer open programmable access to the wireless infrastructure, adopting various controller modules, abstraction layers and enhanced northbound/southbound interfaces able to be fully integrated within the open SDN-based solutions designed and operated in wired networks.

5.6.1.2 Standardisation work

Important standardisation effort has been spent in the last years, related to SDN/NFV, including aspects on wireless networks and devices. Usually, the standards are more oriented on architectures, interfaces and management and less on technologies. Several standardisation entities cooperates, aiming to specifications convergence.

The ONF [10] published several documents on SDN. The wireless networks are addressed in the white paper 'OpenFlow-Enabled Mobile and Wireless Networks'. Use cases are discussed, including mobile traffic management and inter-cell interference management. Flexibility is shown to be a benefit in 4G technology multi-vendor scenario.

ETSI works on NFV standardisation with its ISG for NFV (ETSI NFV SIG). Currently, there are four working groups (WG), two expert groups (EGs) and four root-level work items (WIs). Among them, one can mention WG1 – infrastructure architecture, WG2 – management and orchestration, WG3 – software architecture, WG4 – reliability and availability, EG1 – security and EG2 – performance. The document of ETSI GS NFV-INF 001 is focused on wireless (and specifically mobile BSs) as a possible domain for virtualisation, and specifies standard interfaces and use cases; however, it does not specify how virtualisation is to be realised.

The International Telecommunications Union – Telecommunications Standardization Sector (ITU-T) elaborated standards for SDN applied in future networks. An example is ITU-T Rec. Y.3300 (2014) – Framework of SDN describing the SDN framework (objectives, definitions, capabilities and architecture), but not explicitly addressing the wireless case.

The IETF started work on SDN and network virtualisation. It introduced the concept of service function chaining (SFC), SFC architecture (draft-ietf-sfc-architecture-01) and SFC use cases in mobile networks (draft-ietfsfc-use-case-mobility-01). Architecture and related use cases are described for usage of SFC, that is a carrier-grade process for continuous delivery of services based on NF associations in mobile networks (e.g. in 3GPP).

The IEEE SDN initiative standardisation WG and research groups on virtualisation in wireless networks are also active in the framework of the research group on software defined and virtualised wireless access and the research group on SDN/NFV – structured abstractions.

5.6.2 Examples of early SDN approaches in wireless networks

OpenRoads [22] is an early attempt/experiment to use SDN in wireless network infrastructure based on WiMAX; it uses OpenFlow to separate control from the data path through open APIs. By using FlowVisor software [14], isolated network slices can be created. Therefore, multiple experiments can run simultaneously in a production wireless network. The SNMPVisor mediates device configuration access among different experiments. The architectural protocol stack contains in the CPl: applications on top of FlowVisor and/or SNMPVisor and OpenFlow/SNMP at lower layer of the CPl. The data plane contains OpenFlow switches and BSs. Using the above components, one can virtualise the underlying infrastructure in terms of decoupling mobility from physical network (OpenFlow) and allowing multiple SPs to concurrently control (with FlowVisor) and configure (with SNMPVisor) the underlying infrastructure. However, OpenRoads mainly targeted to Wi-Fi networks with little support for cellular networks [34].

OpenRadio [23] proposes a novel programmable wireless data plane that provides modular programming capability for the entire wireless stack. The wireless protocols are split into processing and decision planes. The OpenRadio system can be built around a commodity multi-core hardware platform, while the core component is a software abstraction layer that exposes a modular and declarative interface to programme the PHY (baseband) and MAC layers. This decoupling provides a declarative interface to programme the platform while hiding all underlying complexity of execution. Such an approach assures flexibility at the PHY and MAC layers and provides modular I/Fs able to process traffic subsets using Wi-Fi, WiMAX, 3GPP long-term evolution (LTE)-Advanced and so on. The processing plane includes algorithmic actions expressed as directed graphs of (e.g. data plane processing for 54 Mbps OFDM Wi-Fi, FFT or special encoding and decoding for video). The decision plane contains the logic which dictates which directed graph is used for a particular packet (e.g. selection between data and video graphs).

The advantage of the approach is that an operator only defines decision plane rules and therefore the corresponding processing plane action graphs to assemble a protocol; the declarative interfaces allow the operator to programme the platform while hiding all underlying complexity of execution. Rules are logical predicates on parameters of packets such as header fields, received signal strength, channel frequency and other fields that may be programmed. Actions describe behaviour such as encoding/decoding of data and scheduling of traffic on the channel.

The system is capable of realising modern wireless protocols (Wi-Fi and LTE) on off-the-shelf digital signal processing (DSP) chips, while providing flexibility to modify the PHY and MAC layers to implement protocol optimisations. OpenRadio can provide programmable BSs for cellular infrastructure (which are more flexible than fixed-function hardware), for example 'software-upgradable' platforms for HSPA/WCDMA/LTE. OpenRadio can be used to specify both the underlying protocols as well as optimisations. Some use cases are cell-size-based optimisation, co-existence of heterogeneous cells, application-specific wireless service and evolving standards. However, OpenRadio does not provide any network controller that takes advantage of its programmable data plane.

5.6.3 Integrated SDN/NFV architectures

This section presents several examples and proposals of architectures trying to integrate SDN and NFV concepts. Emergent SDN and NFV technologies are powerful means to develop and operate networks/services, reducing costs and boosting performance. It is predicted [24,32] that transition from 4G to 5G and also 5G itself can benefit from SDN/NFV approach. Enhancements are proposed in 4G long-term evolution/system architecture evolution (LTE/SAE) architectures, while exploiting SDN and NFV in implementing network nodes. For instance, evolved node Bs (eNBs), MME, S/P-GW, home subscriber server (HSS), etc. can be realised in a virtualised edge cloud environment. An example of implementation of 3GPP compliant EPC, OpenEPC is presented in [33], where upper layer NFs (network applications) runs on SDN/OpenFlow protocol.

5.6.3.1 CellSDN: a software-defined cellular core network

CellSDN [24] aims to achieve a centralised CPl for cellular core networks. The work proposes four main extensions to controllers, switches and BSs: (1) *flexible policies on subscriber attributes*: the controller, application apply policies based on the properties of cellular subscribers (network provider, device and subscriber type and recent usage). The controller translates policies based on subscriber attributes; (2) *scalability through local switch agents*: in order to reduce the signalling between the switches and controllers, switches run software agents performing simple local actions (such as polling traffic counters and comparing against thresholds), at the controller request; (3) *flexible switch patterns and actions:* cellular networks would benefit from support for deep packet inspection (DPI), header compression and message-based control protocols (e.g. stream control transmission protocol); (4) *remote control of virtualised BS resources:* virtualising the BS by time slot and subcarriers. The open API between the controller and BS can enable remote control of radio resource allocation, admission control (AC), handoff and paging.

The CellSDN solution has some limitations: Actually there is no concrete solution for an SD architecture but only some ideas for core network (CN) without the incorporation of RAN. Some topics are only partially discussed or even not touched, for example: network virtualisation functionalities, scalability design for software-defined core network (SD-CN), specific and concrete SD traffic engineering solutions. The authors in [24] focus mainly on radio virtualisation to provide effective resource virtualisation; the approach can compromise overall system performance.

5.6.3.2 A generic 5G architecture based on cloud and SDN/NFV

A major challenge in defining a 5G architecture, while taking advantages from modern cloud concepts and SDN/NFV architectures and technologies, is how to split the functionalities between core and access part, between hardware and software and how to separate CPl and DPl as to finally meet the strong requirements summarised at the beginning of this chapter.

Several works [2,29,34] propose generic architectures, consisting from a RAN part coupled with a core part, where the core could be seen as a cloud.

A general architecture is presented in [2], based on two logical network layers – a network cloud performing higher layer functionalities and a *radio network (RN)* performing a minimum set of lower layers L1/L2 functionalities. Three main design concepts are considered and integrated: NFV and SDN with control/user plane split, to provide flexible deployment and operation; ultra-dense small cell deployments on licensed and unlicensed spectrum, to support high capacity and data rate challenges; the network data are intelligently used in the cloud, to optimise network resources usage and for QoS provisioning and planning. Within the network cloud, different functions could be dynamically instantiated and scaled on the basis of SDN/NFV approach. A redesigned protocol stack eliminates the redundant functionalities and integrates the *access stratum* (AS) and NAS. The architecture enables provisioning of required capacity and coverage based on splitting the control/user (data) planes and using different frequency bands for coverage and capacity. Relaying and nesting configuration are used in order to support multiple devices, group mobility and nomadic hotspots. The NI is data-driven, allowing optimisation of the network resource planning and usage. Connectionless and contention-based access is proposed with new waveforms for asynchronous access of massive numbers of

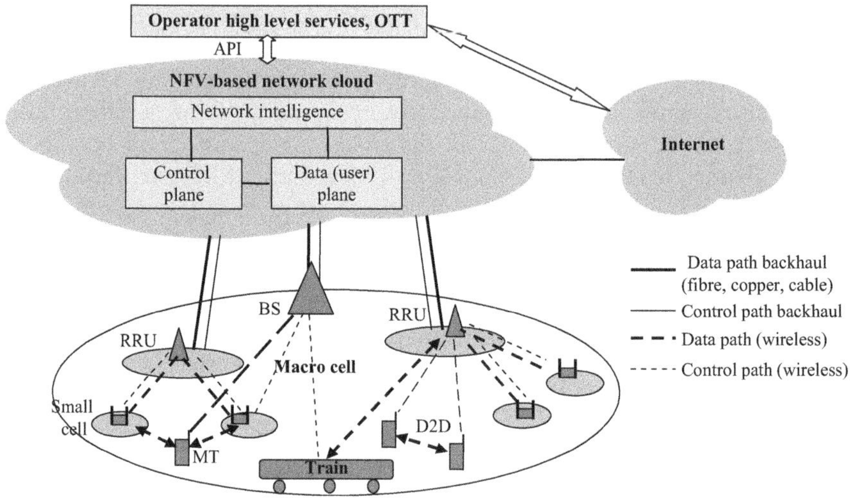

Figure 5.5 Example of 5G network generic architecture [2]. RRU – remote radio unit; D2D – network-controlled device to device; MTC – machine type communication; OTT – over the top; MT – mobile terminal; NFV – network function virtualisation; API – application programmer interface

machine-type communications (MTC) devices like connected cars, connected homes, moving robots, and sensors. Figure 5.5 presents a simplified high level view of this architecture.

The NFV-based network cloud is split into CPl and DPl (following the SDN principle) and a *NI* layer could be put on top of them. The CPl can perform tasks as MM, radio resource control, NAS–AS integration and security functions (e.g. authentication, etc.). The DPl (user plane) assures the data flow paths between different RANs and to/from Internet. Specifically, the DPl contains gateway functions, data processing functions, mobility anchors, security control on the air interface, etc. The NI performs the services orchestration, that is, makes traffic optimisation, QoS provisioning, caching control, and so on. In addition, the NI can analyse the big data collected from the different components (core, RAN) and infer appropriate actions.

In Figure 5.5, the RAN might have macrocells, covering cells and small cells. The SDN principle of CPl/DPl (or C/U) split is also applied in the RAN. All elements in RAN may have a set of lower layers L1/L2 functions. In addition, the main BS may have low carrier frequencies (CF) for respectively non-orthogonal multiple access – as fall back for coverage and high CF for Massive MIMO wireless backhaul. The small cells stations may have high CF and/or unlicensed spectrum for local capacity and switch-on-demand capabilities. The remote radio units (RRU) have allocated high CF with massive MIMO for capacity. Network-controlled direct communication D2D is possible, between terminals, while applying different D2D variants: inband or outband, in underlay or overlay style, thus saving significant radio resources [28]. The mobile terminals may have dual connectivity and independent DPl/CPl mobility.

The proposed architecture allows a flexible deployment and management. The network cloud realisation could be also flexible; CPl and DPl instances can be seen as 'data centres' having high amount of resources. Each data centre can control one or several macrocells and/or RRUs. The DPl and CPL entities could be located close to BSs and also to RRUs, if some latency-critical services requirements should be met. Therefore, the operator can deploy both large and small data centres to support specific service needs. On the other hand, BSs are simpler and more efficient in energy consumption than in conventional 4G case. The network cloud allows for resource pooling, reducing over-provisioning and under-utilisation of network resources. By employing SDN and NFV, the architecture allows a dynamic deployment and scaling on demand of NFs. The local data centres can borrow resources from each other, should the traffic load conditions require this; they also can be enriched (installing new software) to support other applications. The cloud-computing model flexibility is present in the network cloud: when the traffic demand is low, the available cloud resources can be lent out, whereas additional resources can be rented through IaaS when the demand is high.

The business model supported by the generic architecture described above can be enriched to provide specific network functionalities as a service (i.e. Everything as a Service, XaaS) to customers (e.g. NOs, over the top (OTT) players, enterprises) having some specific requirements. Examples of such services are 'mobile NaaS', 'radio NaaS' and even CPl or DPl entities can be offered as a service. Third parties (e.g. like OTT players) might rent defined parts of the platform, for example, to serve applications having low latency requirements.

The proposed architecture has been preliminary evaluated [2], by using a real-time simulator to assess the system-level gains, when part of the candidate 5G technologies are considered for downlink transmission. The results demonstrate, as an example, the gains from the hybrid usage of macrocells at lower frequency bands and small cells at higher frequency bands, together with mMIMO. The authors considered a combination of dense deployment of small cells, using large bandwidths at higher frequency bands and employing massive MIMO techniques at small cells. Promising results show more than 1,000 times throughput gain, compared to a macro-only 3GPP Release 8 LTE.

However, when considering detail functioning of the proposed system, some mutual conflicts of the conceptual components have been identified [2]. Further work is necessary to decide on balancing between different allocation of functions and specifically how to incorporate small cells with NFV and SDN in a cost effective manner use in small cells in different frequency regimes. The problem of constructing intelligent algorithms is open, to better use the available network resources and provide a consistent end-user QoS/QoE.

5.6.3.3 Industrial 5G SDN/NFV platform – examples

SK telecom proposes [25], as an industrial solution, a software-based 5G platform, providing a software-oriented framework Telco-oriented. Both *software defined*

RAN (SD-RAN) and SD-CN are considered, where CPl and DPl are decoupled. The architecture can provide NaaS, where a core function in software framework allows configuration/change of telecommunication and service functions. The platform also provides Telco API for service utilisation, enabling implementation of analytics-based services, that is multi-service carrier Ethernet 2.0 and MPLS edge solution. SD-RAN contains basically radio hardware units and antennas, whereas data plane in SD-CN contains SDN switches with flow tables. A single coherent CPl is proposed.

The platform makes use of three important technologies based on SDN/NFV: NFV-based virtualisation CN operation, virtualised RAN and SDN enriched with integrated orchestration. The platform builds a cloud by virtualising the hardware and operate a range of network/service functions on the software-based network. It centralises and virtualises the digital units of a BS into a standard hardware-based cloud and processes RAN signals in real time. It also provides the life-cycle management of the software-based networks services from a centralised and unified NS orchestrator. The network infrastructure considered in the SK Telecom platform contains the radio hardware, an edge cloud, transport network and centralised cloud. The latter can be connected to Internet. The general functional architecture contains the hardware resources (computing, storage and network). On top of these, there is the software of CPl and data plane including an abstraction and middleware at the bottom part, then virtualised Telco functions (network-related and IT-related). These are virtualised in order to build NaaS. Abstract Telco APIs can offer NaaS to the service layer containing different applications.

While considering an integrated solution (i.e. comprising both RAN and CN), the platform has some limitations [34,35] related to the coarse-grained BS decoupling as CRAN and control traffic unbalancing. It does not carry details for control and data plane decoupling and is supposed to support a high amount of *I–Q* transmissions related to radio processes. The network virtualisation should be still developed in a way related to wireless resource slicing scheme.

SoftRAN [38] starts from the observation that in RAN a major problem is how to use and manage limited spectrum in the best way, as to achieve a good connectivity. In a dense wireless deployment with mobile nodes and limited spectrum, several difficult tasks should be solved, that is to allocate radio resources, implement handover, manage interference, balance load between cells, etc. Therefore, SoftRAN proposes a SDN-like CPl for RANs, which abstracts all BSs in a local geographical area as a *virtual big-BS* comprising a central controller and radio elements. This is an attempt to make more efficient the wireless connectivity of the current LTE-distributed CPl.

The SoftRAN architecture defines a controller supporting different control algorithms to operate on, and ensuring that the delay between the controller and the radio element is acceptable. The centralised controller receives periodic updates from all radio elements, indicating local network state in a given area. So, the controller updates and maintains the global network state in the form of a database, called *RAN information base* (RIB). The RIB conceptually consists of the following elements: (i) *interference map* – a weighted graph, where each node represents a radio element or an active client in a geographical area and the weight of the edges represent the channel strength between the two nodes; (ii) *flow records* – a record of the relevant parameters of an ongoing flow, for example, number of bytes transmitted, average transmission rate, number of packets queued, etc.; (iii) *network operator preferences* – if the NO needs to prioritise certain flows over others, then it can enter his preferences into the RIB. Through this abstraction and architecture, an environment is created that enables efficient and dynamic management of increasingly scarce and strained radio resources.

SoftRAN addresses limited NFV over the antennas in BSs for big-BS abstraction. Also, more detailed consideration about the interaction with SD-CN is missing in the SD-RAN design.

5.6.3.4 A cloud approach for 5G: CRAN

The works [29,30], propose the so-called CRAN solution, consisting in *centralised processing, cooperative radio, cloud and clean (green) infrastructure RAN* (i.e. CRAN) trying to solve many challenges of the 5G networks. A distributed system of base transceiver station (BTS) is defined, composed of the baseband unit (BBU) and remote radio head (RRH). Different function splitting can be between BBU and RRH: (a) *full centralisation*, where baseband (i.e. layer 1) and the layer 2, layer 3 BTS functions are located in BBU; (b) *partial centralisation*, where RRH integrates radio function and also the baseband function, while all other higher layer functions are still located in BBU. For the second solution, although the BBU does not include the baseband function, it is still called BBU for the simplicity. The different function partition methods are shown in Figure 5.6.

The CRAN architectures contains three main parts: distributed radio units RRU composed of *RRH* and antennas which are located at the remote site; high bandwidth low-latency optical or microwave transport network which connect the RRHs and BBU pool (the connection is realised in hub-style from several RRUs to a single BBU); the BBU composed of high-performance programmable processors and real-time virtualisation technology. In the central hardware units, several virtual BSs can exist. The RRUs can be placed in the centre of overlapping cells. In the centralised solution, the white-depicted FBs (see Figure 5.6) are placed in the distributed RRU, whereas the grey ones are placed in the BBU (L1/L2/L3/O&M). In the distributed solution, the BBU has L2/L3/O&M functions, whereas the RRHs have L1 functions and the baseband processing.

Each variant of the CRAN architecture has some advantages and limitations. The 'fully centralised' CRAN is easily upgradable and allows network capacity

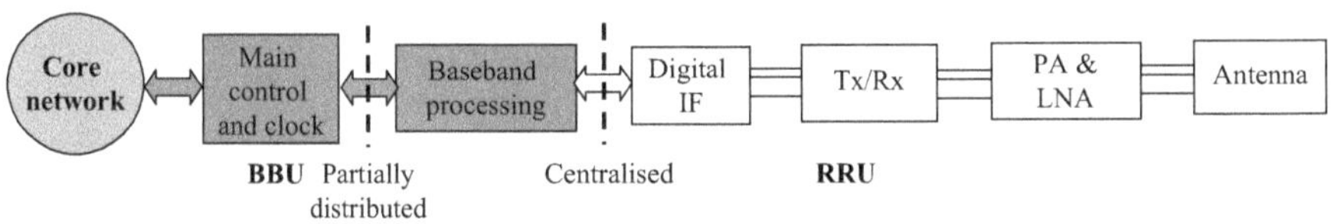

Figure 5.6 Two separation methods of BTS functions [29]

expansion; it can support multi-standard operation, maximum resource sharing and multi-cell collaborative signal processing. However, it has high bandwidth requirement between the BBU, to carry the baseband I/Q digital modulation-related signals (e.g. a TD-LTE 8 antenna with 20-MHz bandwidth will need a 10-Gbps rate). The 'partial centralised' CRAN requires much lower transmission bandwidth between BBU and RRH, by integrating the baseband processing into RRH. The BBU–RRH connection only needs to carry demodulated data, which is only 1/20–1/50 of the original baseband I/Q sample data. On the other side, if the baseband processing is integrated into RRH, then it determines less flexibility in upgrading, and less convenience for multi-cell collaborative signal processing. Only RAN virtualisation functionalities are touched by CRAN solution. Traffic engineering solutions are proposed to be placed mainly in the PHY layer.

CRAN has been deployed in several countries in the last years. However, today it is considered that current fronthaul interface CPRI (*common public radio interface*) has become an obstacle towards CRAN large-scale deployment, especially in the context of 4.5G and 5G technologies which have additional challenges with respect to existing CPRI. For instance, in LTE, the CPRI interfaces have high bandwidth and low latency requirements; whereas wavelength-division multiplexing (WDM) technology can resolve the fronthaul problem and save fibre; it also introduces additional transmission equipment, with higher costs [31]. A flexible, low-bandwidth fronthaul network should be designed, to solve CRAN transmission problems. In addition, the need for protection resulting from CRAN centralisation also necessitates flexible routing between BBU pools and RRUs. In short, traditional CPRIs will try to support the future networking demands of centralised CRAN deployment. To compensate the CRAN centralisation, a flexible routing between BBU pools and RRUs is also necessary.

In [31], (elaborated jointly by several equipment manufacturers) it is proposed the *next generation fronthaul interface* (NGFI), seen as more flexible with respect to function splitting. The requirements, design principles, application scenarios, potential solutions and other NGFI technical aspects are discussed. NGFI is a novel open interface between the BBU and RRHs, redefining their functions. The BBU and RRU architecture is modified, given that some baseband processing functions are shifted to the RRU. The BBU is redefined as the *radio cloud centre* (RCC), and the RRU becomes the *radio remote system* (RRS). The NGFI defines actually a packet switched, multiple-to-multiple fronthaul network. The NGFI should comply with principles [31] as adaptive bandwidth changes responsive to statistical multiplexing and dynamic payload, support for high-gain coordinated algorithms; interface traffic volume decoupled from the number of antennas at the RRU; neutrality with respect to air interface technology; optimisation of RRS–RCC connections and so on. NGFI specifications will determine changes within radio equipment (RE) architecture and impose new requirements on NGFI transport networks.

Bernardos *et al.* [36] proposed a high level SDN-based architecture for future mobile networks. The focus is most on RA issues. It has proposed a 5G mobile network architecture spanning over two layers: a *Radio Network*, performing a basic set of layer L1, L2 functionalities, and a *network cloud* for all upper layer functionalities. A lean protocol stack is proposed by consolidating the redundant functionalities of *AS* and *NAS* signalling. Numerous procedures for MM, session management (SM) and security management can be simplified or potentially removed. On the data plane, dynamic network deployment and ability to scale are achieved, by merging RAN L2 and gateway functionalities in the core network. However, this work lacks requirements and details for the real deployment of the proposed architecture.

The paper [37] describes an all-SDN network architecture featuring hierarchical control capability. It focuses on a 5G CPl aiming at providing *connectivity management as a service,* with a so-called unified approach to mobility, handoff and routing. According to the authors, 'unified' relates to the merger of RAN and CN functions, which are implemented as applications running on one or more hierarchical controllers.

5.6.3.5 A framework for cellular software-defined networking

The work [26] proposes architecture and provides a high-level description of a *Cellular SDN* (CSDN) using the SDN and NFV principles. The overall goal is to optimise the dynamic resource orchestration (RO), by performing real-time context data gathering, analysis and then making intelligent decisions. The network and user information are collected from the mobile edge networks and could be used locally, or exported/ shared to other SPs, to enrich the set of services. The CSDN contains forwarding, control and network application architectural planes.

In addition, a novel *knowledge plane* is added, to co-operate with network application plane, allowing the *mobile services provider* (MSP) to construct an intelligent vision upon its network and users' environment. New applications or virtual functions can be implemented and instantiated (e.g. optimised content distribution and caching, IoT, location-based services, etc.) and linked to the controller northbound interface. The work [26] shows, as an example, a CSDN mainly oriented towards the 4G LTE, whose several functions are proposed to be implemented as VNFs at the CSDN application level, in a centralised cloud-based infrastructure. The subsystems included in the architecture are the LTE EPC and eNB. The LTE virtualised network functionalities interact at the M&C level with the CSDN switches via the controller. The forwarding plane contains CSDN switches of the ePC, and its boundary is placed at the eNB. Figure 5.6 presents the CSDN architecture, instantiated for LTE.

The DPl contains switches corresponding respectively to eNBs, S-GW and P-GW. The CPl has as a main component the NOS and an abstraction/VL.

The three typical SDN interfaces are seen for CPl: north I/F – to the application plane, south I/F to the DPl and east–west I/F towards other controllers. VNFs are defined in the application plane, to execute the functions of UTRAN and EPC. The VNFs, named VeNB, virtual mobility management entity (VMME) VS–GW, VP–GW and Virtual Policy Control and Charging Rules Function, together with their corresponding switches in DPl (e.g. VeNB plus its CSDN switch correspond to eNB functionalities), perform the equivalent of LTE UTRAN and EPC functionalities. Other applications could be added to the application plane.

The DPl OpenFlow-enabled switches provide typical functions (similar to the wire-line SDN) like network traffic measurements, thus allowing NOs to evaluate the traffic load, perform subscriber's usage statistics, billing, assess the QoS, etc. The operator can then flexibly change the path of the flows in order to optimise the transport and enforce QoS network policies. However, specific requirements arise for DPl, from the wireless nature of the network, where a high number of mobile users exist, radio resources are scarce and real-time response is required. In such conditions, centralising all functions in the SDN controller is not scalable. A solution is to 'go back' (partially) to a distributed control solution, by proactively instructing the switches about some actions to be performed. In other words, the switches will execute more functions than in 'pure' SDN case, that is only the forwarding itself. In CSDN approach, the switches could notify the controller if the traffic exceeds a certain threshold, or tag some packets to be redirected to a transcoder, DPI – to help intrusion detection, etc. However, the functions allocation balancing between a switch and a controller is still an open research issue.

The CSDN controller NOS is composed of FBs for topology auto-discovery, topology resource view and network resource monitoring. The NOS produces a consistent updated network global view, while hiding the distributed characteristic. The abstraction/VL (on top of NOS) presents to the application plane an abstract network view. The controller receives via its southbound interface, network measurements or data packets (when a flow-table match-miss event is detected for a flow of data packets).

The work [26] does not analyse the problem of the number of controllers needed in CSDN; one controller for each public land mobile network is suggested. However, the actual number of controllers would depend on the network dimensions and in the case of multi-controllers, inter-controller communication is needed via east–west interfaces. Also the geographical placement of the controllers is still an open issue. The controller placement problem is a NP-hard one [47,48]. Therefore, different solutions can be considered, with specific optimisation criteria, targeting performance in failure-free or more realistic scenarios. However, some criteria could lead to different solutions; therefore, multi-criteria global optimisation algorithms could be attractive. Some examples of specific criteria are:

1. maximise the controller–forwarder or inter-controller communication throughput, and/or reduce the latency of the path connecting them;
2. limit the controller overload (load imbalance) by avoiding to have too many forwarders per one controller;
3. find an optimum controller placement and forwarder-to-controller allocation, offering a fast recovery after failures (controllers, links, nodes).

The VNFs in the Application plane (Figure 5.7) performs the respective functions defined for LTE. For instance, the radio resource management function is allocated to VeNB. Following the SDN principle, the RE is centrally controlled, allowing more consistent (than in the current LTE) radio resource allocation and control of fronthaul and backhaul parts of the network. Other functions can be realised, for example, load balancing between several BSs, cooperative MIMO, sleep mode

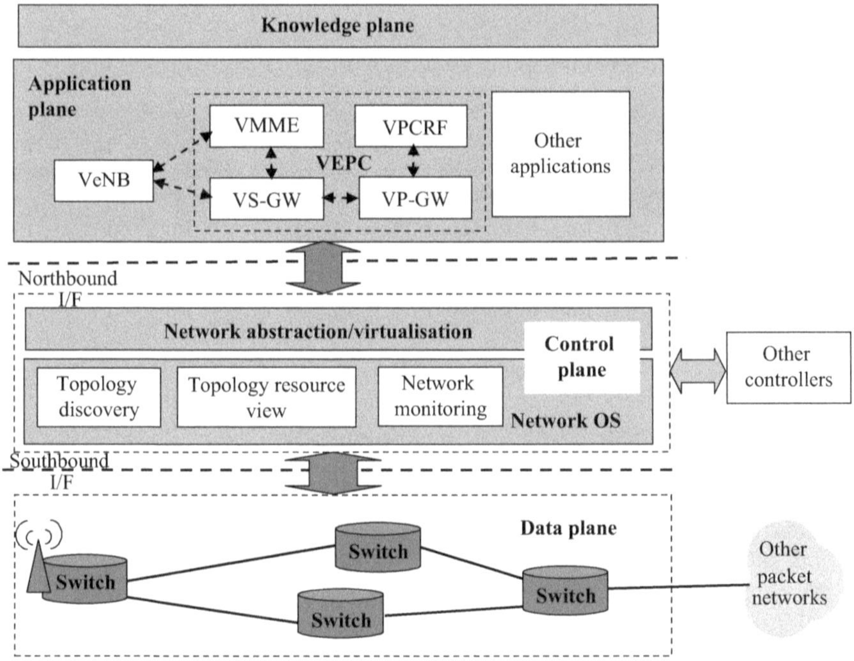

Figure 5.7 CSDN architecture [26]

control of some nodes in order to reduce energy consumption, etc. Similar considerations can be made about some other EPC functions realised as VNFs. To manage the mobility, the VMME executes the functions of the LTE MME. The controller also performs functionalities related to data tunnelling inside EPC, QoS, metering and routing. All these are transformed into packet flow rules, used to configure the CSDN switches.

The work of Guerzoni *et al.* [32] presents a 5G architecture based on SDN, NFV and edge computing. Three control levels are proposed: *device*, *edge* and *orchestration controllers*, fully decoupled from the DPl and implementing a unified security, connection, mobility and routing management for 5G networks. The solution also preserves backward compatibility to 3GPP releases. SDN-based connectivity between VNFs (applications) is proposed, enabling carrier grade communication paths, by avoiding tunnelling. The solution is appropriate for mission critical communications, by realising low E2E latency; it is flexible, reliable and dependable. The implementation could be either 'centralised' or 'distributed at the edge', depending on functional and non-functional requirements of the supported services. Both CPl and DPl logical network elements are decomposed into sets of applications or modules, which can be dynamically instantiated in the cloud infrastructure according to network operation or service requirements. Figure 5.8 (adapted after [32]) shows the unified CPl, including both AS/NAS CPl as well as management plane.

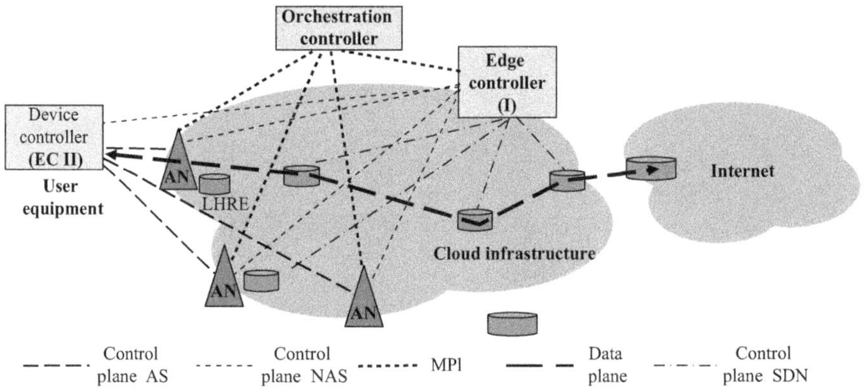

Figure 5.8 The 5G unified control plane and data plane [32]. AN – access node

The *device controller* (DC) (located in the device) controls the physical layer connectivity to the 5G network. The DC handles AS functions such as access/network selection.

Two types of *edge controllers, (EC, (i)* and *(ii)),* implement the network access control, packet routing and transfer, radio resource management, mobility and connection management and security. The EC has similar functions to the AS/NAS 4G functions performed by eNodeB and MME. The EC implementation is distributed over the cloud infrastructure being composed of several interconnected control applications (C-Apps), where each one performs a subset of functions, like RA, Authorisation and Authentication, AC, flow management (FM), MM, connection (session) management (CM) and security (Sec). To fully separate the DPl/CPl also on the radio link, the RA App is split respectively into RAD/RAC applications. The DPl could be instantiated on a different PoP. For some mission critical communications, the mobile devices might be required to support some AS/NAS functions; that is why two types, that is, EC (i) – with C-Apps instantiated in the edge cloud infrastructures and EC (ii) [39] – implemented temporarily or permanently on a mobile device.

The *orchestration controller* (OC) – composed by the RO and *topology management* (TM) modules – has network management functions (similar to those of 4G). It coordinates the utilisation of cloud resources (computational, storage and networking), allocating and maintaining the required resources, to instantiate both CPl and DPl. The RO allocates physical resources to instantiate EC C-Apps that is, it determines the embedding solution for the virtual CPl and DPl to be instantiated.

The TM directly manages the physical resources. It is composed by TM-A (TM – Apps) and TM-L (TM – Links) modules, which handle VMs and virtual links respectively, required to instantiate and connect EC C-Apps. The RO is centralised, having as scope the whole cloud infrastructure. The TM-L and TM-A are distributed, interacting respectively with SDN-based control platforms and cloud management platforms. The OC modules (similar to EC C-Apps) are VNFs embedded in data centres, thus assuring flexibility and adaptability. The CPl can be dynamically reconfigured if the requirements defined by the network administration will change.

In [32, Table 1], a complete mapping of different AS/NAS M&C functions to the ECs and OC is given. Generally, the functions treating the entire DPl or CPl are allocated to OC, whereas some specific functions (e.g. network access and control packet routing, MM, etc.) are allocated to EC (i) and (ii).

A SDN data plane clean-slate architecture has been adopted in [32], which did not define neither dedicated DPl network elements (e.g. 4G SGW and PGW), nor unique logical elements (e.g. mobility anchor points). When a device initiates a network attachment, it is allocated an address and a last hop routing element (LHRE); the latter connects the AP to the backhaul infrastructure. At attachment time, a forwarding path for the device is established by the FM-App, allowing packets coming from the device (or going to) to be forwarded to a network entry point – NEP (or to the device LHRE).

The NEP defines the boundary beyond which the physical infrastructure is no more under the OC and EC control. The NEPs for different attached devices may be also different. The FM-App has to select available links to embed a virtual link between LHRE and NEP; appropriate SDN flow tables should be installed on switches belonging to the SDN-based cloud infrastructure. The RA App of EC (i) should manage the wireless connection between the AP and the device. Optionally, QoS control could be enforced over both the wireless connection and the forwarding path. The architecture also supports D2D communication, managed by the RA App, located in the EC (i) for the in-coverage case or in the EC (ii), in out-of-coverage case.

The main advantages of the architecture proposed in [32] are reconfigurability-operators can dynamically instantiate logical architectures, implementing NFs, services and corresponding states in the optimal location within the cloud infrastructure; the tunnelling protocols (common in 3GPP) are not used any more (reducing the latency of forwarding information installation from ~40 ms (in 4G) to ~20 ms); latency of forwarding paths could be reduced to almost zero by pro-actively configuring the SDN-based infrastructure (thus realising naturally the ‘always-on’ concept already present in 4G EPC). The DPl latency can be additionally reduced by implementing ad-hoc virtual link embedding algorithms in the FM modules. The proposed architecture, functions and procedures have the potential to become the ‘de facto’ solution for 5G.

5.6.3.6 SoftAir architecture

The papers [34,35] introduce a new software-defined 5G architecture called SoftAir, targeting NFV and cloudification and aiming to scalability, flexibility and resilience. The approach includes fine-grained BS decomposition, OpenFlow interfaces, mobility-aware control traffic balancing, resource-efficient network virtualisation and traffic classification. Figure 5.9 presents a high-level view of the SoftAir architecture, composed of SDN CPl and data plane. The DPl contains a complete network infrastructure: SD-CN and SD-RAN. The OpenFlow and SNMP protocols link the two planes. Below, a summary of this architecture is shortly presented, whereas details can be found in [34,35].

The SoftAir SD-CN is composed by SD-switches, under CPl coordination. Development of customised SDN applications, for example, MM, QoS-based routing

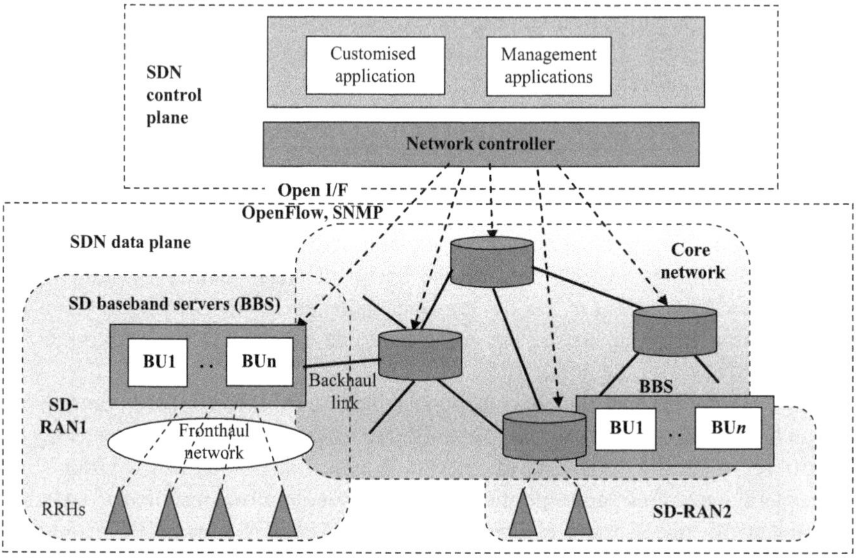

Figure 5.9 Overall architecture of the SoftAir [34]

and billing policies is done in CPl and also global management tools and network virtualisation. Experiments in the field deployment have already shown the SDN advantages (B4 – Google, SWAN – Microsoft, ADMCF – Huawei, etc., [35]). Important increase in link utilisation in SD-CN can be obtained via SDN from 30–40 per cent to over 70 per cent [35]. The scalability in terms of controller-to – [SD-CN] forwarders communication can be solved by using high performances controllers [47] and/or by using multi-controller clusters and multi-threading technologies. Some recent research has shown that in large-scale SDN networks with in-band control channels, the controller–forwarder communication delay can be minimised by using traffic balancing schemes, based on parallel optimisation theories. In addition, the SoftAir adopted a mobility-aware and proactive control traffic balancing scheme, minimising the CPl–DPl delay by exploiting the SD-RAN mobile feature (the control traffic issued by the SD-RAN is following some spatial and temporal patterns).

The SoftAir SD-RAN is flexible, realising layer L1–L3 function virtualisation in a distributed architecture (see Figure 5.9). The SD-BS is split into hardware-only RRH and software-implemented BBUs (these two components could be also remotely located). A fronthaul network (fibre/microwave) connects the RRH to *baseband processing servers* (BBS) using standardised interfaces like *open BS architecture initiative* [49], or CPRI [50]. The distributed SoftAir SD-RAN has some similarities to CRAN [29]. Efforts have been spent by standardisation entities and industry to make the RRH–BBS technology independent.

The CPRI [35,50] interface is based on industry cooperation to define a publicly available specification for the key internal interface of radio BS between the *radio equipment control* (REC) and the RE (Ericsson AB, Huawei Technologies

Co. Ltd, NEC Corporation, Alcatel Lucent and Nokia Siemens Networks GmbH & Co. KG). The CPRI enables flexible and efficient product differentiation for radio BSs and independent technology evolution for REC and RE. The specification includes M&C Plane and DPl transport mechanisms, and means for synchronisation. A focus has been put on hardware-dependent layers: L1, L2. This ensures independent technology evolution (on both sides of the interface), with a limited need for hardware adaptation. The CPRI scope specification is restricted to the link interface only, which is basically a point-to-point interface. Such a link shall have all the features necessary to enable a simple and robust usage of any given REC/RE network topology, including a direct interconnection of multi-port REs. CPRI can enable high-speed (up to 10 Gb/s), low-bit error rate (10^{-12}), and long-distance (up to 40 mi) data exchange between RRHs and BBS, while providing the high-resolution synchronisation.

How it is here SD-RAN versus CRAN [29]? Recall that distributed CRAN architecture is focused on high-performance computing of baseband processing functions (mostly for L1 operations) at remote servers or data centres. However, CRAN cannot achieve scalable PHY/MAC-layer cloudification and does not support network-layer cloudification as SD-CN does. On the other side, SD-RAN offers scalability, evolvability and cooperativeness through fine-grained BS decomposition that overcomes fronthaul traffic burden. In SD-RANs partial baseband processing is done at the RRH (e.g. modem), whereas the remaining baseband functions (e.g. MIMO coding, source coding and MAC) are executed at the BBS.

This split is convenient, given that CPRI, which is not only defined for I–Q sample transport, can still be adopted without designing new interfaces and can lead to considerably reduced data rate requirements between BBS and RRHs. Figure 5.10 illustrates the SoftAir functional split. The SD-RAN (given its reduced data rate requirements) offers scalability. It also can support cooperative gain and is evolvable by allowing the aggregation of a large number of technology-evolving RRHs at BBS and CPRI-supported fronthaul solutions.

In SD-BSs of the SD-RAN OpenFlow interfaces can be implemented (e.g. with *OpenvSwitch* [6]). So, the BSs can be managed in SDN style via a unified interface, even in different wireless standards (multi-technology capability). Seamless vertical mobility is possible, when mobile users roam among BSs having different wireless standards. This is done by rerouting the traffic through CN to different BSs, via OpenFlow interface (enabled on CN switches and also on BSs). In addition, the common OpenFlow interface used in SD-BSs and SD switches assures a transparent interconnection between SD-CNs and SD-RANs under unified management.

SoftAir supports network virtualisation, where multiple VNets can share simultaneously the same physical infrastructure and each VNet (slice) may independently adopt its L1/L2/L3 protocols. In advanced solutions they can be deployed on demand and dynamically allocated. The SoftAir network virtualisation enables a wide range of applications. Each *mobile virtual network operator (MVNO)* may use different wireless technologies (Wi-Fi, WiMAX, LTE, small-cells, HetNets, etc.). RAN sharing may lead to significant capital expenditure (CAPEX) reduction. The virtual slices can be customised for different services and

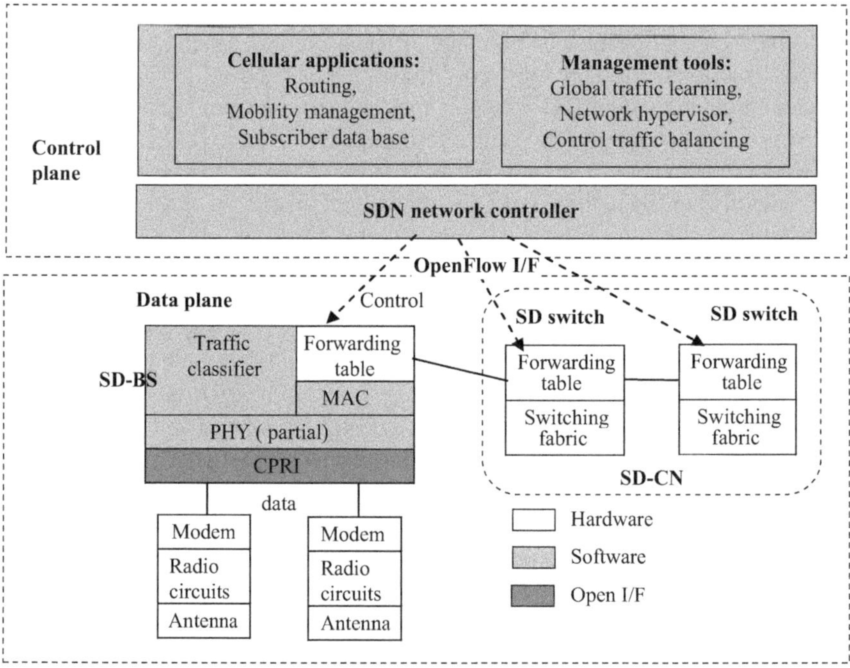

Figure 5.10 Function cloudification in SoftAir system [34]

types of traffic flows – for example, for QoS routing, E2E-controlled performances, etc. The slices isolation might accelerate the innovation, given that in a slice one can experiment novel protocols, without interfering with other slices. In order to realise virtualisation SoftAir proposes, three types of hypervisors: network hypervisor for high-level virtualisation; for low-level virtualisation a wireless hypervisors and switch hypervisors are defined. Thus SoftAir enables the end-to-end network virtualisation traversing both SD-RAN and SD-CN, realising a truly multi-service converged network infrastructure.

The *network hypervisor* distributes non-conflicting network resource blocks among virtual NOs based on their demands. It should maximise the global resource utilisation and guarantee the data-rate requirements requested by each virtual operators. For instance, within the coverage area of each SD-BS, for a given RAT, the VNet should offer a certain average data rate for its users with certain spatial distribution and density. At each SD-BS, one can assign the wireless resource blocks to the VNet.

The achievable average data rate seen by an user will depend on the effective *signal to interference noise ratio* (SINR) within the SD-BS coverage area and the distance from this user to each RRH within the considered SD-BS. The *wireless hypervisor* is a low-level resource scheduler executing the network hypervisor policies, while guaranteeing isolation among VNets and optimising the resource utilisation (e.g. spectrum). It can use different wireless resource dimensioning schemes, for example, orthogonal frequency-division multiple access (OFDMA) or wireless scheduling. Each VNet can use its own and customised NET/MAC/PHY layer protocols. The *switch hypervisor* makes bandwidth partitioning in a single SD-switch, by provisioning predefined bandwidth for specific traffic flows or VNet. It is also responsible for the isolation among slices of the virtualised infrastructure; it can employ traditional techniques like leaky-bucket scheme for bandwidth provisioning.

The SoftAir architecture has several advantages detailed in [34,35]. The architecture allows evolution and is adaptive, due to DPl/CPl separation and DPl programmability (SDN characteristics). The DPl/CPl separation allows both hardware and software infrastructures to evolve independently. For instance, novel RATs (e.g. mm-waves, full-duplex, massive MIMO, THz) can be adopted in hardware. Traffic engineering and network management optimisation solutions can be applied in the CPl. The DPl programmability allows one to dynamically allocate network resources, adapted to highly variable traffic patterns, unexpected network failures and/or required QoS.

The cloud style and network virtualisation creates possibility to offer IaaS on top of the same physical network; this is useful for emerging different NSs, for example, M2M, smart grid, MVNOs, over-the-top content services like Netflix video streaming, etc. Distinct SP can independently control, optimise and customise the underlying infrastructure without owning it and without interfering with other SPs. Network resources (e.g. spectrum) can be dynamically shared among SPs, for example, MVNOs.

A good SE can be achieved, due to cooperativeness (SD-BSs are implemented and aggregated at a BBS). The control information, mobile data and *channel state information* (CSI) associated with different active users can be shared. Inter-cell interference can be reduced on the basis of collaborative processing algorithms. The system can coordinate RRHs equipped with massive MIMO and mmWave (highly directional communications) for ubiquitous coverage. Based on the physical infrastructure (BBS, fronthaul network and RRH clusters with overlapped or non-overlapped coverage areas), SoftAir can provide cooperation and/or coordination mode (mm-Waves).

SoftAir realises convergence of HetNets due to its open and technology-independent interfaces, which enable a smooth transition among different RATs: Wi-Fi, WiMAX, LTE, LTE-A, etc. The management is unified for RAN and core network. End-to-end QoS management is possible.

Last but not least, SoftAir is energy-saving, due to DPl programmability (e.g. SD-BSs can be dynamically scaled – according to traffic patterns, idle BSs, cooperativeness – due to SD-BSs implemented and aggregated at a BBS) number of physical sites for BS is lower than in other solutions.

5.6.3.7 Content and media delivery

Content, media and especially video traffic have become significant part of Internet and integrated networks traffic, including mobile data traffic and will still grow in the next years. Estimation show [40–42] that in 5G networks, the data rate required for a mobile user equipment (MUE) will increase to 10 Mbps or more for high-definition video service and 100 Mbps for ultra-high-definition TV, in various mobility scenarios. Other applications (e.g. 3D video conferences) may require even higher transmission rates up to 10 Gbps. Some forecast [41] show that video traffic (e.g. TV, video on demand, Internet video streaming, peer to peer) is expected to rise between 80 and 90 per cent among overall consumer traffic.

The heterogeneous cloud access networks are recognised as a main evolution trend of the future 5G cellular system, with multiple hybrid coexisting RATs. A centralised baseband processing unit pool can be adopted for high-performance video delivery, to control all the RATs and facilitating efficient video encoding and transmission, compared to its basic baseband processing unit counterpart without central control functions. However, the central control raises other typical centralisation issues; therefore, different solutions for centralised/distributed function allocation are studied.

CRAN [30,31] is a recent solution proposed for 5G, consisting of large number of low-cost RRHs, randomly deployed and connected to the BBU pool through the fronthaul links. The advantages of CRAN are the following: RRHs can be installed closer to the users, thus offering higher system capacity and lower power consumption; the baseband processing centralised at the BBU pool enables cooperative processing techniques to mitigate interferences; exploiting the resource pooling and statistical multiplexing gain provide efficiency in both energy and cost. However, the fronthaul constraints have high impact on worsening CRAN performance and the scale size of RRHs; accessing the same BBU pool is limited and could not be too extensive due to the implementation complexity.

The heterogeneous CRANs (H-CRAN), [40] takes into account the HetNets. The RAN components are *low power nodes (LPN)* (e.g. pico BS, femto BS, small BS, etc.) are key components to increase capacity in dense areas with high traffic demands. *High power node* (HPN – e.g. macro or micro BS) are defined, which can be combined with LPN to form a HetNet. One major problem is that too dense LPNs infrastructure produces high interferences; therefore, it is needed to control the interference degree. Some advanced DSP techniques are applied. In 4G technology, a solution could be to introduce a coordinated multi-point (CoMP) having such tasks; however, in real networks the performance of the CoMP is highly depending on the backhaul constraints. Therefore, cooperative processing capabilities are needed in the practical evolution of HetNets. Note that in 1G, 2G, 3G technologies the inter-cell interference can be avoided by utilising static frequency planning or CDMA, so cooperative processing is not demanded. On the other side in 4G – OFDM-based, inter-cell interference is severe; hence, inter-cell or inter-tier cooperative processing through CoMP is critical.

In H-CRAN-based 5G systems, cloud computing based cooperative processing and networking techniques are proposed to tackle the 4G challenges, alleviating inter-tier interference and improving cooperative processing gains. Such techniques enhance the HPNs capabilities with massive multiple antenna techniques and simplify LPNs through connecting them to a ‘signal processing cloud’ via high-speed optical fibres. The baseband data path processing and LPNs radio resource control are moved to the cloud server. Important advantages result as follows: cloud computing based cooperation processing and networking gains are fully exploited; operating expenses are lowered; energy consumptions of the wireless infrastructure are decreased.

The system H-CRAN architecture [40,43] can include a central entity which is the BBU pool, containing baseband processing units (the architectural layers are L1-baseband, MAC and network). The BBU pool is linked via gateway to the external Internet. Several peripheral 'islands' realised with different technologies are linked to the BBU pool in hub-style, via two types of links: backhaul (BBU – HPNs), or fronthaul links (BBU pool – LPN): 2G/3G/LTE islands containing BS controllers (for 2G/3G), MBS seen as HPNs and LPNs, that is, RRHs (the latter can be linked directly to the BBU pool via fronthaul links); 5G MBSs (as HPNs) and RRHs; WiMAX BS (HPN) and RRHs; IEEE 802.11 HPN AP and RRHs. Each peripheral island can be seen as an alternative path connected to Internet via gateways.

The 5G HetNet solution can increase the capacity of cellular networks in dense areas with high traffic demands. The key components in HetNets are LPNs which primarily serve for the pure 'data-only' service with high capacity. HetNets decouple the CPl and user plane, which can naturally lead to a SDN-type control. LPNs only have a very simple CPl, whereas the control channel overhead and cell-specific reference signals of LPNs can be fully shifted to MBSs. Some drawbacks of the solution appear if an underlaid structure exists, where MBSs and LPNs reuse the same spectral resources; this produces severe inter-tier interferences; therefore, it is critical to suppress such interferences through advanced DSP. This is the reason to adopt the advanced CoMP transmission and reception technique to suppress both intra-tier and inter-tier interferences.

In H-CRANs, there exist a high number of RRHs with low-energy consumption, which perform only the front radio frequency (RF) and simple symbol processing. Other important baseband PHY processing and procedures of the upper layers are executed jointly in the BBU pool. The RRHs perform relaying (by compressing and forwarding) the received signals from user equipments (UEs) to the centralised BBU pool through the wired/wireless fronthaul links. The joint decompression and decoding are executed in the BBU pool. HPNs are still critical in CRANs to guarantee backward compatibility with the existing cellular systems and support seamless coverage, since RRHs are mainly deployed to provide high capacity in special zones. The HPNs help the convergence of multiple heterogeneous RNs and all system control signalling is delivered wherein. The BBU pool is interfaced with HPNs to mitigate the cross-tier interference between RRHs and HPNs through centralised cloud computing-based cooperative processing techniques. The BBU pool–HPNs interfaces mitigate the cross-tier interference RRHs–HPNs through centralised cloud computing based cooperative processing techniques. The data and control interfaces BBU pool–HPNs can be S1 and X2, respectively, imported from 3G/4G technologies.

H-CRAN can offer support for voice and data services, where the HPNs manage the voice services, whereas high data packet traffic is mainly served by RRHs. An extension of H-CRAN can also support video services. The participation of HPNs offers the advantage that H-CRAN alleviates high-fronthaul requirements. The control signalling and data symbols are decoupled in H-CRANs, and this favours a SDN-like approach. All control signalling and system broadcasting data are delivered by HPNs to UEs, which simplifies the capacity and time-delay constraints in the BBU pool–RRHs fronthaul links and makes RRHs active or sleep efficiently to decrease energy consumption. Burst traffic or instant messaging service with a small amount of data can be supported efficiently by HPNs.

In [40], the components of an H-CRAN are developed. A new communication entity Node C (Node with cloud computing) is proposed (seen as a 3GPP BS evolution). It has the task to converge different RANs for the existing legacy/ancestral communication entities (ACEs, i.e. MBSs, micro BSs, pico BSs, etc.) and performs processing of the networking functionalities in physical and upper layers for the RRHs. It can be seen as a convergence gateway to execute three sets of functionalities: cooperative multiple-radio resource managements (CM–RRM), media-independent handover functionalities, and those of traditional RN controller and BS controller. The node C can manage the RRHs; in such a case it acts as the BBU pool (inherited from CRANs). Node C has sufficient computing capabilities to run the large-scale cooperative signal processing in the PHY and large-scale cooperative networking in the upper layers.

The RRHs mainly provide high speed data transmission without the CPl in hot spots; the CPl messages (e.g. cell-specific reference signals) for the whole H-CRAN are delivered by ACEs. The system is flexible with respect of serving UEs; those UEs which are closer to ACEs than RRHs are served by ACEs and called HUEs. The work [40] states that the node C can serve hundreds of RRHs and several tens of ACEs. The RRH PHY layer may have different technologies (e.g. IEEE 802.11 ac/ad, millimetre wave communication, and even optical light communication).

Three H-CRAN architectural planes are user/data plane (U) which carries the actual user traffic, related traffic processing; CPl (C) – which control the signalling and makes resource allocation and traffic processing to improve spectrum usage efficiency and energy efficiency; MPl (M) – making administration and operation (add, delete, update and modify the logic and interactions for the U plane and the C plane). The H-CRAN can make use of SDN and NFV technologies. The SDN part can be co-located with the Internet/IoT network entities and decentralised RRHs/ACEs closer to the desired UEs. The adaptive signalling/control mechanism between connection-oriented and connectionless is supported in H-CRANs, which is more efficient than a pure connection-oriented mechanism.

The H-CRAN can support efficiently video and media delivery services [45].

Recall that in conventional delivery solutions, the video packet encoding and scheduling is done at head-end station (HS). Data will flow on predetermined paths (via assigned RATs) to MUE. The equivalent transmission model is a parallel pipeline; each chosen RAT corresponds to a pipeline, with a packet queue and a server. The drawback is that the path from HS to MUE has a long delay for the feedback represented by the *network state information (NSI)*; so, only certain quasi-static info is accessible to the HS and this determines a low performance for adaptive flow control and video encoding techniques. On the other hand, in the case of the multi-RAT, the 'out-of-order' issue at the MUE constitutes a multi-RAT bottleneck which illustrates the importance of such NSI. Note that delivery delays that happen in different RATs are usually unknown to the HS, thus involving reordering at MUE; the demultiplexing problem at MUE appears for video packets and cause out-of-order events. This can cause retransmission for out-of-order packets and therefore create overhead on the network traffic. The apparent solution to increase the MUE buffer size can create additional problems due to the limit of transmission control protocol (TCP) window size adjustment. The conclusion is that out-of-order issue is severe in the conventional het-nets without central control, due to the lack of perfect NSI in the RATs.

The H-CRAN can offer an efficient solution for video delivery in the upcoming 5G systems. It can jointly and efficiently process, cache and transmit various videos, based on centralised baseband processing unit pool (BBU pool), which controls multiple RRHs and multiple HPNs. The BBU pool and RRHs are inherited from the CRAN. A powerful centralised BBU has the advantage of caching video, scheduling data packets and understanding the statistics of video traffic. Smart content caching (BBU is close to multiple RATs) thus can release the traffic burden. Centralised coordination in a BBU creates the possibility that video packets can be sent to MUEs in parallel, via multiple RATs (the resulting effect is an overall rate increase). The BBU could schedule the video packets into the matched RATs according to the required QoS. The BBU pool can be integrated with basic gateway functions, to control and schedule the video packets across multiple RATs; therefore, improved performance can be obtained, by globally managing the available resources across different RATs.

Several solutions can be developed for H-CRAN video delivery. An initial solution assumes that each RAT usually has its own gateway (GW). An enhanced BBU (eBBU) pool can be composed from a BBU pool and a gateway. At its turn, the GW might cover $n \times$ RATs, performing basic functions like packet buffering and inspection and routing/scheduling for multi-RAT (2G, 3G, 4G, WLAN, RRH, etc.). A possible evolution is that such a GW might replace the related network units, such as the EPC in 4G. H-CRAN supports the configurations where in one cell can exist various coexisting RATs and one eBBU pool.

Caching is important [44–46] in content delivery systems by optimising content placement and significantly improving the QoE perceived by the users. In the case of H-CRAN several variants of caching can be used. When no eBBU Pool caching is applied, the eBBU pool is directly connected to the RATs and can easily obtain their online NSI and utilise it in the packet scheduling (multi-RAT scheduler). Consequently, the delivery performance is better (e.g. addressing the previously discussed out of order issue). Note that the priorities of different video packets (e.g. those generated by scalable video coding) or QoS requirements from multiple MUEs may also affect the scheduling at the eBBU pool. The H-CRAN with packet scheduling exposes better delivery performance than conventional HetNets with only HS scheduling.

The eBBU Pool can have also caching role. The demanded video can be cached at the local eBBU pool, based on the technology of content awareness caching for 5G networks. This solution will significantly reduce the traffic amount from original HS. Both the video encoding and transmission can be adapted to the online NSI of multiple RATs. The eBBU pool works as a SP with the units encoding the source video, controlling the frame rate and managing the pre-caching content and buffering in MUEs. More accurate online NSI can determine the encoding redundancy and the size of pre-caching content could be minimised, thus saving the scarce spectrum resource. More accurate NSI at the eBBU pool may lead to decisions to reduce encoding redundancy and therefore increase the efficiency.

5.6.4 Fog/edge computing approach

As shown in Section 5.6.3, the cloud concepts have been included in several 5G architectures. However, the excessive centralisation brings its own problems, especially for RAN efficient implementation. As an alternative to pure cloud architecture, *Fog computing*-based networks, cooperating with cloud technology, have been recently proposed and investigated, to respond better to challenging 5G needs and traffic demand. The cloud RAN functions adopted centralised management in order to achieve optimal resource utilisation, whereas the Fog network can takes advantage of social information and edge computing to improve the end-to-end latency.

The emergent Fog or Edge computing ('fog' is a term coined by CISCO) [51,52] extends the cloud computing paradigm by bringing services to the network edge – in proximity of the users (e.g. network edge points, APs or even end devices). Fog computing nodes are typically located away from the main cloud data centres, that is, at the network edge and are geographically distributed and available in large numbers. Fog nodes are typically accessed by devices over wireless networks. The proximity-to-users naturally allows low and predictable service latency, and therefore, better QoS can be expected. Fog application code runs on fog nodes as part of a distributed cloud application. Fog computing nodes provide applications with awareness of device geographical location and device context. Mobility of devices is supported, that is, if a device moves far away from the current fog node, the mobile device application can be instructed to associate itself with a new application instance on a new fog node closer to the device.

Fog computing provides to end-users, data, compute, storage and application services. Services can be hosted at the network edge, APs or end devices such as set-top-boxes. *IoT* and, more generally, *Internet of everything* (IoE) applications, requiring real-time/predictable latency (e.g. transportation, industrial automation, networks of sensors and actuators) are well served by fog computing. Densely distributed and big data are other areas to which fog computing can offer support, through its collection points. Wide geographical distribution allows for real time big data and real time analytics.

Fog computing offers significant advantages [53] which are important also for 5G networks, such as data movement reduction across the network resulting in reduced congestion and latency. It eliminates the bottlenecks resulting from centralised computing systems, allows lower costs, improved security of encrypted data since it stays closer to the end user, thus reducing exposure to hostile elements. Fog computing improves scalability arising from virtualised systems. It actually eliminates the core computing environment, thereby reducing a major block and a point of failure; additionally it provides sub-second response to end users, provides high levels of scalability, reliability and fault tolerance. The overall bandwidth consumption is lower than in core cloud computing case.

5.6.4.1 Cloud-RAN limitations

A serious CRANs problem consists in its strong requirements imposed to the fronthaul network, in order to access the centralised BBU pool. A high bandwidth and low latency inter-connection fronthaul is necessary; however, in practice the fronthaul is frequently constrained in terms of capacity and time delay. This could have negative impact on SE and also energy efficiency [54].

The heterogeneous cloud radio access networks (H-CRANs) [43,54,55] tries to solve some of the CRAN disadvantages (Figure 5.11). The control and user/data planes are decoupled; HPNs are mainly used to provide seamless coverage and execute CPl functions and many RRHs mainly execute DPl functions, that is, provide high-speed data rate for the packet traffic transmission. HPNs are connected to the BBU pool via the backhaul links for interference coordination. In H-CRAN the RRHs can provide short-distance communication for mobile terminals (MT) to improve transmission rate and HPN can provide ubiquitous connection to achieve seamless coverage.

H-CRANs still present some issues, having impact in practice. For instance the location-based social (popular) applications generate significant traffic, thus overloading the fronthaul between RRH and centralised BBU. Two major problems in both CRANs and H-CRANs are the high transmission delay and heavy burden on the fronthaul. The H-CRANs do not take benefit from processing and storage capabilities in edge devices, such as RRHs and even 'smart' mobile terminals/UEs, which could be a potential mean to save the burden of the fronthaul and BBU pool. On the other side [54], a high number of fixed RRHs and HPNs should be installed by the operators to offer enough capacity to accommodate peaks of traffic but have low average utilisation.

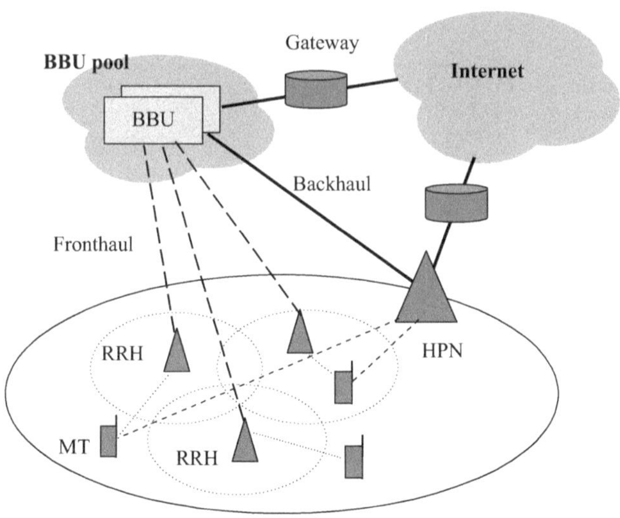

Figure 5.11 Simplified H-CRAN architecture

5.6.4.2 Fog-based radio access network

Fog computing technology applied to RANs , that is, fog-based radio access networks (F-RANs) is seen as a promising solution to increase efficiency in H-CRANs. In [54,55], fog network solutions cooperating with centralised cloud are proposed.

Two sets of functions previously executed in BBUs for H-CRAN solution, that is, *collaboration radio signal processing* (CRSP) and *cooperative radio resource management* (CRRM) – could be, with fog computing not only executed in a centralised BBU pool in H-CRANs but also can be hosted at RRHs and even wearable 'smart' UEs. The UEs can download some content packets not from BBU pool but from neighbours RRHs – should they are available in adjacent RRHs. Real-time CRSP and flexible CRRM performed in the edge devices allows to F-RANs to be adaptive to the dynamic traffic and radio environment and create lower burden on the fronthaul and BBU pool. User-centric objectives can be also achieved supported by several factors like adaptive technique among D2D, wireless relay, distributed coordination and centralised large-scale cooperation. In F-RANs, the traditional RRHs are enriched with caching, CRSP and CRRM capabilities and become *fog-based access point* (F-AP). The F-RAN architecture can be software defined, thus taking benefits of SDN concepts.

Four kinds of clouds are defined in [54]: *global centralised communication and storage cloud,* (the same as the centralised cloud in CRANs); *centralised control cloud* (located in HPNs and intended to complete functions of CPl) *distributed logical communication cloud* (located in F-APs and fog-based user equipment (F-UE), performing local CRSP and CRRM functions) and *distributed logical storage cloud* (for local storing and caching in edge devices).

The model of the proposed system can be split into three layers (Figure 5.12): cloud computing, network access and terminal layer. The fog computing network is actually composed of F-APs (residing in the network access layer) and F-UEs (placed in the terminal layer). Any two terminals F-UEs can communicate with

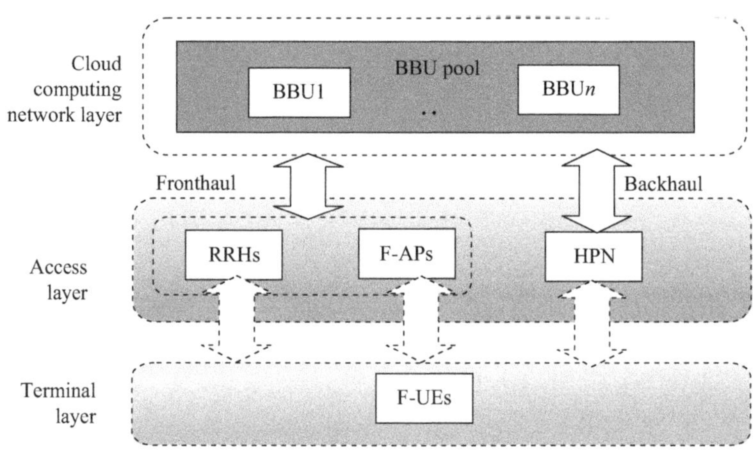

Figure 5.12 System model for implementing F-RAN [54]

each other through the direct D2D mode or through additional F-UEs playing the role of mobile relays. The network access layer is composed by F-APs, HPNs and RRHs., The F-UEs have access the HPN (also in H-CRANs this solution is applied) to perform system-related signalling (this is equivalent to CPl functionality).

The F-APs are used in the data plane to forward and process the traffic data. The F-APs communicate with BBU pool through the fronthaul links and HPN through backhaul links. The signals over fronthaul links are large-scale processed in the BBU pool, whereas over the backhaul links only control information is exchanged between the BBU pool and HPN. The BBU pool in the cloud computing layer plays a similar role as in H-CRANs. It can make also centralised caching. The fog approach alleviates the tasks of the BBU pool and fronthaul links, given that a large number of CRSP and CRRM functions are shifted towards F-APs and F-UEs. Also limited caching can be made by F-APs and F-UEs.

The CRAN, H-CRAN and F-RAN architectures can be compared on the basis of several criteria. The burden on BBU pool and fronthaul is highest for CRAN, medium for H-CRAN and lowest for F-RAN. The latter also offers the lowest latency. Decoupling between the CPl and DPl is only present in H-CRAN and F-RAN. The caching and CRSP functions are centralised in CRAN and H-CRAN while can be mixed, that is, centralised/distributed in F-RAN. The CRRM functions are only centralised in CRAN, whereas in the other, two mixed solution can be used. From the implementation point of view, CRAN or H-CRAN put high complexity in BBU pool and low complexity in RRHs and UEs, whereas the fog approach exposes medium complexity in BBU pool, F-APs and F-UEs. The H-CRANs and F-RANs can better serve real-time flows (e.g. voice).

An important feature of the F-RAN is the possibility of making caching in the edge devices. This can significantly decrease burden on the fronthaul, improve performances of CRSP and CRRM and also can relax the traffic burden at the cloud server. Faster content access and retrieval at F-UEs is possible than in CRAN or H-CRAN. The caching in F-RANs reduces the burden on the fronthaul, backhaul and even backbone, reduces the content delivery latency and increases the implementation flexibility in relationship with object-oriented or content-aware techniques. However, compared with the traditional centralised caching, the space for caching space at each F-AP and F-UE is small. Consequently a low-to-moderate hit ratio can be seen, which leads to the necessity to study intelligent caching techniques to be applied in F-RAN context. Resource allocation strategies and cooperative caching policies among edge devices are needed.

5.6.4.3 SDN and NFV support for F-RANs

The control and data plane decoupling in F-RAN naturally leads to the idea of applying SDN-like control in F-RAN context, with SDN as the core network. The SDN control can be extended to the physical layer, whereas CRSP and CRRM procedures are incorporated into the edge devices. In this way, more flexible and efficient network control can be achieved in F-RANs.

However open research issues still exist related to use SDN style of control in F-RAN environment. The combination of the MAC functions and physical layer functions for edge devices in F-RANs is still not yet clarified. On the other side, SDN is centralisation-based (for control), whereas the F-RAN has a distributed characteristic, based on edge devices. If using SDN control for F-RANs, then one needs to define slices to isolate the CRSP and CRRM in edge devices, so as to provide non-interfering networks to different coordinators. Supposing that SDN controllers are located in cloud computing network layer, then the inherent SDN problem appears, of control traffic overhead (between the controller and forwarding plane) to be transported over fronthaul links and thus decreasing the advantages of F-RANs.

NFV technology is also a promising candidate to support F-RAN, in cooperation with SDN. Programmable connectivity between VNFs can be provided and managed by the orchestrator of VNFs which could play the role of the SDN controller. NFV can support SDN controller virtualisation if installed on the cloud server. So, it could migrate to fit different locations according to the network needs. Not clear yet is how to virtualise the SDN controller in F-RANs, given the distribution characteristic in edge devices. Last but not least, other problems should be further studied, related to VNF interconnection, security, computing performance, portability and backward compatibility with legacy RANs.

5.7 Conclusions

The strong and diversified requirements imposed on 5G networks could be met, provided that powerful architectures and implementations are developed. Novel concepts, architectures and technologies, like cloud computing, SDN, NFV, working in cooperation, could contribute to solve the high challenges imposed to 5G. This chapter performed a partial overview of the Cloud/SDN/NFV-based solutions when applied to wireless networks and in particular to cellular 5G networks.

Given the limited space of this chapter, several aspects related to 5G technologies and services have not been discussed: security and privacy, details on scalability, reliability, mobility, IoT services, M2M and D2D communications and CRN aspects.

The focus of this chapter has been most on the architectures and functional split among the RAN and core networks. The most promising candidates are the integrated cloud/SDN/NFV solutions, trying to take benefit from the most powerful properties of SDN (CPl and data plane decoupling and programmability of the data plane, together with logical centralisation of management) and of NFV (software realisation of NFs and VNFs chaining). However, some important aspects are still research open issues – like scalability (if real-time functions migrate from the radio periphery to the centre), managing the RATs heterogeneity, mobility, unification of the CPl – to solve jointly the radio and core resource management, seamless horizontal and vertical mobility, routing, interference problem solving in dense cells environments, etc.

A short introductory text is inserted for RAN architecture incorporating fog computing into H-CRANs. Compared with the traditional centralised cloud computing based CRANs/H-CRANs, some important functions like cooperative radio signal processing and CRRM procedures in F-RANs can be adaptively implemented at the edge devices and are closer to the end users.

Certain limitations in the above-discussed technologies still should be further investigated. For instance, NFV where all network elements run on the cloud and rely on virtualisation, might not provide the necessary reliability and robustness. Real-time aspects inherent to the mobility-enabled networks put additional requirements to SDN/NFV-based technologies. Moving the VMs among network elements (e.g. MME or S/P-GW) because of HW failure or when there is need for additional processing resources are still open research subjects. Reliability and robustness need to be addressed in the proposed virtualisation platform.

An overall conclusion of this chapter is that 5G networks architectures and implementations can be significantly supported by *cloud/SDN/NFV* concepts and technologies. However, important open research issues still remain to be studied and clarified in the future, on architectural but also design, implementation, deployment and inter-operability aspects. The heterogeneity, dynamicity and extreme density characteristics of the 5G networks make its challenges even stronger. Some of them are summarised below, but limiting the essentially to the topics discussed in this chapter. More discussion on such issues can be found in [1–4,34,35,54–56].

The full advantage of adopting SDN and NFV into 5G mobile networks is not yet completely understood and hence needs further research. A common understanding and trade-off is still needed, taking into account several aspects as network control and management, access network performances, backhaul network overheads of distributed and centralised programmable networks. Related to SDN and NFV architectures and technologies applied to 5G several open research issues can be stated:

- The centralised nature of the conventional SDN approaches creates bottlenecks and thus can reduce the resilience and scalability. Therefore, a balance between centralised logical control and actual distributed infrastructure of controllers can be found. Flat or hierarchical architecture of SDN CPl – with multiple controllers should be adapted to 5G dynamic environment, both in core and RAN. Generally one can say that flat organisation of the conventional SDN controllers does not provide an effective and flexible management solution for 5G networks that can meet the requirements for resilience and scalability.
- Different reconfiguration policies should be applied to the network elements in a dense environment, at different time scales, due to the dynamicity and density of this network, and this can also result in high signalling overhead.
- Wireless link quality in RAN is usually unreliable and unstable, interrupting temporarily the communication between the controller and its forwarders (if the controller communication channel uses in-band signalling), finally leading to isolated wireless networks.
- 5G network might have cells with particular configuration policies, which should be considered in a differentiated way by the SDN controllers.
- The partition of functions to be implemented in each plane is still an open issue in the SDN/NFV/5G, particularly in the RAN area.

- The edge heterogeneity (including D2D, M2M and V2V communications) determines very dynamic topologies; this leads to complexity in SDN and NFV functions planning, increased by several distinct mobility models and hardware constraints (e.g. the SDN controller should instruct the switches or network hypervisor which terminal node should forward packets).
- When integrating SDN and NFV, the SDN programmability needs standardising the northbound and southbound interfaces between physical and VNFs that form a single NS chain.
- Virtualisation might a negatively impact the virtual LTE and Wi-Fi services; therefore, the virtualised NFs performance should be carefully analysed in order to decide about physical/virtual implementation option.
- Standardisation is still in-progress, a unified cellular programmable interface for implementing SDN and NFV is under development, including a service chain through the integration of SDN and NFV.

Related to CRAN, H-CRAN and F-RAN proposals one should mention:

- CRAN has strong requirements imposed to the fronthaul network, in order to access the centralised BBU pool. A high bandwidth and low latency interconnection fronthaul is necessary; however, in practice the fronthaul is frequently constrained in terms of capacity and time delay. This could have negative impact on SE and also energy efficiency.
- H-CRANs try to solve some of the CRAN disadvantages. The control and user/data planes are decoupled; HPNs are mainly used to provide seamless coverage and execute CPl functions and many RRHs mainly execute DPl functions, that is, provide high speed data rate for the packet traffic transmission.
- Two major problems in both CRANs and H-CRANs are the high transmission delay and heavy burden on the fronthaul. The H-CRANs do not take benefit from processing and storage capabilities in edge devices, such as RRHs and even 'smart' mobile terminals/UEs, which could be a potential mean to save the burden of the fronthaul and BBU pool.
- Open research issues still exist related to use SDN style of control in F-RAN environment. The combination of the MAC functions and physical layer functions for edge devices in F-RANs is still not yet clarified.
- SDN is centralisation-based (for control), whereas the F-RAN has a distributed characteristic, based on edge devices. If using SDN control for F-RANs, then one needs to carefully define slices to isolate the signal processing from resource management in edge devices, so as to provide non-interfering networks to different coordinators.
- If SDN controllers are located in cloud computing network layer, then inherent SDN problem appears, of control traffic overhead (between the controller and forwarding plane) to be transported over fronthaul links and thus decreasing the advantages of F-RANs.

Although significant research has been done and is still under progress on 5G/SDN/NFV/cloud, much additional work should be performed to achieve a significant level of practical demonstrations of this very promising technology.

References

[1] Andrews J.G., *et al.* 'What Will 5G Be?'. *IEEE Journal on Selected Areas in Communications*, Vol. 32, No. 6, 2014, pp. 1065–1082.

[2] Agyapong P.K., Iwamura M., Staehle D., Kiess W., Benjebbour A. 'Design Considerations for a 5G Network Architecture'. *IEEE Communications Magazine*, November 2014, pp. 65–75.

[3] 'Analysis Understanding 5G: Perspectives on Future Technological Advancements in Mobile'. 2014. Available from https://gsmaintelligence.com/research/?file=141208-5g.pdf&download [Accessed October 2015].

[4] Panwar N., Sharma S., Singh A.K. 'A Survey on 5G: The Next Generation of Mobile Communication'. *Physical Communication*, vol. 18, part 2, pp. 64–84, March 2016. Available from http://arxiv.org/pdf/1511.01643v1.pdf [Accessed December 2015].

[5] Checko A., *et al.* 'Cloud RAN for Mobile Networks – A Technology Overview'. *IEEE Communications Surveys & Tutorials*, Vol. 17, No. 1, 2015, pp. 405.

[6] Kreutz D., Ramos F., Verissimo P., Rothenberg C.E., Azodolmolky S., Uhlig S. 'Software-Defined Networking: A Comprehensive Survey'. 2014. Available from http://arxiv.org/pdf/1406.0440.pdf [Accessed September 2015].

[7] Astuto B., Mendonca M., Nguyen X.N., Obraczka K., Turletti T. 'A Survey of Software-Defined Networking: Past, Present, and Future of Programmable Networks'. 2013. Available from http://hal.inria.fr/hal-00825087/ [Accessed September 2015].

[8] Hakiri A., Gokhale A., Berthou P., Schmidt D.C., Gayraud T. 'Software-Defined Networking: Challenges and Research Opportunities for Future Internet'. *Computer Networks*, Vol. 75, 2014, pp. 453–471.

[9] Mendonca M., Obraczka K., Turletti T. 'The Case for Software-Defined Networking in Heterogeneous Networked Environments'. 2012. Available from https://hal.inria.fr/hal-00687002 [Accessed October 2015].

[10] 'Open Networking Foundation Overview'. Available from https://www.opennetworking.org/about/onf-overview [Accessed September 2015].

[11] 'OpenStack: The Open Source Cloud Operating System'. Available from https://www.openstack.org/software/ [Accessed September 2015].

[12] Han B., Gopalakrishnan V., Ji L., Lee S. 'Network Function Virtualisation: Challenges and Opportunities for Innovations'. *IEEE Communications Magazine*, February 2015, pp. 90–97.

[13] ETSI GS NFV 003 V1.2.1 (2014-12), 'Network Functions Virtualisation (NFV): Terminology for Main Concepts in NFV'. Available from http://www.etsi.org/deliver/etsi_gs/NFV/001_099/003/01.02.01_60/gs_NFV003v010201p.pdf.

[14] ETSI NFV 001 V1.1.1, 2013-10, 'NFV Use Cases'. Available from http://www.etsi.org/deliver/etsi_gs/nfv/001_099/001/01.01.01_60/gs_nfv001v010101p.pdf [Accessed September 2015].

[15] ETSI GS NFV 002 v1.2.1 2014-12, 'Network Functions Virtualisation: Architectural Framework'. Available from https://www.etsi.org/deliver/etsi_gs/NFV/001_099/002/01.02.01_60/gs_NFV002v010201p.pdf [Accessed October 2015].

[16] ONF, 'OpenFlow-Enabled SDN and Network Functions Virtualisation'. Available from https://www.opennetworking.org/images/stories/downloads/sdn-resources/solutionbriefs/sb-sdn-nvf-solution.pdf [Accessed September 2015].

[17] *SDN and OpenFlow World Congress*, Frankfurt, Germany, 2013.

[18] Rouskas G. '*Network Virtualization: A Tutorial*'. 2012, http://rouskas.csc.ncsu.edu/ [Accessed September 2015].

[19] Marcondes C. 'NFV-WPEIF2014'. Available from www2.ic.uff.br/~michael/EngRedes/Cesar-Marcondes-NFV-WPEIF2014.ppt 19 [Accessed August 2015].

[20] Kolias C. 'Bundling NFV and SDN for Open Networking'. *NetSeminar*, Stanford, USA, May 2014. Available from http://netseminar.stanford.edu/seminars/05_22_14.pdf [Accessed October 2015].

[21] Matias J., Garay J., Toledo N., Unzilla J., Jacob E. 'Toward an SDN-Enabled NFV Architecture', *IEEE Communications Magazine*, April 2014, pp. 187–193.

[22] Yap K.K., *et al*. 'Blue Print for Introducing Innovation into Wireless Mobile Networks'. *VISA10: Proceedings of the Second ACMSIGCOMM Workshop on Virtualized Infrastructure Systems and Architectures*, New York, USA, 2010, pp. 25–32.

[23] Bansal M., Mehlman J., Katti S., Levis P. 'Openradio: A Programmable Wireless Dataplane'. *Proceedings of the First Workshop on Hot Topics in Software Defined Networks (HotSDN'12)*, Helsinki, Finland, 2012, pp. 109–114.

[24] Li L.E., Mao Z.M., Rexford J. 'Toward Software-Defined Cellular Networks'. *European Workshop on Software Defined Networking (EWSDN)*, 2012.

[25] SK Telecom, 'SK Telecom 5G White Paper: SK Telecom's View on 5G Vision, Architecture, Technology, and Spectrum'. Technical Report, 2014. Available from http://www.sktelecom.com/img/pds/press/SKT_5G%20White%20Paper_V1.0_Eng.pdf [Accessed August 2015].

[26] Bradai A., Singh K., Ahmed T., and Rasheed T. 'Cellular Software Defined Networking: A Framework'. *IEEE Communications Magazine – Communications Standards Supplement*, June 2015, pp. 36–43.

[27] Granelli F., Gebremariam A.A., Usman M., *et al*. 'Software Defined and Virtualized Wireless Access, in Future Wireless Networks: Scenarios and Standards'. *IEEE Communications Magazine – Communications Standards Supplement*, June 2015, pp. 26–33.

[28] Feng D., Lu L., Yuan-Wu Y., Ye Li G., Li S., and Feng G. 'Device-to-Device Communications in Cellular Networks'. *IEEE Communications Magazine*, April 2014, pp. 49–55.

[29] China Mobile Research Institute, 'CRAN White Paper: The Road Towards Green RAN'. October 2011. Available from http://labs.chinamobile.com/cran/wp-content/uploads/CRAN_white_paper_v2_5_EN.pdf [Accessed August 2015].

[30] China Mobile Research Institute, 'CRAN White Paper: The Road Towards Green RAN'. June 2014. Available from http://labs.chinamobile.com/cran/

wp-content/uploads/2014/06/20140613-CRAN-WP-3.0.pdf [Accessed August 2015].

[31] China Mobile Research Institute 'White Paper of Next Generation Fronthaul Interface'. Version 1.0', October 4, 2015. Available from http://labs.chinamobile.com/cran/2015/06/03/white-paper-of-ngfinext-generation-fronthaul-interface-version-1-0/ [Accessed August 2015].

[32] Guerzoni R., Trivisonno R., Soldani D. 'SDN-Based Architecture and Procedures for 5G Networks'. 5GU 2014, November 26–27, Levi, Finland, DOI:10.4108/icst.5gu.2014.258052.

[33] Mueller J., Chen Y., and Reichel B. 'Design and Implementation of a Carrier Grade Software Defined Telecommunication Switch and Controller'. *Network Operations and Management Symposium (NOMS)*, IEEE, 2014.

[34] Akyildiz I.F., Wang P., Lin S.C. 'SoftAir: A Software Defined Networking Architecture for 5G Wireless Systems', *Computer Networks*, Vol. 85, 2015, pp. 1–18.

[35] Akyildiz I.F., Lin S., Wang P. 'Wireless Software-Defined Networks (W-SDNs) and Network Function Virtualization (NFV) for 5G Cellular Systems: An Overview and Qualitative Evaluation'. *Computer Networks*, Vol. 93, 2015, pp. 66–79. Available from http://dx.doi.org/10.1016/j.comnet.2015.10.013 [Accessed September 2015].

[36] Bernardos C.J., Oliva A., Serrano P., Contreras L.M., Banchs A., Jin H. 'An Architecture for Software Defined Wireless Networking', *IEEE Wireless Communications Magazine*, Vol. 21, No. 3, 2014, pp. 52–61.

[37] Yazici V., Kozat U.C., Sunay M.O. 'A New Control Plane for 5G Network Architecture with a Case Study on Unified Handoff, Mobility, and Routing Management', *IEEE Communications Magazine*, Nov. 2014, pp. 76–85.

[38] Gudipati A., Perry D., Li L., Katti S. 'Softran: Software Defined Radio Access Network'. *Proceedings of the First Workshop on Hot Topics in Software Defined Networks (HotSDN'13)*, Hong Kong, 2013, pp. 25–30.

[39] Neal A., *et al.* 'Mobile-Edge Computing', White Paper to ETSI ISG MEC, September 2014. https://portal.etsi.org/portals/0/tbpages/mec/docs/mobile-edge_computing_-_introductory_technical_white_paper_v1%2018-09-14.pdf

[40] Yong Li Y., Zhao Z., and Wang C. 'System Architecture and Key Technologies for 5G Heterogeneous Cloud Radio Access Networks'. *IEEE Network Magazine*, Vol. 29, No. 2, 2015, pp. 6–14.

[41] 'Cisco Visual Networking Index: Forecast and Methodology 2013–2018'. White Paper, June 2014. Available from: http://www.cisco.com/c/en/us/solutions/collateral/service-provider/visual-networking-index-vni/complete-white-paper-c11-481360.html [Accessed October 2015].

[42] Stockhammer T. '3GPP Content Delivery Efforts'. Available from http://dashif.org/wp-content/uploads/2015/08/6e-3GPP-Content.pdf [Accessed September 2015].

[43] Peng M., Li Y., Jiang J., Li J., and Wang C. 'Heterogeneous Cloud Radio Access Networks: A New Perspective for Enhancing Spectral and Energy

Efficiencies', *IEEE Wireless Communications*, December 2014, [online] Available from: http://arxiv.org/abs/1410.3028 [Accessed October 2015].

[44] Hachem J., *et al.* 'Content Caching and Delivery over Heterogeneous Wireless Networks'. Available from http://arxiv.org/abs/1404.6560 [Accessed September 2015].

[45] Sheng M., Han W., Huang C., Li J., and Cui S. 'Video Delivery in Heterogenous Crans: Architectures and Strategies', *IEEE Wireless Communications*, June 2015, pp. 14–21.

[46] Dramitinos M. 'Video Delivery over Next Generation Cellular Networks'. Workshop SETM 2013. Available from http://www.cnsm-conf.org/2013/documents/papers/SETM/p386-dramitinos.pdf [Accessed September 2015].

[47] Yeganeh S., Tootoonchian H., Ganjali Y. 'On Scalability of Software-Defined Networking', *IEEE Communications Magazine*, February 2013, pp. 16–20.

[48] Heller B., Sherwood R., McKeown N. 'The Controller Placement Problem'. *Proceedings of the HotSDN*, 2012, pp. 7–12.

[49] 'Open Base Station Architecture Initiative BTS System' Ref. Document Version 2.0. Available from http://www.obsai.com/specs/OBSAI_System_Spec_V2.0.pdf [Accessed September 2015].

[50] 'CPRI Specification V6.0, Common Public Radio Interface (CPRI); Interface Specification', 2013. Available from http://www.cpri.info/spec.html [Accessed September 2015].

[51] Bonomi F., Milito R., Zhu J., Addepalli S. 'Fog Computing and Its Role in the Internet of Things'. August 2012. Available from http://conferences.sigcomm.org/sigcomm/2012/paper/mcc/p13.pdf [Accessed October 2015].

[52] Abdelshkour M. 'IoT, from Cloud to Fog Computing'. Available from http://blogs.cisco.com/perspectives/iot-from-cloud-to-fog-computing [Accessed September 2015].

[53] Klas I.G. 'Fog Computing and Mobile Edge Cloud Gain Momentum – Open Fog Consortium-ETSI MEC-Cloudlets'. November, 2015. Available from www.yucianga.info [Accessed September 2015].

[54] Peng M., Yan S., Wang C. 'Fog Computing based Radio Access Networks: Issues and Challenges'. Available from http://arxiv.org/abs/1506.04233 [Accessed November 2015].

[55] Hung S.C., Hsu H., Lien S.Y., Chen K.C. 'Architecture Harmonization Between Cloud Radio Access Networks and Fog Networks'. *IEEE Access*, December 2015. Available from http://ieeexplore.ieee.org/xpls/abs_all.jsp?arnumber=7359093 [Accessed September 2015].

[56] Borges V.C.M., Cardoso K.V., Cerqueira E., Nogueira M., Santos A. 'Aspirations, Challenges, and Open Issues for Software-based 5G Networks in Extremely Dense and Heterogeneous Scenarios'. *EURASIP Journal on Wireless Communications and Networking*, 2015, p. 164, DOI:10.1186/s13638-015-0380-8.

Chapter 6

Towards a FOG-enabled navigation system with advanced cross-layer management features and IoT equipment

Y. Nikoloudakis, S. Panagiotakis, E. Markakis, G. Mastorakis, C.X. Mavromoustakis and E. Pallis

Abstract

In this chapter, we present a cross-layer fog-enabled framework that offers visitors of small venues; such as museums, malls, convention centres, hospitals, and so on; enhanced context-aware experience and navigation services over 5G small-cell infrastructure. Distributed fog-enabled devices provide 5G networking throughout the surrounding establishment. The visitor, after signing into the network, is able to view various information and multimedia content concerning the narrow points of interest (POIs). The infrastructure also provides the ability to navigate the visitor throughout the establishment, using well-known positioning techniques. The positioning takes place with the mobile device receiving and juxtaposing the signal strength of small RF beacons sculling the local area. Finally, the network proposes other nearby POIs, depending on the user's preferences, based on the meta-data information stored inside the user's mobile device. The framework logic and calculations are transferred and sent back to the user through the cloud.

6.1 Introduction

We are currently experiencing the era of informational revolution where everyone can have access to any information from anywhere. The emergence of 5G networks, along with the cloud networking paradigm, has played a significant role for the enrichment in volume and diversity of information and services provided to the user. The realisation of 5G networks will also facilitate the seamless in terms of speed and ubiquitous access to information and services.

The use case presented in this chapter displays some of the possibilities presented by pervasive context-aware architectures and paradigms in combination with 5G networking. It merely depicts some of the potential evolution of real-time personalised user-centric services. We confer on technologies and architectures that can bring services and information, tailored to the user's needs and preferences, down to the user's fingers. We propose a cloud-based context aware tour guide for users browsing through a certain establishment, delivered via 5G-serving small cells scattered around each establishment.

The chapter is organised as follows. Sections 6.2–6.7 are a theoretical introduction to some of the elements discussed in the presented use case. In Section 6.8, the proposed use case scenario is presented and explained. In Section 6.10, the use case scenario is presented, and in Section 6.10, conclusions are presented.

6.2 State of the art

6.2.1 5G networks

Mobile networks and the incremental growth of subscribers and data transferred through them have been the main topic of discussion for a long time [1]. The ever-growing demand for mobile network capacity and network efficiency along with the emergence of the Internet of Things (IoT) has led the discussion towards the creation of an infrastructure capable of accommodating the modern network that needs offering ubiquitous, ultra-broadband and super high-speed user experience.

There has not been any official specification concerning 5G networks, but rather a general idea of what 5G networks should be and what it should be able to deliver [2]. There has been an ongoing research from many initiatives on a global level [3]. Their goal is to address the requirements that fifth-generation networks will have to achieve and the challenges an infrastructure of such proportion will have to face, with the ultimate goal being the delivery of official specifications.

5G will play a significant role in next-generation networks, as user experience will be enriched with fully immersive services, also referred as 'anything or everything as a service' [4], where context information will be bundled with various services and delivered through the 5G infrastructure. The later poses a number of challenges and arouses the need for redesigning services, interfaces, architectures, algorithms and so on.

As described in [3,5,6], since network traffic and the bandwidth demand are expanding exponentially, one of the main goals of 5G networks is to maximise the network throughput and capacity. The network throughput is addressed with various techniques as described earlier; yet, reducing the cell size has been the traditional approach towards the growth of the network capacity. Therefore, a denser distribution of access points also referred to as small cells, femtocells or picocells depending on the size of the cell, which they cover, was introduced. An immersive approach such as small cells can pave the way towards seamless provision of services, guaranteeing high speed and throughput. Those distributed devices are stationed at the network edge, providing 5G networking, with minimum latency.

6.2.2 Internet of Things and the fog

IoT has been given many definitions [7,8], but the most suitable is the one that states, 'Internet of Things is a network that inter-connects ordinary physical objects with identifiable addresses, so that it provides intelligent services' [8]. Basically, the definition implies that every day non- or low-intelligence devices (thin) can be connected to the internet or any kind of network with a unique address and interact in such a way that will produce some intelligent and generally useful outcome.

6.2.2.1 Advantages

The internet itself does not have any sensing ability. The internet of things provides that ability and enables the interconnection amongst non-intelligent devices. It may be difficult for someone to understand what advantages the IoT brings. But 'things' are already in our lives for some time. Such devices are smart phones or smart watches and devices with limited intelligence, which make our lives easier or more interesting. IoT will first feature in our houses with smart home implementations that make a house more energy-efficient and tailored to the household needs and preferences. Another field that IoT will bring revolution to is healthcare. A doctor will be able to assess a patient's vital measurements remotely and predict or prevent certain situations. Many issues and challenges concerning the IoT are yet to be addressed and dealt with. Some issues are discussed in the following sub-section.

6.2.2.2 Issues

There have been many issues tantalising the IoT from the very beginning of its birth. Initially, IoT needs to inter-connect a large number of heterogeneous devices with weak capabilities and resources. This implies that every device may have a different connection interface and computation ability [8]. Furthermore, the underlying network, where the devices will connect to, may be multi-hop, intermittent or susceptible to the surrounding environment [8,9]. Therefore, a universal architecture should be designed to tackle the integration of the system and surpass the connectivity issues. Such architecture that tackles many of the challenges mentioned above is the 802.15.4 standard (ZigBee).

Having a swarm of devices connected and continuously transmitting data makes the collection, forming and preparation of those data more and more tedious. Therefore, a middle-layer system, placed at the edge of the network could play the mediator role. That system will be collecting the data, maybe pre-processing it and uploading it to the cloud. Cloudlets placed closer to the ground that perform such operations are referred to as 'The Fog' [10–12] in a seamless collaboration with the cloud could be the solution for that issue. The above-mentioned small cells can act as dedicated fog nodes, since they are placed at the network edge, providing 5G networking. Complementary, those devices can perform a number of actions, such as computations, data analytics and so on. Fog nodes will bring data, compute, storage and application services closer to the end user, enabling the realisation of heterogeneity, geographical distribution and low-latency features.

6.2.3 Positioning methods

There has been an increasing interest in context-aware or location-based systems over the last few years [13,14]. One of the factors that provide context awareness to a service is location. Positioning is classified into two main categories, outdoor and indoor. These two categories, along with some of the methods commonly used to calculate position in each case, will be discussed in later sections.

6.2.3.1 Outdoor positioning methods

Outdoor positioning is mainly used by navigation services or military applications in order to calculate the position of a node or a user device in an open environment. Many different techniques and technologies have been researched and employed to achieve optimum accuracy.

6.2.3.2 Global positioning system and assisted GPS

The global positioning system (GPS) project was initialised in 1973 by the US government. It is a system that provides time and position information in any weather conditions, anywhere on earth where there is an unobstructed line of sight to four or more GPS satellites [15]. The receiver calculates position, using measurements received from the satellites, based on geometrical properties of triangles. In addition to the former, as a 911 requirement [16], the development of the assisted GPS (A-GPS) was accelerated in order to make cell phone location available to any emergency call dispatcher. The A-GPS leverages the cellular network to acquire external data to improve the start-up performance and overall accuracy of the receiver in exceptionally poor GPS signal conditions [17].

6.2.3.3 Time of arrival and time difference of arrival

Positioning methods based on triangulation can be divided into two sub-categories – lateration and angulation. Lateration methods employ various measurements in order to calculate the distance between a base station and the mobile device, whereas angulation method measures the angles of signals received and calculates distances based on geometric properties.

Due to signal attenuation, the time needed for the signal to travel from the transmitter to the receiver will vary depending on the distance. In that respect, time of arrival (TOA) can be transformed to distance and by performing geometrical computations or even least-squares approximations, position is obtained [13]. This technique has several drawbacks. One of them is signal multipath due to the complexity of surrounding environment, which alters the TOA, causing dubious results. Another problem is that all nodes, receivers and transmitters must have synchronised clocks and the signals must carry a timestamp of the time the signal was broadcasted. To reduce the effects of the problems mentioned above, since TOA is a two-dimensional approach, the time difference of arrival (TDOA) method, is a three-dimensional hyperbolic positioning approach, that measures the time difference between the signals received from different base stations. That technique is more effective, yet it requires the receiver to store a number of measurements.

6.2.3.4 Indoor positioning methods

The emergence of novel wireless communication technologies and mobile devices with sensing abilities has given birth to various pervasive applications and systems that offer users a very different and enhanced experience in a given environment [14]. Many techniques have been engaged to calculate position in indoor environments. Some of these techniques were initially used for outdoor positioning. Nevertheless, these techniques, albeit some alterations, were successfully applied in indoor environments as well. In the following sub-sections, some commonly used techniques are presented.

6.2.3.5 Proximity

The proximity approach is most commonly used among use cases, since it is a very simple method for assuming position [18]. In that method, distributed thin base stations on known locations, each periodically broadcast a unique signature. Mobile devices, when near a base station, receive that signature. That signature identifies a unique base station, therefore coarsely placing the mobile device to that location. That method has no apparent accuracy; yet, dense distribution of base stations increases precision of the technique.

6.2.3.6 Triangulation

Triangulation, as the term suggests, is the calculation of position using measured angles from a known base station to a mobile device. That method relies on basic trigonometric properties to calculate distance between the base station and the receiver. The cartographer Gemma Frisius first introduced triangulation method in his 1533 pamphlet 'Libellus de Locorum describendorum ratione', as a method for accurately positioning faraway places for map-making. In recent applications, a base station located in a known position broadcasts a signal containing a unique signature and various other information concerning the base station. The mobile device receives the signal and measures the received angle. By applying basic trigonometric formulas, the mobile device is able to calculate its position with respect to the base station.

6.2.3.7 Trilateration – multi-lateration

The trilateration method is the process of determining relative locations by utilising the geometry attributes of circles, spheres or triangles. In this method, three measurements of base stations in known coordinates are needed for the algorithm to calculate the position [19]. In a two-dimensional environment, the method is accurate. In a three-dimensional environment, though, the position calculated is a rather coarse estimation. By inserting multiple measurements of base stations in the algorithm (multi-lateration), the calculation error is reduced [20].

6.2.3.8 Fingerprinting

The fingerprinting method, also known as 'scene-analysis' [13], consists of two phases: the offline phase where the measurements of distributed base stations are taken and tagged with the location, wherein the measurements were taken and stored in a database. The second phase is the online phase where a mobile device browses through the area of interest, takes measurements and compares them to the measurement stored in the database. The measurements that are more similar to the stored values determine the devices coarse location.

To improve accuracy of the algorithm, a large number of measurements have to be taken and stored in distributed databases. That increases the size of those databases, and higher computation capabilities are needed.

6.2.3.9 Dead reckoning

Dead reckoning is the method that calculates the current position of a mobile device by constantly measuring velocity, direction and so on, given that the starting position is known [21]. The method requires an accurate measurement of speed and direction at all times for the position to be calculated. Since a large number of factors can conduce to measurement accuracy degradation, the method is prone to cumulative error that can result in entirely incorrect calculations.

6.2.4 Related technologies

Various existing and emerging technologies have contributed and continue to contribute to the research concerning mainly indoor positioning methods. Such technologies are nodes that enable distance measurement or proximity awareness and so on, or protocols that enable the inter-connection between nodes and/or mobile devices.

6.2.4.1 Bluetooth low energy

Bluetooth low energy (BLE) is an emerging wireless communication protocol stack that operates in the 2.4-GHz unlicensed band and employs frequency hopping to avoid interferences from other 2.4-GHz technologies [22]. It was developed by Bluetooth special interest group and is optimised for devices that require maximum battery life rather than high data rates [23]. A large number of existing commercial sensors and nodes utilise the BLE protocol stack for transmitting their measured values or IDs.

6.2.4.2 Beacons

A beacon, as the name implies, is a node that broadcasts small pieces of information to the surrounding environment, usually utilising the BLE communication protocol to provide contextual awareness to the vicinity. Beacons are usually short-range nodes. Dense distribution of beacons is required to cover large places of interest.

6.2.4.3 Radio-frequency identification

Radio-frequency identification (RFID) technology is used for low-range identification over radio frequency [24]. The RFID technology consists of tags, readers and specific software that juxtaposes the ID read from the tag to a database in order to match the ID with specific information related to it. The tags are printed antennas attached to a memory unit. There are two basic methods by which the tags communicate with the reader – the 'inductive coupling' and the 'electromagnetic waves' methods. In the first method, the antenna coil of the reader induces a magnetic field in the antenna coil of the tag. The tag utilises that induced energy to send the data stored in the memory unit back to the reader. In the second method, the reader sends the tag an amount of energy in the form of electromagnetic waves. The tag uses some portion of the energy received to turn on its circuit and then uses the rest of the received energy to send the stored data back to the reader. There are three basic frequencies that RFID systems operate. Low (100–500 kHz), intermediate (10–15 kHz) and high (850–950 MHz, 2.4–5.8 GHz).

6.2.4.4 Geo-fencing

Geo-fencing is the technology used to track and monitor mobile objects such as vehicles, persons, packages and so on located by GPS [25]. The coordinates of the mobile object are continuously sent to a control centre that positions the object on a map. Another set of coordinates is used to form a virtual fence around a certain area of interest. The system produces a pre-configured action when the tracked device enters or exits the marked area. A major disadvantage of that method is the continuous use of the GPS system, which leads to extensive energy consumption [26].

6.2.5 Content delivery networks

The internet could be characterised as a living organism that continuously expands and evolves. Nevertheless, even though the internet infrastructure bandwidth is constantly improving, providing users with last-mile high-speed connections, still it remains a best effort network due to the lack of central supervision and overall administration, making it impossible to ensure appropriate quality of services [27,28].

The content delivery networks (CDN) paradigm is an effective approach that tackles the issues discussed above by alleviating internet congestion by bringing content closer to the end user, thus making the routing path from the user to the content server as short as possible. In more detail, CDN replicates selected content from the origin servers to widely distributed replica servers close to the edge of the network. The users request is delivered from the origin server to the most geographically suitable replica server, and finally the content is delivered to the user [29–31].

6.2.6 Recommender systems

A recommender system (RS) is a complementary service that collects information concerning a target user in order to build a certain profile consisting of likes and dislikes about certain items such as movies, songs books, travel destinations and so on. That information can be acquired explicitly by simply collecting target user's ratings. It can also be acquired implicitly by monitoring user's behaviour, such as websites visited or music listened to, movies watched, book read and so on [32]. After building the user profile, the system can then predict and recommend items of possible interest to the user.

Many algorithms have been developed to materialise that endeavour. Some of them have reached commercial utility like Internet Movie Database (IMDB), Netflix or Amazon. However, there are two main types of algorithms used in RS – content-based methods and collaborative filtering. The former method compares the attributes of the proposed item to similar items the target user's likes or dislikes. The later gathers opinions from users with tastes and preferences similar to the target user, and based on their past ratings concerning the proposed item, makes a decision whether the proposed item is a suitable proposal for the target user [33].

6.2.7 Software-defined networking and virtualisation

6.2.7.1 Software-defined networking

Software-defined networking (SDN) [34] is a novel emerging approach that provides a centralised administration to the underlying network. The SDN follows a

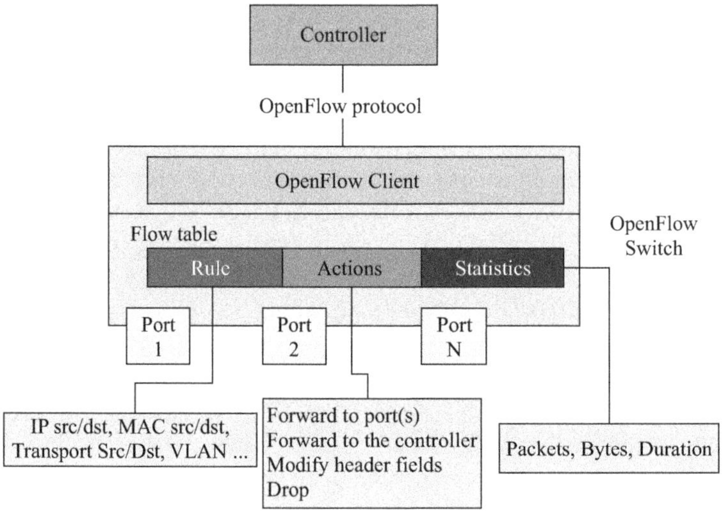

Figure 6.1 SDN architecture

basic principle separating the control from the data plane, meaning that the switch no longer performs in a non-intelligent manner. On the contrary, it operates rather as thin client of a separate network orchestrator located elsewhere [34]. The communication between the SDN-enabled switch and the network orchestrator has been a subject of research that has yielded a number of protocols with the most successful being the OpenFlow protocol [35,36]. The SDN paradigm provides the network with a certain amount of intelligence, thus enabling centralised administration, dynamic provision of network resources along with vertical and horizontal scalability. Figure 6.1 illustrates the SDN basic architecture.

6.2.7.2 Network function virtualisation

As the internet grows, so does the diurnal demand on bandwidth and services. Thus, service providers and network operators face great challenge in scaling up due to the excessive cost of hardware. Moreover, it is rather difficult and expensive to provide personalised services to users since dedicated vertical systems must be utilised. Network functions virtualisation (NFV) tackles those issues by deploying network functions such as routers or firewalls, services or even whole operating systems (OS) in a virtual manner. Functions can be deployed and administered individually or in bulk inside one hardware appliance and share memory and computational resources. Because of their virtualised state, functions can dynamically initialise and terminate, scale upwards or downwards in resources or even migrate or get instantiated to another distributed appliance as required. NFV is highly complementary to the SDN paradigm, yet it does not depend on it to operate. Although should SDN and NFV be implemented together, the overall outcome can be potentially greater [37–39].

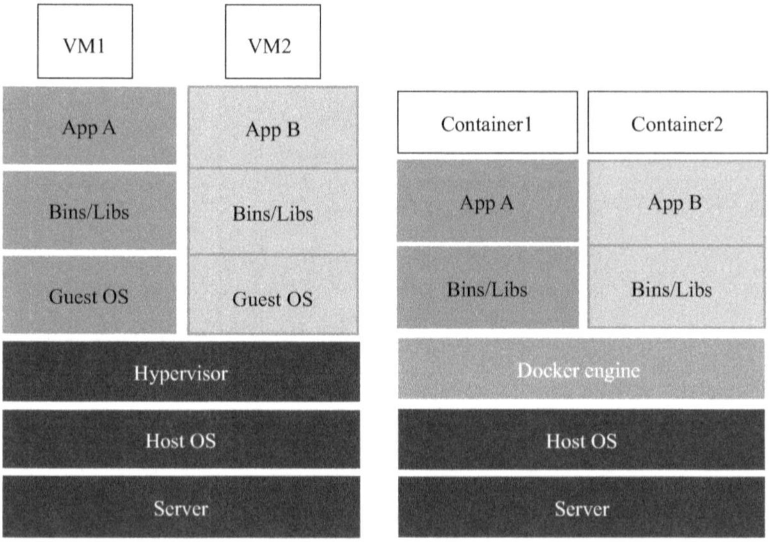

Figure 6.2 Comparison between standard VMs and dockers

6.2.7.3 Containers

A virtual machine (VM) is applied on top of a hypervisor that provides each VM a complete emulation of the underlying hardware. This way, it fools the VM into assuming exclusive access on that hardware. In practice, each VM is exclusively provisioned the hardware resources needed for it to operate. This fact provides limited scalability due to hardware constraints. Yet, a novel approach to virtualisation has emerged that aims to alleviate system administrators from that burden. Containers do not emulate any of the underlying hardware. On the contrary, the virtualised OS or service in the container communicates with the host OS, which by its turn makes the appropriate system calls to the hardware [40]. The container only virtualises the functions needed for the VM or service to operate. It operates like a 'sandbox', therefore offering a very lightweight solution for virtualisation. Figure 6.2 depicts the differences between the standard VMs and the dockers.

6.3 Beyond state of the art – use case

In the presented use case scenario, a user enters the premises of an establishment, and is immediately prompted with a URL via BLE, from the beacon stationed at the entrance. This URL leads to the web application that will guide the user throughout the establishment. If the user owns a device with no BLE capability, as soon as the device connects to an internal access point, the user is re-directed to the web application's URL by the captive portal setting of the access point. As soon as the application is initialised, it starts collecting advertisements from the surrounding beacons and periodically sends them to the user-service container in the cloud,

Figure 6.3 Navigation illustration

spawned by the cloud orchestrator as described in Section 6.8.2.3. The service calculates the user's current position and sends it back to the user. Consequently, the user is presented with recommendations of preference and location-relevant POIs. The user is then navigated towards the selected POI with the use of illustrated instructions (Figure 6.3). During the navigation, the user is also presented with recommendations of other relevant nearby POIs. Given the user chooses another destination, the navigation instructions dynamically change to navigate the user towards the new-chosen location. In addition, the user is provided with further information, location-related temperatures and crowdedness, gathered from the sensors scattered around the premises. When the user passes by or arrives to a POI, location-relevant content is optionally presented.

Upon exiting the premises, the user is prompted with further recommendations of other nearby preferable establishments. The service then only provides the geographical location of the POI of the user's choice, as those POIs are located outside the premises, wherein our infrastructure is situated. Figure 6.4 illustrates the basic use case of our infrastructure.

6.4 Position-aware navigation system with recommendation functions

The proposed system is an infrastructure that is able to determine the current position of visitors of a certain establishment and navigate them throughout the surrounding premises leading them to POIs of their preference and presenting them with location-relevant multimedia content and information, stored in geographically distributed fog-enabled CDN servers. Exiting the premises, visitors are prompted with recommendations of different nearby POIs that relate to their interests.

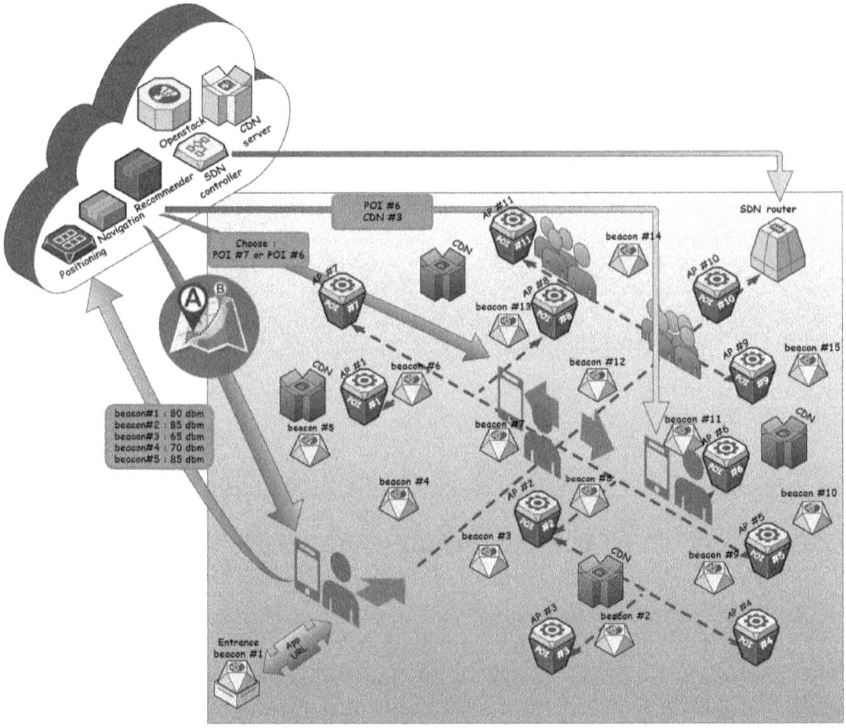

Figure 6.4 Infrastructure use case

6.4.1 System architecture

The discussed infrastructure can be divided into three layers containing the essential ontologies and functionalities. Figure 6.5 depicts the organisation of layers. In the following sub-sections, we describe every layer individually and confer on the architecture [41].

6.4.2 Real-world plane

At the bottom layer of our infrastructure, the equipment used to host the client application, which is the main medium for the user–server interaction, plays a very significant role. It is rather facilitative for the users to use their own hand-held portable devices, such as cell phones, tablets or even wearable devices, than any other device.

The exclusive use of such equipment, on one hand allows us to avoid the use of info-kiosks and other equipment statically stationed around the premises. This is a powerful characteristic since it enhances the system's agility and allows the user to

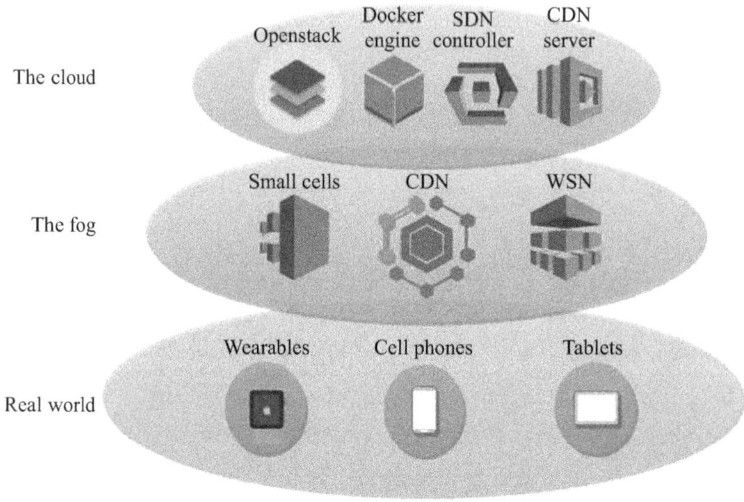

Figure 6.5 Infrastructure layers

freely browse through the establishment, exploiting the system's capabilities to the fullest, constantly receiving guided navigation instructions and location-relevant content. On the other hand, due to the physical constraints of such devices, for example CPU and power, we are dictated towards a centralised architecture, moving the computation burden and the overall 'intelligence' to the cloud, thus alleviating the end devices from the computational cost and energy consumption.

The user's mobile device hosts the client service, yet all the computation takes place on a private cloud server. As a front end, an HTML5 web application that graphically illustrates the user's current position, by collecting the signal strength measurements of nearby beacons and sending them back to the server, is used to navigate the user. The server computes the user's location using the multi-lateration positioning method and sends the location back to the user. The location is constantly updated and depicted on a floor plan inside the web application.

Lastly, the web application used in our infrastructure was developed using lightweight, client-side frameworks to avoid unnecessary latency and reduce payload exchange. The python Flask framework was used as the core of the web application, and angular Java Script (angular JS) along with the Twitter Bootstrap Frameworks were used for the implementation of the dynamic functionality and the responsive graphical user interface (GUI), respectively.

6.4.3 The fog plane

The fog computing paradigm as described in [11,10,42] is a dispersed version of the cloud, also referred as cloudlet, which shifts a number of cloud functionalities at the edge of the network. In the proposed scenario, the second layer plays the fog role, and hosts the small calls, the wireless sensor network and the CDN servers.

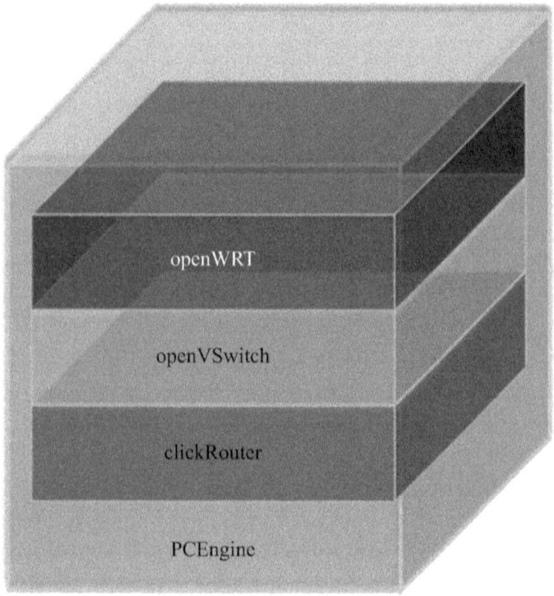

Figure 6.6 Small cell

The dense distribution of small cells paradigm was approached for the realisation of the proposed infrastructure in order to achieve optimal utilisation of the network and meet the 5G demands. The small cells scattered around the premises utilise the IEEE 802.11ac 5 GHz only Wi-Fi technology, providing up to 1 Gb/s data rates to the end user. For our use, case implementation, the SDN-enabled, Virtual-Network-Function-ready (VNF-ready) EmPOWER platform [43] was used due to its dynamic capabilities and seamless scalability potential. Figure 6.6 depicts the EmPOWER small cell (access-point) architecture. The EmPOWER platform is based on the PCEngines-ALIX platform, equipped with two mikrotik 802.11ac miniPCIe wireless modules. The ALIX board exploits the open-source router platform openWRT as an operating system. A virtual instance of OpenVSwitch is deployed in each access point to provide SDN capabilities along with a virtual instance of click modular router [44] to provide programmable control capabilities.

For each user, a virtual instance of an access point, with a unique service set identifier (SSID), is created and applied on the top layer of the platform by the SDN controller, creating, thus, to the user the illusion that is served by a single dedicated access point. The SDN controller also installs a policy for the flow connecting that particular virtual Instance and the border Router. In that way, separate or group quality of service (QoS) policies and load balancing can be easily applied to ensure the performance of each connection. Each access point, in collaboration with the SDN controller, is able to hand-off the user connection to the next access point. The virtual instance is then migrated to that access point, so that seamless roaming from one access point to another can be achieved.

As an attempt to meet 5G bandwidth expectations, we must ensure that the backbone inter-connection of the network devices is also a high-speed connection and can deliver the data rates that are expected, avoiding bottlenecks. In this respect, the G.fast-based distribution system is used, which is specified in [45] and also proposed in [5] as the preferred medium for the radio over cable distribution. G.fast can deliver up to 1 Gb/s data rates in a range of 500 m using the existing copper cable infrastructure, providing maximum cost-effectiveness since existing copper cables can be used instead of fibre optic cables to achieve the expected high data transfer rates. In more detail, all access points, the router and the CDN relay servers are inter-connected with plain copper Ethernet cables, yet all devices implement the G.Fast protocol, thus achieving speeds up to 1 Gb/s.

A swarm of sensor nodes densely scattered around the premises enhance the infrastructure with the provision of information such as proximity, temperature and location, thus elevating the level of the system's contextual awareness. Such information, correlated with the meta-information collected from the end user's device, can result in the formation of a relatively precise recommendation service. The sensor nodes broadcast (advertise) data packets containing the signal strength, proximity, telemetry values or even URLs, exploiting the Google's Eddystone beacon protocol. The Eddystone beacon protocol is an open source protocol, developed by Google, designed to enable beacons seamlessly provide diverse real-time, contextual and non-pervasive information to the end user. Aspects of that protocol are discussed in [18] and elaborated on its role in the physical-web project in [46]. The beacon advertisements are periodically broadcasted with a preconfigured interval. The Estimote product family [47] is used for the population of the wireless sensor network swarm.

In the proposed infrastructure, the user is constantly served with POI-related multimedia content. In order to ensure that the user is served with the requested content, with the least possible latency, reducing the routing path to the minimum, CDN nodes are distributed throughout the premises. The content delivery nodes are Debian-based Samba servers, hosted on the Raspberry pi 2 B platform [48]. The nodes are assigned the same IP address. The access points, obeying the ANYCAST methodology directives [49], route the content requests to the node closer to them according to their routing table as described in Section 6.5. Figure 6.7 presents an overview of the ANYCAST paradigm. The presented content is stored in distributed nodes mapped with the same IP address. All content requests are routed to the nearest node, through the access points with the use of the ANYCAST methodology. In more detail, each access point populates its routing table using the open shortest path first (OSPF) routing protocol. As a result, every access point creates a database describing the network topology. Then, the protocol will calculate all the routes considering the distance based on metrics depending on the network topology and populate the routing table with the shortest paths. The service's request for content will be resolved to the matching entry with the longest prefix inside the routing table on the access point. That way, the requests for content are routed to the node closest to the access point, and any changes in the network topology will not affect the underlying network since the access points' routing tables are dynamically updated.

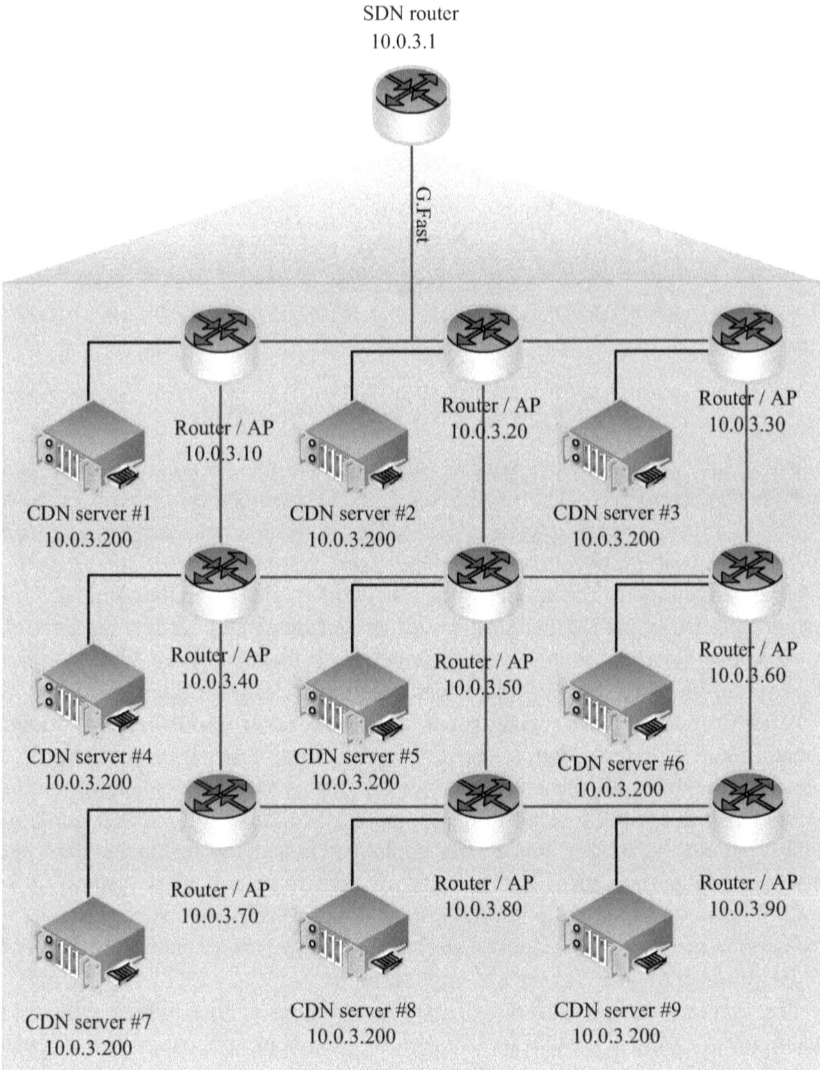

Figure 6.7 ANYCAST

6.4.4 The cloud plane

The cloud server is the most important element of our infrastructure owing to the fact that it hosts and implements the overall functionality and intelligence of the system. The cloud hypervisor, the SDN controller, the web services, the positioning and the RS are all implemented on the cloud server [50,51].

The cloud orchestrator is a framework that provides centralised management of large pools of compute, storage and networking resources. It dynamically

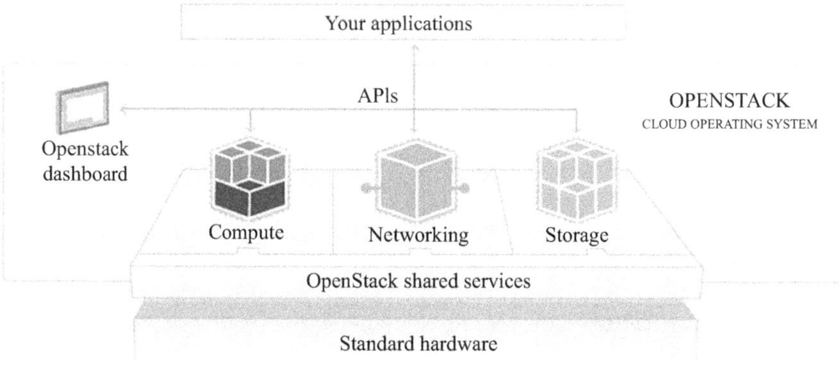

Figure 6.8 OpenStack architecture

provisions or retracts those resources according to the current demands of the underlying infrastructure. The OpenStack [52,53] open-source *framework*, hosted on CentOS Linux operating system, is used as the cloud orchestrator. Figure 6.8 depicts the generic architecture of the OpenStack framework.

Complementary to the orchestrator's network node, an SDN controller manages and provisions all the network resources. As described earlier, the controller creates a new flow for each new user, since each user is served with a separate virtual router instance, thus efficient load balancing can be applied.

Apart from the system management, the cloud server also provides ositioning, navigation and recommendation services to the users. A per user dockerised container that integrates all the above services is deployed. Each user is served by a unique docker container instead of deploying a VM that would make it impossible for the system to scale out, since after deploying that many instances, the system would eventually run out of resources. A docker container, as described in Section 6.7.3, is not a separate operating system. It is basically an abstraction of a VM that is applied on top of the docker engine. Therefore, the docker paradigm is the optimal solution for the current use case scenario.

The core of the client service is the positioning service which collects the measurements of the nearby beacons taken by the client application and calculates the user's position. The service gathers as many measurements as possible and performs the multi-lateration method to calculate distance between the user and the beacons. The algorithm calculates the position, with the height variable set to zero. The height is not required since the positioning is applied on a two-dimensional floor plan. This method uses as many measurements as possible in order to export a relatively accurate result. Nevertheless, the overall accuracy of the algorithm will vary depending on the spatial diversity of the area, wherein the beacons are scattered. Physical obstacles and signal reflections may lead to inaccurate results; however, the selected algorithm significantly reduces that effect. The beacons are positioned in pre-stored locations around the premises. Therefore, since every beacon has a unique ID, the user position can be placed relatively accurately on the floor-plan depending on the beacons' IDs from whom the measurements were extracted.

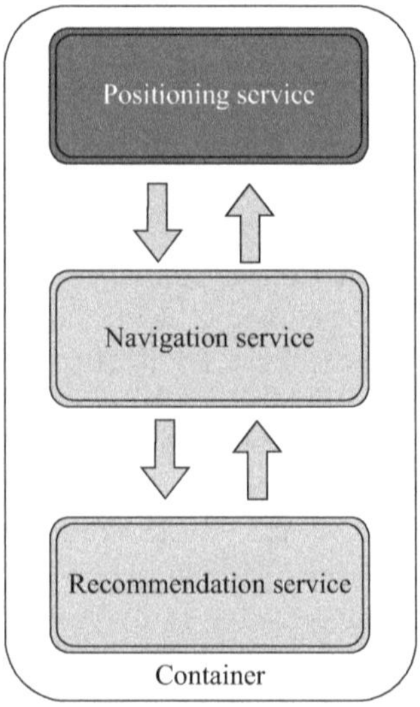

Figure 6.9 Container layers

The navigation service uses the user's position as an input to calculate the route from the initial user's position to the final destination. The positioning and navigation system interaction is bi-directional in the notion that as the user moves around, the user position is updated by the positioning system and the route on the floor plan is updated by the navigation system.

The recommendation service is the most complex of the services. It is an intelligent system which performs personal, location-based recommendations of POIs, depending on the user's interests and preferences. That intelligence stems from the explicit collection and classification of the user's meta-information, stored in the device used to host the client web application. The service, based on a content-based algorithm, initially recommends a number of POIs, and after the navigation begins it proposes new nearby POIs that may interest the user. The interaction of the recommendation service with the navigation service is also bi-directional. Should a user change the desired destination during the navigation, the navigation system dynamically changes the route and navigates the user to the new destination. Figure 6.9 graphically depicts the layer structure of the container described.

6.5 Conclusion

In this chapter, we proposed an immersive cloud infrastructure that offers navigation throughout a given establishment, recommends preference-related POIs to the user and provides multimedia content and information concerning the recommended POIs. The infrastructure utilises 5G networks, employing 5G-serving small cells to achieve minimum latency and maximum quality of user experience. The proposed infrastructure employs a number of novel emerging technologies, including cloud computing, fog computing, SDN networking, virtualisation and so on.

In this chapter, we elaborated on a number of indoor positioning techniques, yet our project mainly focuses on the multi-lateration method as an attempt to diminish calculation errors. As an alternate approach, the proximity method could also be used. The proximity method, in opposition to the multi-lateration method, does not require complex mathematical computations for the calculation of the user's location. Also, due to its nature of implementation, it could majorly decrease the cost of the pre-requisite Opex of the infrastructure. Nevertheless, it would not provide the same amount of positioning precision.

As a future endeavour, our goal is to create a generic framework, able to target any variety of venues such as museums, historic sites, monuments, commercial stores, malls and so on, employing the crowd-sourcing paradigm as an attempt to enforce the recommending mechanism and enrich the multimedia content concerning each POI, based on previous visitors' comments, likes and dislikes. In that respect, we will be able to provide a ubiquitous real-time guiding-as-a-service framework, which will be able to navigate a user throughout any kind of establishments utilising cloud and 5G networks.

Acknowledgement

The presented work was undertaken in the context of the "nExt generation eMergencY commuNicatiOnS – (EMYNOS) project" with contract number 653762. The project has received research funding from the H2020-EU.3.7 European Framework Programme.

References

[1] O. Shiakallis, C. X. Mavromoustakis, G. Mastorakis, *et al.*, "Traffic-based S-MAC: A novel scheduling mechanism for optimized throughput in mobile peer-to-peer systems," *Int. J. Wirel. Networks Broadband Technol.*, vol. 4, no. 1, pp. 62–80, 2015.
[2] M. Fiorarli, P. Monti, B. Skubic, J. Martensson, L. Valcarenghi, P. Castoldi, and L. Wosinska, "Challenges for 5G transport networks," in *2014 IEEE International Conference on Advanced Networks Telecommunication Systems (ANTS 2014)*, vol. 1, 2015, pp. 1–6.
[3] E. Hossain and M. Hasan, "5G cellular: Key enabling technologies and research challenges," *IEEE Instrum. Meas. Mag.*, vol. 18, no. 3, pp. 11–21, 2015.
[4] D. Soldani, K. Pentikousis, R. Tafazolli, D. Franceschini, N. E. E. Nd, T. O. Nd, I. Nfrastructure, and D. Soldani, "End-To-End Architecture and Infrastructure," *IEEE Commun. Mag.*, vol. 52, pp. 62–64, Nov. 2014.
[5] K. Chandra, R. V. Prasad, and I. Niemegeers, "An architectural framework for 5G indoor communications," in *IWCMC 2015 – 11th International*

Wireless Communications and Mobile Computing Conference, 2015, pp. 1144–1149.

[6] D. Bercovich, L. M. Contreras, Y. Haddad, A. Adam, and C. J. Bernardos, "Software-defined wireless transport networks for flexible mobile backhaul in 5G systems," *Mob. Netw. Appl.*, vol. 20, no. 6, pp. 793–801, 2015.

[7] M. Botterman, "Internet of Things in 2020: A Roadmap for the Future," in *European Technology Platform on Smart Systems Integration*, 2008, pp. 1–27.

[8] H. D. Ma, "Internet of Things: Objectives and scientific challenges," *J. Comput. Sci. Technol.*, vol. 26, no. 6, pp. 919–924, 2011.

[9] J. Cubo, A. Nieto, and E. Pimentel, "A cloud-based Internet of Things platform for ambient assisted living," *Sensors (Basel)*, vol. 14, no. 8, pp. 14070–14105, 2014.

[10] F. Bonomi, R. Milito, J. Zhu, and S. Addepalli, "Fog computing and its role in the Internet of Things," in *Proceedings of the First Edition of the MCC Workshop on Mobile Cloud Computing*, 2012, pp. 13–16.

[11] L. M. Vaquero and L. Rodero-Merino, "Finding your way in the fog: Towards a comprehensive definition of fog computing," *ACM SIGCOMM Comput. Commun. Rev.*, vol. 44, no. 5, pp. 27–32, 2014.

[12] I. Stojmenovic and S. Wen, "The fog computing paradigm: Scenarios and security issues," in *Proceedings of 2014 Federated Conference on Computer Science and Information Systems*, vol. 2, 2014, pp. 1–8.

[13] T. Ruiz-López, J. L. Garrido, K. Benghazi, and L. Chung, "A survey on indoor positioning systems: Foreseeing a quality design," *Adv. Intell. Soft. Comput.*, vol. 79, pp. 373–380, 2010.

[14] D. H. Stojanović and N. M. Stojanović, "Indoor localization and tracking: Methods, Technologies and Research Challenges," *Facta Univ. Ser. Autom. Control Robot.*, vol. 13, no. Iii 43007, pp. 57–72, 2014.

[15] Wikipedia, "GPS," *Wikipedia*, 2016. [Online]. Available: https://en.wikipedia.org/wiki/Global_Positioning_System. [Accessed: 31-Oct-2016].

[16] Wikipedia, "911 Enforced," 2016. [Online]. Available: https://en.wikipedia.org/wiki/Enhanced_9-1-1. [Accessed: 31-Oct-2016].

[17] Wikipedia, "A-GPS," 2016. [Online]. Available: https://en.wikipedia.org/wiki/Assisted_GPS. [Accessed: 31-Oct-2016].

[18] Y. I. Daradkeh and D. Namiot, "Network proximity and physical web," *Math. Comput. Sci. Ind.*, 2015. [Online]. Available: http://www.inase.org/library/2015/books/bypaper/MCSI/MCSI-32.pdf. [Accessed: 20-Nov-2015].

[19] B. Li, J. Salter, A. G. A. Dempster, and C. Rizos, "Indoor positioning techniques based on wireless LAN," *October*, vol. CD-ROM pro, no. January 2016, pp. 13–16, 2006.

[20] Wikipedia, "Multi-Lateration Method," 2016. [Online]. Available: https://en.wikipedia.org/wiki/Multilateration. [Accessed: 31-Oct-2016].

[21] E. Wikipedia, "Dead Reckoning," *Wikipedia*, 2014. [Online]. Available: http://en.wikipedia.org/wiki/Dead_reckoning. [Accessed: 31-Oct-2016].

[22] C. Gomez, J. Oller, and J. Paradells, "Overview and evaluation of Bluetooth low energy: An emerging low-power wirelcss technology," *Sensors (Switzerland)*, vol. 12, no. 9, pp. 11734–11753, Aug. 2012.

[23] Bluetooth SIG, "A Look at the Basics of Bluetooth Technology," BlueTooth SIG, 2012. [Online]. Available: http://www.bluetooth.com/Pages/Basics.aspx. [Accessed: 24-Jul-2015].

[24] H. D. Chon, S. Jun, H. Jung, and S. W. An, "Using RFID for accurate positioning," *J. Global Positioning Syst.*, vol. 3, no. 1, pp. 32–39, Dec. 2005.

[25] F. Reclus and K. Drouard, "Geofencing for fleet & freight management," in *2009 Ninth International Conference on Intelligent Transport Systems Telecommunications (ITST 2009)*, pp. 353–356, 2009.

[26] D. Namiot and M. Sneps-Sneppe, "Geofence and network proximity," in *Lecture Notes in Computer Science (including Subseries Lecture Notes in Artificial Intelligence and Lecture Notes in Bioinformatics)*, vol. 8121 LNCS, 2013, pp. 117–127.

[27] G. Peng, "CDN: Content distribution network," *arXiv Prepr. cs/0411069*, p. 26, 2004.

[28] A. Bourdena, C. X. Mavromoustakis, G. Kormentzas, E. Pallis, G. Mastorakis, and M. B. Yassein, "A resource intensive traffic-aware scheme using energy-aware routing in cognitive radio networks," *Future Gener. Comput. Syst.*, vol. 39, pp. 16–28, 2014.

[29] A. Vakali and G. Pallis, "Content delivery networks: Status and trends," *IEEE Internet Comput.*, vol. 7, no. 6, pp. 68–74, 2003.

[30] C. X. Mavromoustakis, G. Mastorakis, E. Pallis, *et al.*, "On the perceived quality evaluation of opportunistic mobile P2P scalable video streaming," in *International Wireless Communications and Mobile Computing Conference*, 2015, pp. 1515–1519.

[31] C. X. M. E. Markakis, D. Negru, J. Bruneau-Queyreix, E. Pallis, and G. Mastorakis, "A P2P home-box overlay for efficient content distribution," *Emerg. Innov. Wirel. Netw. Broadband Technol.*, vol. 37, no. 3, IGI Global, 1AD, pp. 199–220, 2016.

[32] J. Bobadilla, F. Ortega, A. Hernando, and A. Gutiérrez, "Recommender systems survey," *Knowledge-Based Syst.*, vol. 46, pp. 109–132, 2013.

[33] J. He and W. W. Chu, "A social network-based recommender system (SNRS)," *Data Min. Soc. Netw. Data*, vol. 12, pp. 47–74, 2010.

[34] K. Kirkpatrick, "Software-defined Networking," *Commun. ACM*, vol. 56, no. 9, pp. 16–19, Sep. 2013.

[35] B. A. A. Nunes, M. Mendonca, X.-N. Nguyen, K. Obraczka, and T. Turletti, "A survey of software-defined networking: past, present, and future of programmable networks," *IEEE Commun. Surv. Tutorials*, vol. 16, no. 3, pp. 1617–1634, 2014.

[36] T. Limoncelli, "OpenFlow: A radical new idea in networking," *Commun. ACM*, vol. 55, no. 8, pp. 42–47, 2012.

[37] N. McKeown, T. Anderson, H. Balakrishnan, *et al.*, "OpenFlow," *ACM SIGCOMM Comput. Commun. Rev.*, vol. 38, no. 2, p. 69, 2008.

[38] E. N. I. S. Group, "Network functions virtualisation (NFV): White paper," *Course.Ipv6.Club.Tw*, vol. 1, no. 1, pp. 1–15, 2015.

[39] C. X. Mavromoustakis, G. Mastorakis, A. Bourdena, and E. Pallis, "Energy efficient resource sharing using a traffic-oriented routing scheme for cognitive radio networks," *IET Netw.*, vol. 3, no. 1, pp. 54–63, 2014.

[40] G. Mastorakis, C. X. Mavromoustakis, E. Markakis, E. Pallis, and G. Skourletopoulos, "Virtual network functions exploitation through a prototype resource management framework," in *2014 International Conference on Telecommunications and Multimedia (TEMU)*, 2014, pp. 24–28.

[41] S. J. Vaughan-Nichols, "New approach to virtualization is a lightweight," *Computer (Long Beach, CA)*, vol. 39, no. 11, pp. 12–14, 2006.

[42] Y. Kryftis, G. Mastorakis, C. X. Mavromoustakis, J. M. Batalla, E. Pallis, and G. Kormentzas, "Efficient entertainment services provision over a novel network architecture," *IEEE Wirel. Commun.*, vol. 23, no. 1, pp. 14–21, 2016.

[43] R. W. Ahmad, A. Gani, S. H. A. Hamid, M. Shiraz, A. Yousafzai, and F. Xia, "A survey on virtual machine migration and server consolidation frameworks for cloud data centers," *J. Netw. Comput. Appl.*, vol. 52, pp. 11–25, 2015.

[44] R. Riggio, T. Rasheed, and F. Granelli, "EmPOWER: A testbed for network function virtualization research and experimentation," in *SDN4FNS 2013 – 2013 Workshop on Software Defined Networks for Future Networks and Services*, 2013, pp. 1–5.

[45] E. Kohler, R. Morris, B. Chen, J. Jannotti, and M. F. Kaashoek, "The click modular router," *ACM Trans. Comput. Syst.*, vol. 18, no. 212, pp. 263–297, 2000.

[46] J. M. M. Timmers, M. Guenach, C. Nuzman, and J. Maes, "G.fast: Evolving the copper access network," *IEEE Commun. Mag.*, vol. 51, pp. 74–79, August 2013.

[47] G. Inc., "Physical-Web." [Online]. Available: https://google.github.io/physical-web/. [Accessed: 24-Jul-2015].

[48] B. Estimote Inc., Apple, Google, "Estimote Beacons," 2013, 2013. [Online]. Available: http://estimote.com. [Accessed: 31-Oct-2016].

[49] B. Horan, "Raspberry Pi," 2013. [Online]. Available: https://www.raspberrypi.org/. [Accessed: 31-Oct-2016].

[50] E. Basturk, R. Engel, R. Haas, V. Peris, and D. Saha, "Using network layer anycast for load distribution in the internet," *Tech. Rep., IBM TJ Watson Res. Cent.*, vol. 20938, 1997.

[51] C. X. Mavromoustakis, G. Mastorakis, A. Bourdena, *et al.*, "A social-oriented mobile cloud scheme for optimal energy conservation," *Resour. Manag. Mob. Cloud Comput. Netw. Environ.*, pp. 97–121, 2015.

[52] C. X. Mavromoustakis, P. Mousicou, K. Papanikolaou, G. Mastorakis, A. Bourdena, and E. Pallis, *Dynamic Cloud Resource Migration for Efficient 3D Video Processing in Mobile Computing Environments*. New York, NY: Springer, 2015.

[53] OpenStack, "OpenStack." [Online]. Available: http://www.openstack.org/. [Accessed: 31-Oct-2016].

Chapter 7

Internet of Things: a systematic literature review

Ioannis Deligiannis[1], George Alexiou[1], George Papadourakis[1], Evangelos Pallis[1], Evangelos Markakis[1], George Mastorakis[1] and Constandinos X. Mavromoustakis[2]

Abstract

The "Internet of Things" (IoT) is becoming an increasingly growing topic of conversation worldwide that it promises to offer a revolutionary fully connected "smart" world. IoT represents a vision, in which the Internet extends into the real world involving everyday objects equipped with sensors, processing, and communications capabilities that will allow them to interconnect to each other over the Internet to accomplish some objective. This chapter reports on the current status of research on the IoT by examining the literature, identifying trends, exploring issues, challenges, and opportunities associated with IoT.

7.1 Introduction

Although technology advances, society is moving toward an "always connected" reality. The Internet is constantly changing under the influence of new technologies and concepts. One of those concepts is the so-called, Internet of Things (IoT). It is a global Internet-based phenomenon that is widely used for the exchange of services and goods describing a new reality where devices are part of the Internet. The IoT is emerging as the third wave in the development of the Internet. Although the fixed Internet that grew up in the 1990s connected

[1]Department of Informatics Engineering, Technological Educational Institute of Crete, Heraklion 71500, Greece
[2]Department of Computer Science, University of Nicosia, 46 Makedonitissas Avenue, Engomi, 1700 Nicosia, Cyprus

1 billion users via PCs, and the mobile Internet of the 2000s connected 2 billion users via smartphone, the IoT is expected to connect 28 billion "things" to the Internet by 2020, ranging from wearable devices such as smart-watches to automobiles, appliances, and industrial equipment. The combination of IoT with services and intelligence leads to a homogenization of the digital world with the physical one; this will bring a new ubiquitous computing and communication era and change people's life extremely [1] (Figure 7.1).

The term "IoT" was first coined by Kevin Ashton, executive director of the Auto-ID Center, in the context of supply chain management [2]: "I could be wrong, but I'm fairly sure the phrase 'Internet of Things' started life as the title of a presentation I made at Procter & Gamble (P&G) in 1999. Linking the new idea of RFID in P&G's supply chain to the then-red-hot topic of the Internet was more than just a good way to get executive attention. It summed up an important insight which is still often misunderstood." However, in the past decade, the concept got more inclusive covering a variety of applications like healthcare, smart cities, environmental monitoring, etc. Although the definition of "Things" is changing as technology evolves, the main goal of making intelligent machines communicating without the aid of human intervention remains the same. The IoT today consists of many different sensor networks and protocols, connected to dedicated cloud services [3], providing access through smartphone and browser apps. It is rare for these separate "silos" to cooperate or interact with each other.

An example of IoT is a smart house equipped with a smart lock, a smart thermostat, a smart security camera at the front door, and a smart TV. The lock and the thermostat intercommunicate and automatically turn off the heat when there are no traces of residents of the house within it for a specific amount of time. The security camera at the front door transmits a picture to a smart TV to show who is ringing the doorbell.

The vision of IoT can be seen from two perspectives; Internet-centric and Thing-centric [4]. In the Internet-centric architecture, the involved Internet services are the main focus, whereas data are contributed by the objects. The Thing-centric is focused on the capabilities of real-world objects connected to the network or augmented with Information Technology (IT) services as is the case with Radio Frequency Identification (RFID) or smart objects (Figure 7.2).

IoT is getting into all aspects of production and life, and gradually changes society's behavior and thinking. The capabilities of the concept lead to applications in nearly every domain of the modern life. It is involved in industry (manufacturing, logistics, service sector, banking, financial governmental authorities, intermediaries, etc.), in environment (agriculture & breeding, recycling, environmental management services, energy management, etc.), and in society (governmental services for citizens, e-inclusion for aging or disabled people, etc.) [5]. There are also terms of developing new applications and services that apply at inter-domain level. For example, monitoring of the food chain, or dangerous goods, has not only to do with the industry itself but also has effect on the society.

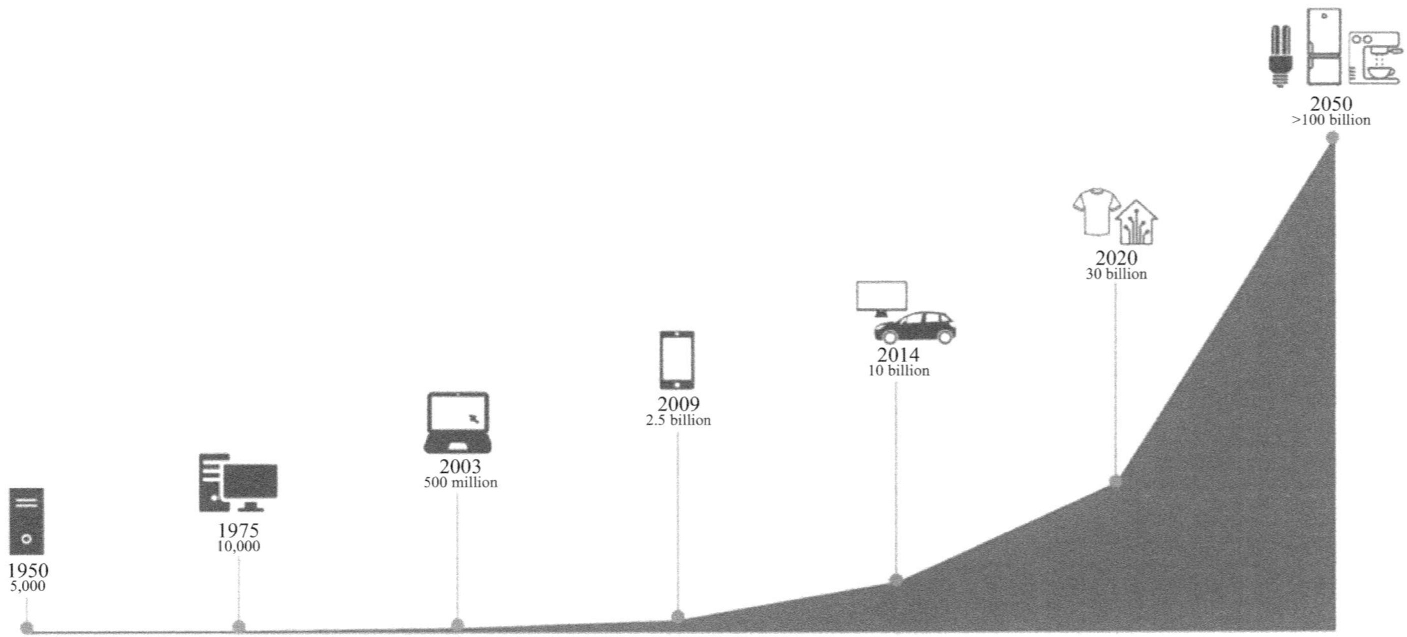

Figure 7.1 Expansion of the IoT devices

Figure 7.2 Most commonly used IoT technologies

7.2 Search methodology

The objective of this research is to report on the current state of IoT research by examining the literature, identifying current trends, describing the challenges that threaten IoT diffusion, presenting open research questions and future directions, and compiling a comprehensive reference list to assist researchers.

The search process covered journal articles and conference papers, excluding surveys and literature reviews, available in four major electronic databases; ACM Digital Library, IEEE Explorer, Springer-Link, and Science Direct. In the initial stage, we identified relevant papers by analyzing publications title and abstract. In the second stage, a full-text analysis was undertaken to discover and record the concrete technologies reported in each of the relevant papers. After the second filtering stage, the primary studies were subsequently divided on the basis of their keywords. One author was responsible for the initial search process stage, and another was responsible for the second.

Table 7.1 Classification scheme

Category	Description
Technology	Hardware, software and architecture
Applications	
Challenges	Security, privacy, legal
Business models	New business models for corporations
Future directions	

Since the review is generally about IoT, we used the search query "IoT" to find articles that focus on every aspect of the concept. The date range of the queries was 2008–16. We searched the four databases mentioned above for articles whose titles matched the search query. Due to conduction of the queries, the provided search engine of each database as well as Google Scholar is used.

Each paper was carefully analyzed and classified into a single category. The classification was mainly performed by the authors who agreed on the classification of each article. The literature was classified according to its content into the following major categories: technology, applications, challenges, business models, future directions, and overview/survey. The search was conducted on August, 2016; therefore, results that were indexed after this date have not been included in this study (Table 7.1).

7.3 The technology behind IoT

The core of the concept of the IoT is the perception that everyday "things" such as vehicles, refrigerators, medical equipment, and general consumer goods will be equipped with tracking and sensing capabilities. When this vision is thoroughly realized, "things" will also include more sophisticated processing and networking capabilities that will enable these smart objects to sense their environments and interact with people. Like any information system, the IoT will rely on a combination of hardware, software, and architectures.

7.3.1 Hardware

7.3.1.1 Radio-frequency identification

RFID is a technology that incorporates the use of electromagnetic or electrostatic coupling in the radio-frequency portion of the electromagnetic spectrum to identify an object, animal, or person uniquely. It is a generic term that is used to describe a system that transmits the identity (in the form of a unique serial number). There is a classification of automatic identification technologies.

Auto-ID technologies include bar codes, optical character readers, and some biometric technologies such as retinal scans. The auto-ID technologies have been used to reduce the amount of time and labor needed to input data manually and to improve data accuracy. Some auto-ID technologies, such as barcode systems, often require a person to scan a label or tag to capture the data manually. RFID is designed to enable readers to capture data on tags and transmit it to a computer system—without needing a person to be involved.

Thousands of companies have used RFID technology for a decade or more. RFID Business Applications spells out some of the ways the technology has been and will be utilized. Until recently, the cost of RFID has limited its use. For many applications, such as tracking parts for just-in-time manufacturing, companies could justify the cost of tags—a dollar or more per tag—by the savings, a RFID system could generate. Moreover, when RFID was used to track assets or reusable containers within a company's four walls, the tags could be reused. Tags may contain different forms of data, but the data form most commonly used for IoT applications are the Electronic Product Code or EPC. An EPC is a universally unique identifier for an object. These unique identifiers ensure that objects tracked with RFID tags have individual identities in the IoT.

This kind of technology has applications in the areas of logistics and supply chain management, aviation, food safety, retailing, public utilities, and others.

7.3.1.2 Near-field communication

A newer technology based on RFID is the near-field communication (NFC). It is a short-range high-frequency wireless communication technology that enables the exchange of data between devices over about a 10-cm distance. This technology combines the interface of a smartcard and a reader into a single appliance. It allows users to share content between digital devices seamlessly, pays bills wirelessly, or even uses their cell phone as an electronic traveling ticket on existing contactless infrastructure already in use for public transportation.

The significant advantage of NFC over Bluetooth is the shorter set up time. Instead of performing the old-fashioned manual configurations to identify Bluetooth devices, the connection between two NFC devices is established at once (under a 1/10 s). Due to its shorter range, NFC provides a higher degree of security than Bluetooth and makes NFC suitable for crowded areas where correlating a signal with its transmitting physical device (and by extension, its user) might otherwise prove impossible. The NFC technology is integrated into smartphones that can exchange data with one another when brought together. NFC devices are also able to make connections with passive, unpowered NFC tags that are attached to objects.

7.3.1.3 Sensor networks

A sensor is a device that detects and responds to some input from the physical environment. The particular input could be light, heat, motion, moisture, pressure, or any one of a significant number of other environmental phenomena. The output is a signal that is converted to human-readable display at the sensor location or transmitted electronically over a network for reading or further processing.

When multiple sensors are used together and interact, they are referred to as a wireless sensor network (WSN). WSNs contain the sensors themselves and may also include gateways that collect data from the sensors and pass it on to a server. Although sensors "sense" the state of an environment or object, actuators perform actions to affect the environment or object in some way. Actuators can affect the environment by emitting sound, light, radio waves, or even smells. These capabilities are one way that IoT objects can communicate with people. Actuators are frequently used in combination with sensors to produce sensor–actuator networks.

7.3.1.4 System-on-chip

The growth of IoT solutions creates vast new opportunities for developers of embedded systems by providing capabilities that can be added to just about any physical object including medical devices, household appliances, home automation, industrial controls, even clothing, and light bulbs [6]. This collection of billions of end devices, from the tiniest ultra-efficient connected end-nodes to the high-performance gateways, creates a continuously growing demand in the embedded systems industry and sophisticated software design for efficiently supporting the demanding applications running on IoT devices.

IoT system-on-chip (SoC) designers have some difficult choices to make on storing data. They usually have to decide how much memory to include for major SoC functions, add on-chip or off-chip memory and whether data programing requirement is one time, a few times, or many times. Usually, these options seem mutually exclusive especially when the system does not provide an efficient memory management algorithm. Due to high-volume and low-price expectations for the IoT-enabled system, the cost is of great concern [6]. Moreover, hardware security support is now a requirement, and forthcoming ARMv8-M microcontrollers will be the new benchmark for security. This new architecture differs from the higher-end platforms, because it is designed to provide low, deterministic latency support. It also does not provide hypervisor support found in platforms like the Cortex-A, because that would also incur overhead that microcontroller applications cannot afford either in timing or hardware overhead.

The implications of the ARMv8-M architecture for IoT are significant. It provides a common security architecture that will be adopted by the wide array of Cortex-M vendors. It is scalable from the Cortex-M0 to the Cortex-M7. This will make it easier for developers to target the microcontroller space that has included a variety of restrictive security measures.

7.3.2 Software

Recent advances in networking, sensor, and RFID technologies allow connecting various physical world objects to the IT infrastructure, which could, ultimately, enable realization of the IoT and the ubiquitous computing visions. Although the IoT may rely upon the existing hardware infrastructure into a large extent, new software must be written to support the interoperability between numerous heterogeneous devices and searching the data generated by them.

The interconnectivity of computing and physical systems could, however, become “the nightmare of ubiquitous computing” [7] in which human operators will be unable to manage the complexity of interactions in the system, neither even architects will be able to anticipate that complexity, and thus to design the system. The IBM vision of autonomic computing [7] proclaims the need for computing systems capable of "running themselves" with minimal human management that is mainly limited to definition of some higher level policies rather than direct administration.

Semantic technologies are viewed today as a key technology to resolve the problems of interoperability and integration within heterogeneous world of ubiquitously interconnected objects and systems. Semantic technologies are claimed to be a qualitatively stronger approach to interoperability than contemporary standards-based approaches [8]. The IoT should become in fact the Semantic Web of Things [9,10].

It seems to be generally recognized that achieving the interoperability by imposing some rigid standards and making everyone comply could not be a case in ubiquitous environments. Therefore, the interoperability requires existence of some middleware to act as the glue joining heterogeneous components together.

The IoT will include vast numbers of heterogeneous devices generating enormous quantities of variable data. The IoT middleware sits between the IoT hardware and data and the applications that developers create to exploit the IoT. Thus, IoT middleware helps one to bring together a multitude of devices and data in a way that enable developers to create and deploy new IoT services without having to write different code for each kind of device or data format. There are a couple of EU FP6 research projects that have, as one of their goals, the development of some middleware for embedded systems. They are Reconfigurable Ubiquitous Networked Embedded Systems, 2004–07 and ongoing Service-Oriented Cross-Layer Infrastructure for Distributed Smart Embedded Devices, 2006–09. However, the middleware needs of the IoT domain have to go well beyond interconnectivity of embedded systems themselves. An approach for accomplishing that is the Global Enterprise Resource Integration, where all different types of resources get seamlessly integrated: physical devices with embedded electronics, web services, software applications, humans along with their interfaces, and other. The components of ubiquitous computing systems should be able not only to communicate and exchange data, but also to flexibly coordinate with each other, discover and use each other, and jointly engage in different business processes [11].

However, semantic web technology ought to be utilized in current browsers and search engines. IoT devices are mobile, dynamic, and will generate huge amounts of ever-changing data. Thus, there is the requirement for an IoT browser that is capable of distinguishing smart objects, discovering their services, and interacting with those objects [12] likewise as an IoT search engine that's capable of looking out the apace ever-changing data generated by IoT-enabled objects (Figure 7.3).

7.3.3 Architecture

Architectures had a need to represent, organize, and structure the IoT in a manner that could enable it to operate efficiently. Specifically, the distributed, heterogeneous character of the application form is necessary by the IoT of hardware/network, software, and process architectures capable of supporting these devices, their services, and the ongoing workflows they'll impact. Many hardware/network architectures have been proposed to support the distributed computing surroundings required by the IoT [13]. The differing designs which may be used to support the IoT also highlight

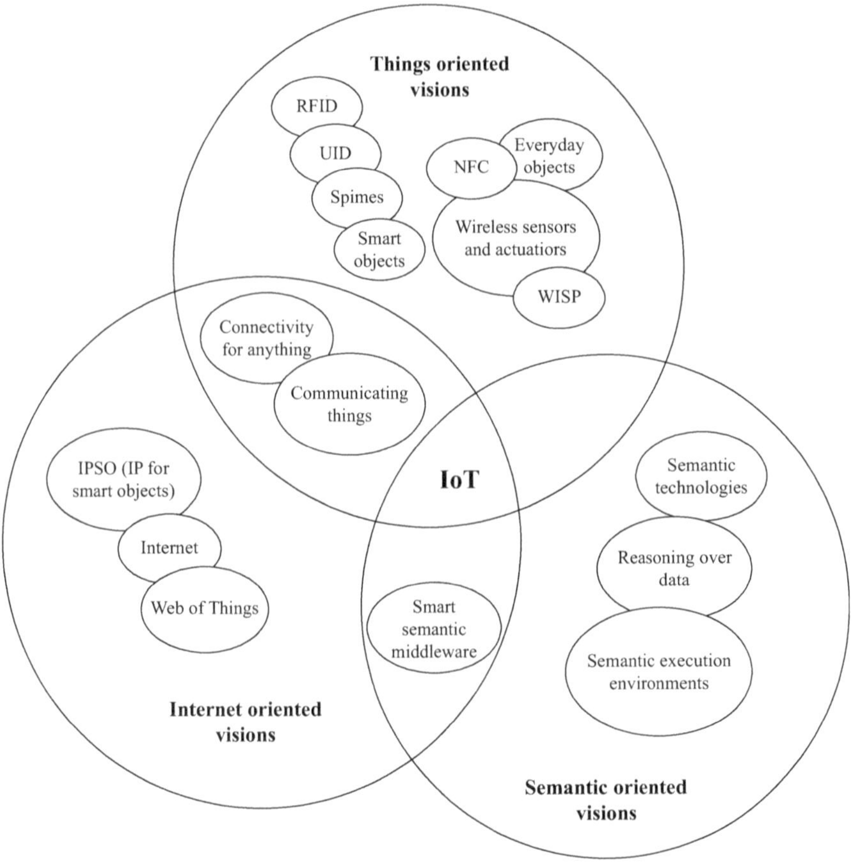

Figure 7.3 Different visions of IoT paradigm

the importance of the problem of standardization. Software architectures are essential to provide access to and enable the showing of services provided by IoT devices. The IoT will certainly have an impact on business processes. Process architectures are essential to structure the business processes that will incorporate the IoT effectively. There is absolutely no agreement on a single architecture that best fits the IoT. Several articles suggested various conceptual structures designs, whereas others suggested requirements for the diagnosis of proposed architectures and conceptual structures to meet the needs of smart objects.

7.4 The Internet of Things

The IoT consists of three stages or waves. The first one referred to things becomes connected to the Internet. Smart meters, Internet refrigerators, and even coffee makers are some cases of ordinary items that are becoming connected and sharing their data to the Internet. The second wave is when smart objects become connected to each other. It brings the potential of automated mundane tasks and removes a part of the necessity of human or even computer supervision, although it is the most challenging one since there is lack of machine-to-machine communication standards. The final stage of IoT is where applications are written to exploit the interconnectivity and automation of devices for the development of a new world where physical effects can be programmed.

As mentioned in the introduction, IoT "reality" comprises a broad area of research and developments, with applications in almost every possible field such as healthcare, industry, agriculture, transportation, cities, military, etc. Considering the limited nature of this review due to constraints in length we will cover five fields; social Internet of Things (SIoT), smart cities, smart health, security, and environmental monitoring.

7.4.1 Social Internet of Things

The notion of SIoT can be seen as the integration of Social Network concepts into the IoT allowing objects to establish relationships autonomously. This paradigm is rapidly evolving due to the awareness it carries, in a world where objects intercommunicate with each other or humans.

An initial concept of socialization between objects was popularized by Holmquist *et al.* [14]. Their work was mainly concerned on establishing relationships between wireless sensors and controlling such processes, although the sensors weren't integrated into social networks due to the lack of them in the time. More recent literature focuses on objects that enter humans' daily activities rather than abstract sensors. The concept of notifying communities by objects was introduced by Kranz *et al.* [15]. They presented a cognitive office that empowers everyday physical objects such as mug temperature sensors or plant moisture sensors, to share states, pictures, and sensor data via social networks. Their work also investigated the implications integration of social networks and IoT will bring. Building upon the previous notion [16] demonstrated awareness of senior citizens activities. Practically, two novel roles that the augmented everyday objects will play were introduced in the following:

1. Mediate the human-to-human communication and
2. Support additional ways for making noticeable and noticing activities in everyday life.

SIoT is said by many to be the next step in the evolution of ubiquitous computing. However, there are still a number of challenges and open issues that should be faced by the research community in order to mature this technology. Network Architecture, design, technologies, and interoperability are some of the many challenges rising. Ortiz [17] makes a step in the right direction in order to face these challenges presenting an overview of SIoT and proposing a general architecture that should be embraced.

7.4.2 Smart cities

The mass shift of the population from rural-to-urban areas is generating several kinds of problems. Difficulty in waste management, scarcity of resources, air pollution, human health concerns, traffic congestions, and inadequate, deteriorating and old infrastructures are only some of the total technical, physical, and material problems [18]. In order to overcome those problems and improve quality of life, our cities need to get smarter [19]. According to Frost & Sullivan, a [20] "smart city" is a city well performing in eight characteristics; smart governance, smart energy, smart building, smart mobility, smart infrastructure, smart technology, smart healthcare, and smart citizen.

7.4.3 Application of the IoT in healthcare

People's demands for health care have been increasing lately challenging the existing public health service, resulting in shortages of health resources and inadequacy of medical resources. These challenges have been constraining its further development [19]. In the forthcoming years, the delivery model of healthcare will transform fundamentally from the present hospital-centric, through hospital-home-balanced in the 2020th, to the final home-centric in the 2030th [21]. Current and emerging developments in IoT, ubiquitous wearable devices, and services will be contributing to this paradigm shift. The principal research agenda of IoT-based personalized healthcare systems are ubiquitous real-time monitoring systems [22].

The most typical type of existing remote monitoring systems (RMSs) is pure software apps on a smartphone or tablet. Their functionality is limited by the hardware [23], and the lack of engagement in contextual information, social networks, and multimedia [24]. Another type of RMS is the binding of software apps and external sensors; their connection is usually made through wireless native interfaces of the mobile terminal [19]. The challenges in these RMSs are the minor compatibility of medical sensors and mobile terminals, and the difficulty in facilitating interoperability between software and wireless sensors [25]. The last type, also the most profound, of solutions customizes the RMS together with biomedical devices, specific communication protocols, and complex application software [26]. These devices measure various vital signs such as apnea, heart rate, blood pressure, respiratory rate, posture, etc. Then the data are transmitted to the RMS through various wireless body area network techniques and finally propagated to a service backend through various communication environments [27].

Although there has been a lot of research work in this field, the use of IoT in healthcare is still in an infantry stage; Limburg *et al.* [28] have thoroughly summarized some of the major challenges the technology development is facing. Further research in this field could provide low-cost solutions and more robust services significantly improving a person's outcomes and quality of life.

7.4.4 Agriculture monitoring

Nowadays, the necessity of supporting agricultural activities is constantly rising due to escalating issues, such as the decline of people engaged in agriculture and their increasing age. On the other hand, quality evaluation and control of agricultural production are crucial to providing high-quality outcomes. The data acquisition of plant condition in an environment and the utilization of the data by using IoT will lead to stable production and improvement in productivity. Also, the acquired data could be utilized for future elimination of cultivation issues. An interesting application for environmental monitoring and management described in [29] proposes a system that connects several different devices on a network and processes the data collected by them. The system contains four layers; perception layer, network layer, the middleware layer, and application layer. The area of study was performed in a natural environment that is vulnerable and sensitive to climate change and human activities. Understanding the intensity of climate change and its ecological responses is critical for the sustainable regional development in the area. Another application, based on precision agriculture, utilizes wireless sensors on an IoT network, in southern Spain, for managing crops in a location where water deficit is a major challenge for local farmers [30]. Their system measured matric potential of the soil water and other crop parameters to keep consistent water levels needed for optimal crop growth. The information provided by the sensor nodes was essential to farmers for control irrigation, during each agronomic stage.

In agriculture monitoring, most of the research is focused on developing systems that monitor underground crop parameters. Optimal crop growth and disease prevention are obtained, nonetheless, by observing external crop parameters such as leaf and trunk condition. However, monitoring external crop condition remains a major challenge for researchers in this field yet.

7.5 Challenges

The challenges facing the emergence of the IoT are numerous. They are both technical and social. These difficulties must be overcome to ensure IoT adoption and diffusion. We subclassify challenges into security, privacy, legal/accountability, and general.

7.5.1 Security

Security is an important part of almost every IoT deployment, yet it is too often neglected in the development of systems. Considering that people become progressively familiar with IoT, security remains a concern and continues to be a challenge. IoT is susceptible to various security issues and has some significant privacy concerns for the end users. Numerous IoT devices such as sensors and RFID tags may have insufficient resources or low power while highly secure measures require many resources and energy consumption since they cost much computation and communication. In other words, security measures and energy efficiency often stand on the opposite side of each other. IoT designers are willing to apply high secure measures so to improve security, but at the lowest cost possible. One of the main challenges in IoT is to achieve security and low-energy consumption at the same time for the sensors. Traditionally security measures of sensor networks are targeting on selecting or proposing energy efficiency security algorithms or protocols. In addition, previous studies have shown that traditional security mechanisms lack resilience, security measures for IoT should be more adaptive to the current context.

There is a need for solutions that provide flexible, secure measures by context-aware computing and dynamic enforcement of policies for a complicated environment [31]. Current developments are divided into four primary layers:

- Perceptual layer: Perceptual nodes have limited computation power, storage capacity, and power. Thus, it is difficult to apply a high-security protection system. Node authentication is important in order to prevent unauthorized node access. Sensor data integrity and authenticity is becoming research focus, whereas lightweight encryption algorithms and cryptographic protocols [32] are necessary to this layer.
- Network layer: Security mechanism in this layer is crucial to the IoT. Identity authentication [33] is an important problem; there is also need to establish data privacy. Man-in-the-Middle and Distributed denial-of-service (DDoS) attacks are the most frequent methods of attack in the network [34]. Thus, DDoS attacks prevention of the vulnerable nodes is another important issue to be solved in this layer.
- Support layer: The main characteristic of this layer is the data processing and intelligent decision of network behavior that is prone to attacks from malicious information [35]. This layer needs a lot of the application security architecture such as cloud computing, strong encryption algorithms, encryption protocols, and virus protection.
- Application layer: Data sharing is the primary concern of this layer, which creating problems of data privacy, access control, and sensitive data exposure. The solution of the problem in this layer requires two aspects. One is the authentication and key management across the network, and the other is privacy protection [36,37].

With the sustained development of IoT, the small networks will merge into a large network. By then it would be more difficult to ensure the security. These security problems would be the key factor to decide the development of IoT [38].

7.5.2 Privacy

As increasingly gadgets come to be traceable via IoT, threats to nonpublic privacy come to be extra-extreme. Further to securing information to make certain that it does not fall into the incorrect palms, troubles of facts possession want to be addressed in an effort to make sure that customers sense at ease participating inside the IoT. For this reason, the ownership of statistics accumulated from smart objects should be simply established. The facts owner has to be confident that the records will not be used without his/her consent, especially whilst the records will be shared. Privacy rules may be one technique to making sure the privacy of statistics. Smart gadgets and reading devices in the IoT can each be prepared with privacy guidelines. Although the item and reader come into touch, they are able to every check the alternative's privacy policy for compatibility before communicating.

7.5.3 Energy

The IoT is a growing innovation for future enterprises as well as the everyday of million individuals, where a plethora of battery depended on objects like sensors, actuators, and (inter)connected smart device through the Internet, which provide variety of crucial services like medicinal services, smart transport systems, environmental monitoring observing, etc. Since energy efficiency is critically important to the batteries, constrained IoT devices, IoT-related standards, and research works have focused on the device energy sustainability issues. The network aspects of IoT using these wireless technologies are different from those for traditional wired or wireless legacy networks as a result of the huge number of devices participating in the communication. Aside from the traffic generation per IoT device that is typically low, each device transfers a small amount of data to a corresponding server, although data generated from a massive amount of devices have severe impacts on the network performance [39,40]. Moreover, IoT networks should operate autonomously for a longer period without the requirement for human interference and with a high degree of quantitative confidence [41]. Another aspect is that gateways may incorporate multiple wireless interfaces for versatile purposes such as throughput, latency, and energy efficiency [42].

Devices in such IoT networks will mainly work on battery-based power sources; therefore, energy efficiency is crucial in IoT device management. Looking into a particular WSN domain, energy efficiency for battery-dependent objects and lifetime extension have been research issues for many years, where medium access control layer protocols mainly focus on adapting the activity cycle for sensor nodes, and routing layers protocols are designed for data aggregation and unicast transmission. Furthermore, since IoT objects operating in IoT network paradigm are also battery dependent, energy consumption should be kept in mind as an important factor during IoT network deployment. Essentially, IoT network aspects and deployment scenarios are much more complex than traditional WSNs in various aspects, for example, the storage capabilities of IoT devices, traffic generated between objects and servers, heterogeneous data from sensors and actuators, usage of heterogeneous wireless access technologies, gateways, etc. [43–45]. Accordingly, some legacy WSN power management strategies like homogeneous data aggregation are not suitable for the most IoT scenarios. Extensive research is being conducted for better energy management for battery-dependent IoT devices from many aspects such as standardization, research, and industry applications.

7.5.4 Business models

Changes in technology clearly require adjustments in enterprise models. For instance, Web 2.0 technologies have pushed new enterprise models that include software program as a service, disintermediation, and an accelerated reliance on online marketing and strategic records aggregation. The IoT will clearly pressure the improvement of latest commercial enterprise fashions that capitalize on its pervasiveness and ubiquity. Researchers have proposed market structures and pricing schemes for the IoT and defined how IoT may want to force competitive advantage through better information and more localized choice making.

7.6 Future directions

In this review, we survey the state of the art in IoT developments in various fields. IoT is shaping the human life with greater connectivity and ultimate functionality, and all this is happening through ubiquitous networking to the Internet. IoT will merge the physical world and virtual world to create a highly personalized and often predictive connected experience [41]. In all the fields as mentioned earlier in this review, a pattern of lack of standardization protocols and difficulties in interoperability is observed. These challenges are ceasing in a way the further development of IoT. If these difficulties are overcome, there is no doubt that real growth in IoT will take place, and more and more companies will invest their future in it. Since the IoT has not yet been realized, it might seem precocious to forecast the future directions of the IoT. However, future visions of the IoT will affect its current development and must, therefore, be considered.

One future vision for the IoT is the integration of even new gadgets into the IoT referred because of the net of nano-matters. The Internet of nano-matters can be described because of the interconnection of nanoscale devices with conversation networks and the Internet. Although these gadgets are proposed to speak thru electromagnetic conversation, numerous technical challenges must be overcome before the concept becomes viable. The net of nano-things might be a fair greater granular approach to ubiquitous computing than the IoT.

References

[1] L. Coetzee and J. Eksteen, "The Internet of Things—Promise for the future?," in *IST-Africa 2011 Conference Proceedings*, 2011.

[2] K. Ashton, "That 'Internet of Things' thing," *RFID Journal*, vol. 22, pp. 97–114, 2009.

[3] C. X. Mavromoustakis, P. Mousicou, K. Papanikolaou, G. Mastorakis, A. Bourdena and E. Pallis, "Dynamic cloud resource migration for efficient 3D video processing in mobile computing environments," in *In Novel 3D Media Technologies*, Springer, New York, 2015, pp. 119–134.

[4] M. Komarov and D. Nemova, "Emerging of new service-oriented approach based on the internet of services and Internet of Things," in *e-Business Engineering (ICEBE)*, 2013.

[5] H. Sundmaeker, P. Guillemin, P. Friess and S. Woelfflé, Vision and challenges for realising the Internet of Things, European Commission, 2010.

[6] I. Deligiannis and G. Kornaros, "Adaptive memory management scheme for MMU-less embedded systems," in *11th IEEE Symposium on Industrial Embedded Systems (SIES)*, 2016.

[7] D. C. J. Kephart, "The vision of autonomic computing," *Computer*, vol. 36, no. 1, pp. 41–50, 2003.

[8] O. Lassila and M. Adler, "Semantic gadgets: Ubiquitous computing meets the semantic web," in *Spinning the Semantic Web: Bringing the World Wide*

Web to Its Full Potential, D. Fensel, W. Wahlste, H. Lieberman, and J. Hendler, eds., pp. 363–376, Cambridge MA: MIT Press, 2003.
[9] S. Heeps, N. Dulay, A. Egon, *et al.*, "The autonomic management of ubiquitous systems meets the semantic web," in *Second International Workshop on Semantic Web Technology for Ubiquitous and Mobile Applications*, http://citeseerx.ist.psu.edu/viewdoc/summary?doi=10.1.1.531.7984, 2006.
[10] D. Brock and E. Schuster, "On the semantic web of things," *Semantic Days*, pp. 26–27, 2006.
[11] Y. Kryftis, G. Mastorakis, C. X. Mavromoustakis, J. M. Batalla, E. Pallis and G. Kormentzas, "Efficient entertainment services provision over a novel network architecture," *IEEE Wireless Communications*, vol. 23, no. 1, pp. 14–21, 2016.
[12] J. Alvarez-Lozano, P. Estrada-Martinez, E. Aviles-Lopez and A. Garcia-Macias, "Browsing the Internet of Things with sentient visors," *Computer*, vol. 44, no. 5, pp. 46–52, 2011.
[13] A. Andreou, C. X. Mavromoustakis, G. Mastorakis, A. Bourdena and E. Pallis, "Adaptive heuristic-based P2P network connectivity and configuration for," in *In Resource Management in Mobile Computing*, Springer International Publishing, 2014, pp. 221–240.
[14] L. E. Holmquist, F. Mattern, B. Schiele, P. Alahuhta, M. Beigl and H.-W. Gellersen, "Smart-its friends: A technique for users to easily establish connections between smart artefacts," *Ubicomp 2001: Ubiquitous Computing*, vol. 2201, pp. 116–122, 2001. http://link.springer.com/chapter/10.1007/3-540-45427-6_10.
[15] M. Kranz, L. Roalter and F. Michahelles, "Things that Twitter: Social networks and the Internet of Things," in *Eighth International Conference on Pervasive Computing*, 2010.
[16] E. Nazzi and T. Sokoler, "Walky for embodied microblogging: Sharing mundane activities through augmented everyday objects," in *13th International Conference on Human Computer Interaction with Mobile Devices and Services*, 2011.
[17] A. Ortiz, D. Hussein, S. Park, S. Han and N. Crespi, "The cluster between Internet of Things and social networks: Review and research challenges," *IEEE Internet of Things Journal*, vol. 1, no. 3, pp. 206–215, 2014.
[18] H. Chourabi, T. Nam, S. Walker, *et al.*, "Understanding smart cities: An integrative framework," in *45th Hawaii International Conference on System Sciences*, 2012.
[19] G. Papadourakis, I. Deligiannis, E. Potetsianakis, *et al.*, "A Guardian Angel for the Extended Home Environment (GUARANTEE) Project: Decision engine," in *11th International Symposium on Ambient Intelligence and Embedded Systems*, 2012.
[20] "Frost & Sullivan: Public-private sector collaboration opportunities key to successful Smart Cities," Available from: http://ww2.frost.com/news/press-releases/frost-sullivan-public-private-sector-collaboration-opportunities-key-successful-smart-cities.

[21] C. E. Koop, R. Mosher, L. Kun, *et al.*, "Future delivery of health care: Cybercare," in *Engineering in Medicine and Biology Magazine*, vol. 27, pp. 29–38, 2008.

[22] M. Simonov, R. Zich and F. Mazzitelli, "Personalized healthcare communication in Internet of Things," 2008. https://scholar.google.com/citations?

[23] I. Plazaa, L. Martína, S. Martinb and C. Medranoa, "Mobile applications in an aging society: Status and trends," *Journal of Systems and Software*, vol. 84, no. 11, pp. 1977–1988, 2011.

[24] P. Klasnja and W. Pratt, "Healthcare in the pocket: Mapping the space of mobile-phone health interventions," *Journal of Biomedical Informatics*, vol. 45, no. 1, pp. 184–198, 2012.

[25] C. Liua, Q. Zhua, K. Holroydb and E. Sengb, "Status and trends of mobile-health applications for iOS devices: A developer's perspective," *Journal of Systems and Software*, vol. 84, no. 11, pp. 2022–2033, 2011.

[26] J. Baek, S. Chung, K. Kim and S. Park, "A smart health monitoring chair for nonintrusive measurement of biological signals," *Information Technology in Biomedicine*, vol. 16, no. 1, pp. 150–158, 2012.

[27] M. S. H. Talpur, W. Liu and G. Wang, "A ZigBee-based elderly health monitoring system: Design and implementation," *International Journal of Autonomous and Adaptive Communications Systems*, vol. 7, no. 4, pp. 393–411, 2014.

[28] M. van Limburg, J. E. van Gemert-Pijnen, N. Nijland, H. C. Ossebaard, R. M. Hendrix and E. R. Seydel, "Why business modeling is crucial in the development of eHealth technologies," *Journal of Medical Internet Research*, vol. 13, no. 4, e124, 2011.

[29] S. Fang, L. Da Xu, Y. Zhu, *et al.*, "An integrated system for regional environmental monitoring and management based on Internet of Things," *IEEE Transactions on Industrial Informatics Journal*, vol. 10, no. 2, pp. 1596–1605, 2014.

[30] J. A. Lópeza, H. Navarrob, F. Sotoa, N. Pavónc, J. Suardíaza and R. Torresd, "GAIA2: A multifunctional wireless device for enhancing crop management," *Agricultural Water Management*, vol. 151, pp. 75–86, 2015.

[31] E. K. Wang, T.-Y. Wu, C.-M. Chen, Y. Ye, Z. Zhang and F. Zou, "MDPAS: Markov decision process based adaptive security for sensors in Internet of Things," *Genetic and Evolutionary Computing*, vol. 329, pp. 389–397, 2015.

[32] G. Panić, T. Basmer, S. Henry, S. Peter, F. Vater and K. Tittelbach-Helmrich, "Design of a sensor node crypto processor for IEEE 802.15.4 applications," in *SOC Conference (SOCC)*, 2012.

[33] F. Chu, R. Zhang, R. Ni and W. Dai, "An improved identity authentication scheme for Internet of Things in heterogeneous networking environments," in *Network-Based Information Systems (NBiS)*, 2013.

[34] P. Kasinathan, C. Pastrone, M. A. Spirito and M. Vinkovits, "Denial-of-service detection in 6LoWPAN based Internet of Things," in *Wireless and Mobile Computing, Networking and Communications (WiMob)*, 2013.

[35] G. Gang, L. Zeyong and J. Jun, "Internet of Things security analysis," in *International Conference on Internet Technology and Applications (iTAP)*, 2011, pp. 1–4.

[36] J.-C. Yang and B.-X. Fang, "Security model and key technologies for the Internet of Things," *The Journal of China Universities of Posts and Telecommunications*, vol. 18, pp. 109–112, 2011.

[37] Q. Jing, A. V. Vasilakos, J. Wan, J. Lu and D. Qiu, "Security of the Internet of Things: Perspectives and challenges," *Wireless Networks*, vol. 20, no. 8, pp. 2481–2501, 2014.

[38] J. T. Kim, "Privacy and security issues for healthcare system with embedded RFID system on Internet of Things," *Advanced Science and Technology Letters*, vol. 72, pp. 109–112, 2014.

[39] A. Bourdena, C. X. Mavromoustakis, G. Kormentzasa, E. Pallis, G. Mastorakis and M. B. Yasseind, "A resource intensive traffic-aware scheme using energy-aware routing in cognitive radio networks," *Future Generation Computer Systems*, vol. 39, pp. 16–28, 2014.

[40] O. Shiakallis, C. X. Mavromoustakis, G. Mastorakis, A. Bourdena and E. Pallis, "Traffic-based S-MAC: A novel scheduling mechanism for optimized throughput in mobile peer-to-peer systems," *International Journal of Wireless Networks and Broadband Technologies*, vol. 4, no. 1, p. 19, 2015.

[41] G. Mastorakis, C. X. Mavromoustakis, A. Bourdena, *et al.*, "Energy-efficient routing in cognitive radio networks," in *Resource Management in Mobile Computing Environments*, vol. 3, pp. 323–340, Springer, 2014.

[42] C. X. Mavromoustakis, G. Mastorakis, A. Bourdena and E. Pallis, "Energy-efficient management using a traffic-oriented routing scheme for cognitive radio networks," *International Journal of Network Management*, vol. 25, no. 6, pp. 418–434, 2015.

[43] E. Markakis, D. Negru, J. Bruneau-Queyreix, E. Pallis, G. Mastorakis and C. X. Mavromoustakis, "A P2P home-box overlay for efficient content distribution," in *Emerging Innovations in Wireless Networks and Broadband*, igi-global, pp. 199–220, 2016.

[44] E. Markakis, C. Skianis, E. Pallis, G. Mastorakis, C. X. Mavromoustakis and A. Antonas, "Peer-to-peer constellations in an IMS-enabled network architecture based on interactive broadcasting," in *Resource Management in Mobile Computing Environments*, Springer International Publishing, 2014, pp. 271–282.

[45] G. Mastorakis, C. X. Mavromoustakis, E. Markakis, E. Pallis and G. Skourletopoulos, "Virtual network functions exploitation through a prototype resource management framework," in *Telecommunications and Multimedia (TEMU)*, 2014.

Chapter 8

Internet of Everything: a survey on technologies, challenges, and applications

Chandu Thota[1], Constandinos X. Mavromoustakis[2], George Mastorakis[3] and Jordi Batalla[4]

Abstract

Internet of Everything (IoE) is an interconnection of individuals, data, method, and devices. It identifies the convergence of numerous environments such as cloud computing, mobility, data processing, and to end with, an explosion in interconnected things. The IoE integrates the various methodologies and techniques, tries to construct a process mechanism, and includes individuals in this method in order to develop additional smart systems. IoE primarily used to collect and examine information from various sources such as instruments, senor devices, payment processing equipment, mobile devices, data stores, and it is also used to find predictions in future. The IoE is creating new challenges and opportunities that will be analyzed during the subsequent years. Large amounts of data will be produced and consumed, so Internet of Things frameworks will need to identify new methodologies and techniques associated to big data analysis, performance, and scalability. We consider that the configuration of local clouds of devices, close to the location where data is produced and consumed, is a good solution to solve these issues that may involve in security as well. This paper studies the definitions, architecture, fundamental technologies, and applications of IoE. In addition, this paper also discusses the emerging techniques such as device-to-device communication, machine-to-machine, and 5G mobile network for the implementation of IoE. Finally, the major applications, open issues, and challenges related to the IoE are investigated.

8.1 Introduction

Internet of Everything (IoE) is an enhanced version of Internet of Thing (IoT) and it includes machine-to-machine (M2M) communications, device-to-device (D2D)

[1]Albert Einstein Lab, Infosys Ltd., Hyderabad, India
[2]Department of Computer Science, University of Nicosia, Nicosia, Cyprus
[3]Technological Educational Institute of Crete, Crete, Greece
[4]National Institute of Telecommunications NIT, Warsaw, Poland

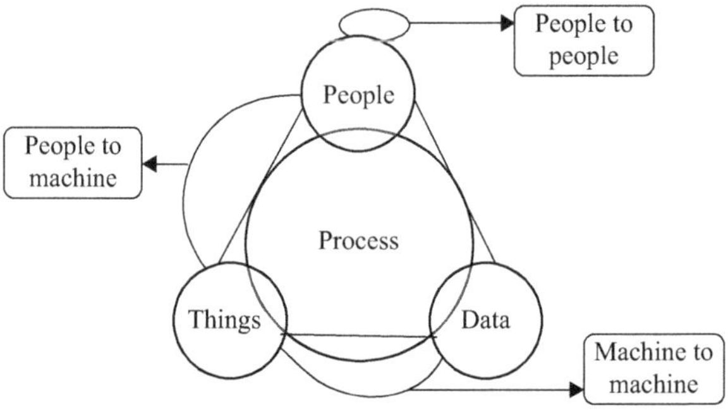

Figure 8.1 People, things, data, and process of IoE

communications, machine-to-people (M2P) communications, and people-to-people (P2P) communications (Figure 8.1). The following advancements in IoE includes D2D, M2M, M2P, and P2P that improve the various environments such as smart health care, smart governance and smart transportation, smart water management system, and so on. It is observed that IoT provides the platform to service with supporting communication among physical objects and virtual representations. IoT consists of various tools and technologies as controllers, sensors, or low-powered wired and wireless services [1]. Wireless devices play an important role compared with wired devices in IoT applications. Storage capacity of the data in these wireless devices, which are connected by Internet, should be worth able. In [2], a stream-oriented modeled scheme is proposed based on each node's self-scheduling energy management. This scheme is taking into account the overall packet loss in order to form the optimal effect for the end-to-end connection-throughput response. The scheme also—quantitatively—takes into account the asymmetrical nature of wireless links and the caching activity that is used for data revocation in the ad-hoc-based connectivity scenario. Through the designed middleware and the architectural layering and through experimental simulation, the proposed energy-aware management scheme is thoroughly evaluated in order to meet the parameters' values in which the optimal throughput response for each device/user is achieved.

The above-mentioned definitions found the following themes: it includes efficient connectivity and interaction between devices. In general, these connected devices communicate with the help of numerous small-scale wireless tools and technologies defined for embedded tools communication. The above-mentioned devices are continuously generating enormous data. Hence, nowadays many organizations have started using cloud computing. Though cloud computing provides possible storage space, there is a need to process such huge amount of data. In order to overcome this issue, Big Data analytics has come into the picture. Nowadays, many organizations such as government and private institutes, healthcare industries [3] and research and development organizations are interested in using Big Data analytics.

These IoT devices generate large amounts of structured and unstructured data continuously. Big data tools and technologies are mostly preferable to supervise the huge amount of data generated from various IoT devices used to measure the body temperature, body glucose level, heart rate, and so on. In the IoT environment, the smart devices and things uses the following tools and technologies: micro-controllers (MCU), sensors, real-time operating system (RTOS), middleware, and connectivity providers to observe the specific value and transmit it to the sensor server. Gateways, protocols, and communications like network technologies are used to transfer the device-generated data into the cloud storage [4]. Nowadays, there is a huge development in network technologies, such as 4G to 5G networks. This advancement in network technology solves many connectivity problems and latency issues, and improves the data transfer speed between the IoT devices and data storage spaces. With the help of 5G mobile network, IoT is enhanced to IoE and change the day-to-day environment of individual's life [5].

8.1.1 Internet of Everything

In general, Internet connections are always used for the laptop, desktop computers, and tablets. Nowadays, many advanced devices such as heart pressure watch, body temperature belt, and so on are also connected to the Internet to transfer the individual's health information continuously not only in health care, but also in more applications like smart city, smart traffic control, and weather monitoring applications. Normally, IoE technologies vary in range from digital sensor devices used for various applications to smarter and numerous interconnected wireless devices, smart industrial applications, and various distributed hardware technologies that have just become more automated and smarter. The work of Mavromoustakis *et al.* [6] proposes a scheme for sharing resources using the opportunistic networking paradigm, whereas it enables EC by allocating real-time traffic-based dissimilar sleep/wake schedules to wireless devices. The scheme considers the resource-sharing process, which according to the duration of the traffic through the associated channel, impacts the sleep-time duration of the node. The paper examines the traffic's backward difference in order to define the next sleep-time duration for each node. In general, features of IoE have been classified into two types, namely input and output. Input function is used to allow the external data into a device, whereas output function is used to transfer the device data into Internet.

Recently, the IoE term plays a vital role in information technology fields. For example, Cisco is one of the leading institute that has focused more in IoE-based technologies. IoE is enhanced from the previous versions of Internet-based technologies such as IoT, Internet of humans, industrial IoT, and Internet of digital. In other words, IoE is a system with end-to-end connectivity among processes, technologies, and concepts engaged across all connectivity use cases.

IoE basically consists of four connection parts such as people, things, data, and process [7].

8.1.1.1 People

Destination or target nodes are interconnected with the Internet to distribute activities and data. IoE enables people to connect to the Internet in incalculable ways. Nowadays, many people connect to the Internet using their own smart devices such as PCs, TVs, tablets, and smartphones. In addition, they also use social networks such as Twitter, Facebook, LinkedIn, and Pinterest. As the Internet grows toward IoE, we will be connected in more related and helpful ways.

8.1.1.2 Things

Things are the most important component in the IoE used to observe the more relevant data from the physical devices. Collected data from IoE devices are used to take valuable decisions in near future and emergency situations. For example, the medical smart devices in IoE health care application are used to observe the individuals' information that efficiently monitor the patient health in emergency situations. This collected information is transferred into the data store to analyze further appropriate and valuable decisions.

8.1.1.3 Data

IoT devices normally collect data and stream it over the Internet to a sensor server, where it is processed and analyzed. Due to the fact that capabilities of things connected to the Internet persist to advance, they will become additionally intellectual by combining data into more valuable information. Unprocessed data after being generated from devices will be processed and analyzed into valuable statistics to provide control mechanisms and intelligent decisions. For example, high and low heart rate measurements are used to find the average heart rate of patient in healthcare industry.

8.1.1.4 Processes

Process plays a significant role in measuring how entities like data, people, and things work with others to bring value to the connected world of IoE. With the accurate process, connections turn into applicable and add value because the exact information is transferred to the specific destination or device in the proper way. In addition, the strong connectivity between the smart devices, data, and individuals is used to gain the high-value insights from the IoE system. For example, use of social networks and smart fitness devices to promote pertinent healthcare offerings to prospective customers.

8.1.2 IoE uses for next generation

The services offered by the IoT make it possible to develop several applications of different industries currently suffering from lot of attributes like cost, maintenance, resources, and so on. In forthcoming days, there will be current applications with intelligence to turn them out as smart telecommunication industry, smart medical and smart healthcare industry, smart independent living, smart pharmaceutical industry, smart retail industry, and smart logistic systems. The following are the different industries: aerospace and aviation industry, automotive industry, telecommunication industry, medical and healthcare industry, independent living, pharmaceutical industry, retail, logistic, supply chain management, transportation, agriculture, and manufacturing industry are looking for to use IoT type of technology term to make our next generation smarter. Here I am discussing advantages of few smart applications.

In smart healthcare application, medical clinic centers and laboratories are shifting from providing test and diagnose of the patients on premise, that is, in hospitals and clinics, to isolate self-monitoring. Self-monitoring profits the patients by providing them with better freedom and individuality in observing the patient's health condition. It will lead to keep patient in home or which place he needed to support and encourage himself/herself, who feel scared about hospital atmosphere.

For IoT-based smart logistic application, it is possible to visualize that goods or products can transport without human resources involvement in certain areas from companies to merchants. This system makes warehouses completely programmed to intelligent decisions with goods moving in and out, based on statistics received via devices and global positioning systems (GPS) to minimize the transiting directions.

8.1.3 Internet of Things

More recently, a report published from the scientific adviser of the United Kingdom states that the overall connected things (devices, mobile phones, personnel digital assistants (PDAs), etc.) are expected to increase from 20 to 100 billion by 2020 [8]. Nowadays, the advancement in the fields includes Wi-Fi, ZigBee, and 4G/5G, which are changing the network connectivity of the globe. For example, smart homes, smart health care, smart grid, and smart cities are some of the examples of IoT [9]. Recently, numerous enabling technologies are identified such as radio frequency identification (RFID) or near field communication, optical tags and quick response code, and Bluetooth low energy (BLE). In general, all IoT objects have been assigned an IP address to communicate with each other. Till recent decade, we used IPv4 and our electronics devices such as personal computers and laptops are becoming more complicated to communicate on Internet protocol version 6 (IPv6), with IPv6 we would be able to take care of the IP addresses to roughly for everyone [10]. The huge variation between IPv6 and IPv4 is improved in address space. Addresses of IPv4 are 32 b, whereas IPv6 addresses are 128 b. Due to the size of IP addresses in IPv6, users are capable of handling IoT kind of upcoming technology [11]. In [12], sensor networks contribute to the interconnection of a large variety of devices (i.e. transducers, sensors, and actuators), thus enabling monitoring and control processes. Although new wireless technologies are emerging, a major issue of interoperability has to be addressed in terms of data communications, controlling, and interfacing in order to confront the heterogeneity of networks and connected devices and enable end-to-end communication, as well as efficient resource management.

There exists no common definition of IoT. Speaking generally, the IoT is a system consisting of communications of different devices as sensors, actuators, and smart objects. The purpose of IoT is a way to make interconnected “all” things intelligent, programmable, and more capable, including daily used objects to engineering objects by interacting with humans and each other. In [13] future, Internet is foreseen to fully handle a wide range of multimedia services allowing their access through diverse computing devices such as laptops, TVs, PDAs, and 3G mobile phones interconnected via different wired and wireless networking technologies. Such a diversification in the computational context reinforces the need of personalized and adaptive media services toward better end-user experience.

IoT ecosystem: The IoT ecosystem enables entities to connect to and control their IoT devices. In the ecosystem, an entity uses a remotely connected device like smartphone, tablet, and so on to send a command or a request for information over a network to an IoT device. The devices then perform the command and analyze the data, then send the information back over the network to be analyzed and deployed on the remote. There are multiple locations where the data generated by the IoT devices can be analyzed and stored—in databases deployed in cloud, a local database on the remote, or locally on the IoT device itself.

IoT Business and marketing: IoT solutions’ uses will ultimately be to achieve business values for the organizations. The following are the three ways through which the IoT can expand organization business: (1) dropping operating costs, (2) growing productivity, and (3) creating new markets. The IoT market is rapidly increasing; initially organizations are active and currently creating products for which they see a market. To create market for these products, these companies need to implement proprietary resolutions. Nowadays, IoT is trending toward vertical smart applications. Verticals presenting early development are advance agriculture, smart health, transportation, energy, and so on. IoT enlargement and positioning are motivated by the aspiration to provide existing cheaper, faster, and better goods and services more efficiently that will drive new revenue streams. Connecting things are able to create huge data, and allowing data to move across different locations will open new markets in software companies that are maintaining virtual server centers and data centers. The governments are mainly concentrated on increasing productivity, decreasing costs, and trying to provide quality life to the citizens. They will be the second-largest adopters of IoT ecosystems. Intel and its ecosystem help businesses use the IoT to solve long-standing industry-specific challenges. Quickly developing IoT solutions connect things, collect data, and derive insights with Intel’s portfolio of open and scalable solutions so one can reduce costs, improve productivity, and increase revenue [14].

8.1.4 Communications

Nowadays, development in wireless communication technologies has changed the traditional communication methods. In last decade, man-to-man communication and man-to-machine communication were most often used in communication environments. The push toward the network communications has increased, and M2M

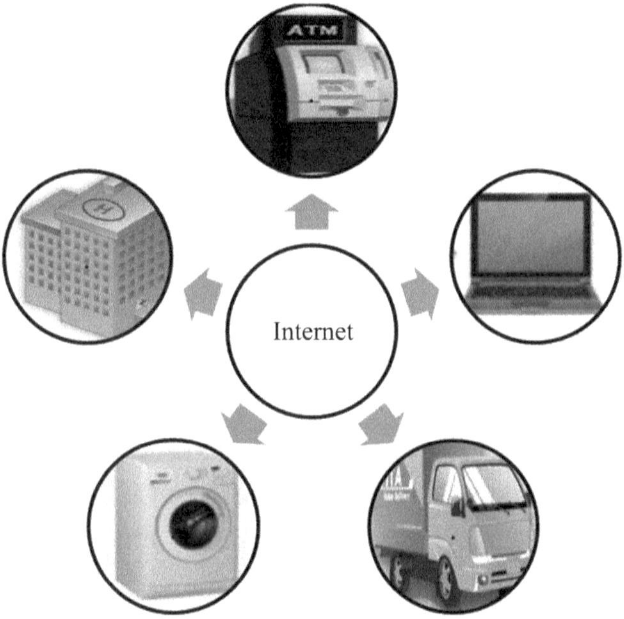

Figure 8.2 Machine-to-machine communication

communications are recently used in many platforms. For example, vehicle-to-vehicle communication or car-to-car is one of the types of M2M mobile communications. In general, all devices and things in the M2M communication systems are in movement. These devices and things can moreover be vehicles or mobile devices [15].

Machine-to-machine (M2M): It is an essential portion of smart transportation, logistic services, smart homecare services, and electronic shopping (Figure 8.2). M2M communication is the backbone of IoT technologies. M2M communication consists of various devices and technologies such as sensors, routers, Wi-Fi, 4G/5G cellular infrastructures, and software platforms to transfer the message between one machine to another. Till last decade, an external environment or platform was needed to communicate the machines. This would cause high delay and overhead. In order to overcome this issue, M2M technologies are introduced to transfer data between one machine to another without any additional software and devices. The most familiar applications of M2M communications include e-Health, continuous traffic management, and robotics and automation [16].

The following applications of M2M envelop many areas and the areas in which M2M is presently used:

- Security and privacy—alarm systems, surveillances and access control
- Tracing & tracking—order management, fleet management, pay as you drive, asset tracking, road tolling, navigation, traffic information, and traffic optimization/steering

- Payment processing—vending machines, point of sales, and gaming machines
- Health care system—supporting the aged or handicapped, monitoring vital signs, remote diagnostics, and web access telemedicine points
- Remote control/maintenance—pumps, sensors, lighting, valves, elevator control, vehicle diagnostics, and vending machine control
- Metering function—water, power, gas, heating, industrial metering, and grid control
- Manufacturing function—automation and production chain monitoring
- Facility management system—home/campus/building automation
- Device-to-device (D2D)—it is considered as a machinery component that develops the spectral bandwidth and competence that provides direct communication between nearby mobile devices, radio communication resources, and PDAs—which is an exciting and novel feature of next invention mobile network systems (Figure 8.3). In addition, D2D communications are being considered and researched for 4G long-term evaluation (LTE) advanced techniques. Nowadays, cellular spectrum is started using the 4G LTE D2D technology to enable the strong connection of a device, user equipment, and so on to another device. D2D technology 2G and 3G systems enables users to transfer large amount of data from one mobile device to another over short distances and enhanced wireless link. D2D technology also provides communication capabilities with less human or external involvement.

Efficient communications between devices: LTE D2D technology is also used to converse nearby devices to deliver high-dependability communication especially

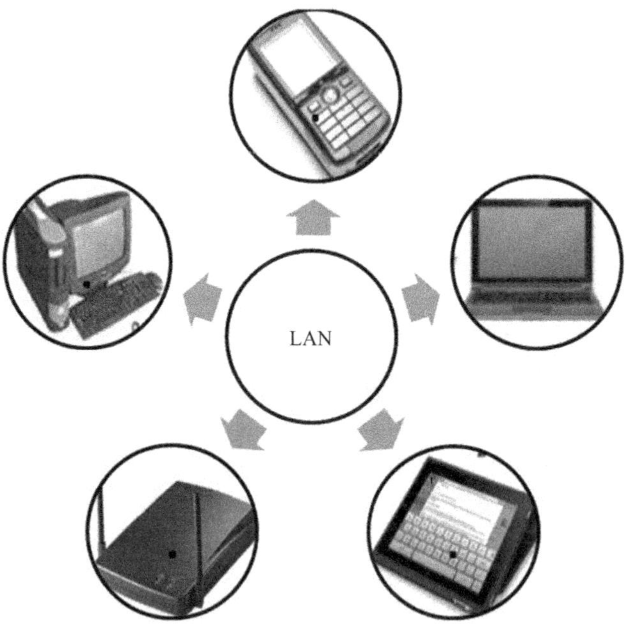

Figure 8.3 D2D communication

if the LTE network has failed for any cause. For example, unfortunately if any natural disaster occurs, the LTE D2D technology efficiently handles the failure and network problems.

Lack of interference: in general, every communication technology uses additional links and process with base station. In the case of D2D communication technology, LTE does not maintain any communication or link directly with a base station. This would cause high availability and reliability. Hence, there is no need to maintain additional link or communication between the source and base station.

Immediate communications: In general, D2D communications do not use any additional link or communication with base station. Hence, the communication speed and performance has improved. As D2D communication does not consist of any connection with base station, the devices work and communicate with each other based on direct communication methods. For example, walkie-talkies are most often connected with direct link and this does not require any installation or communication with source or base station. Hence, D2D communications are most often used for emergency services and real-time applications.

Power consumption: Normally, communication technologies use base station transfer data between the devices. As D2D communication focuses only on direct communication, the power consumption for message transfer in D2D communication is very low. In addition, D2D communication technologies have used only low-power sensors and actuators. Hence, the power taken to transfer the data between the devices is also reduced.

8.1.5 5G mobile network

Earlier, 2G and 3G systems were considered as the backbone of all IoT applications. Evolved 4G and 5G mobile networks with new functionalities assure to build and create new generations of communications that will present even premium capabilities and higher data throughput such as wireless IP-based video, location, and occurrence, which eventually progress efficiencies in functionalities. Mavromoustakis *et al.* [17] present an efficient 3D video processing, dynamic cloud computing scheme for efficient resource migration and 3D media content processing in mobile computing environments. It elaborates on location and capacity issues to offload resources from mobile devices due to their processing limitations toward efficiently manipulating 3D video content. Ciobanu *et al.* [18] witness an explosion in the number of applications being developed for mobile devices. Many such applications are in need or generate a lot of Internet traffic, and so such mobile devices are today equipped with more networking capabilities—from mobile broadband (3G/4G) to Wi-Fi, Bluetooth, and others. The functional idea of 5G is completely designed on IP model for the wireless communication and mobile devices. The operation model of 5G network is typically designed based on the MasterCore architecture that will help to work in parallel manner. Hence, both IP network and 5G network modes work parallel with 5G network without any additional requirement or involvement [19]. 5G mobile network is simple and well suited for the real-time application and streaming data processing [20]. Mavromoustakis *et al.* [21] propose an energy-efficient delay-aware cooperative scheme, exploited for efficient resource management and maximum energy conservation in a 5G mobile cognitive radio network architecture.

The key concepts of 5G and beyond 4G are mentioned below:

- Generally, no more constraint for cellular network world with zone and access concerns.
- Visiting care for mobile Internet protocol address is used in the 5G based on the connected network and location.
- The emerging pervasive computing enables user to shift various environments without need of any additional process and involvement. For example, user can use any of the mobile communication technologies including 2.5G network, 3G network, 4G or 5G networks, Wi-Fi, WPAN, or any advanced mobile networks.
- Cognitive radio technology is used in the advance mobile networks that enable user to use various communication spectrums.

8.2 Cloud computing and Big Data in IoE

Nowadays, Big Data have been playing a vital role in almost all environments such as health care, education, business organizations, and scientific research. There is a strong relationship between Big Data and IoE [22]. In general, IoE applications are used to capture or observe some specific values to find the hidden values and take better decisions. When the device is connected to the Internet, it always senses the specific metric and stores those metrics into a connected data store. This would increase the size of the data stored in a data store. Hence, high-end devices and scalable storage systems are needed to store such huge size of data. The amount of data to be stored and processed becomes an important problem in real life. Relational database management system is generally used to store the traditional data, but day by day the volume, velocity, and variety of sensor data is growing toward the Exabyte [23]. This requires advanced tools and techniques to store, process, and display such large amount of sensor data to the end users. Thus, storing and querying large amount of data require database clusters and additional resources. However, storage and retrieval are not the only problem but also extract useful information from huge data. In order to overcome this issue, cloud computing is used to provide scalable storage systems and high-end devices for computation. The data must be effectively stored and retrieved by the IoT service providers. The solution is to access the data through application programming interfaces. The paper focuses on the problem of designing an effective data storage service for IoT, which will be available through the universal application programming interface.

Wireless sensor network (WSN) is composed of spatially distributed connected sensor nodes with limited computing power and storage. Saleem *et al.* [24] give an introduction to WSN, mobile-sink-based WSN, and cloud computing. After then, we give an overview of state-of-the-art work on wireless-sensor-based cloud computing. Subsequently, integration of WSN and cloud computing is highlighted with some insights on how WSN and clouds can both benefit from each other. Applications of wireless sensors over the cloud are then described. Afterward, we explain incorporation of mobile link between WSN and cloud. Skourletopoulos *et al.* [25] evaluate different cloud-supported mobile services subjected to limited capacity, as the selection of a service may introduce additional costs, such as those that derive from the additional amount of memory required for processing.

Cloud computing is a type of computing, and it is used for the delivery of hosted services over the Internet. In other words, cloud computing relies on sharing computing resources and hardware rather than having personal devices or local servers to manage the real-time applications. In [26], the mobile cloud can be considered as a marketplace, where the mobile services of the mobile cloud-based system architectures can be leased off via the cloud. This context elaborates on a novel fluctuation-based quantification model, which is based on a cost–benefit appraisal, adopting a nonlinear and asymmetric approach. The proposed model aims to predict the incurrence and the risk of entering into a new technical debt in the future and provide insights to inform effective investment decision-making. The lease of a cloud-based mobile service was considered, when developing the formula, and the research approach is investigated with respect to the cost that derives from the unused capacity. Papanikolaou and Mavromoustakis [27] address some of these traditional concepts combined in a "multi-sharing" cloud application environment and discusses how these concepts evolve in the context of cloud computing. In general, cloud providers are called as cloud service providers (CSPs). Amazon simple storage service (Amazon S3) is the first cloud offered by Amazon in 2006. Thereafter, other cloud providers have developed a number of cloud services such as Microsoft, Rackspace, Apple, IBM, Joyent, Google, Cisco, Citrix, Salesforce.com, and Verizon/Terremark. Hence, the IoE devices are interconnected with cloud server to store the device-generated data. Once the data is stored efficiently into the cloud, there is a need for scalable algorithms to process those data. In order to fulfill the requirements, Amazon web services provide Elastic MapReduce to process the device-generated data.

8.2.1 Big Data and analytics

In [28] today, with high volume, high variety, velocity, and value characteristics, big data is playing a key role in the data analytics, data storage, and visualization of the current trending technologies like cloud computing and IoT of current generation. This growing background makes one thing clear: the current terms of networking have changed:

- IoT platform builds on very strong communication and networking infrastructure, needs longer "data transport"; it is about "intelligence" resultant from network data to reach better business and policy results.
- A distributed IoE architecture is evolving, where data can be stored and evaluated in real time at the edge of the network, at the same time in the cloud.
- High-performance computing capability of Big Data analytics is increasingly surrounded in the network to store, sort, and analyze, where data is moving among numerous devices.

8.2.2 Functionality of the proposed architecture

There are a number of cloud services available to store and process Big Data. When user tries to connect with number of cloud services, there is need to select the appropriate cloud service without any delay or additional user involvement. In order to overcome this issue, MetaFog-redirection (MF-R) architecture with grouping choosing (Figure 8.4) is proposed in this paper. Proposed architecture is explained in the following sections such as data flow diagram for collecting sensor data from personal health system, big data storage in MetaFog-redirection architecture, security in MetaFog-redirection architecture, and application of MetaFog-redirection architecture. The proposing grouping & choosing (GC) architecture mainly focuses on the integration of fog to cloud in terms of application integration, data transfer from fog servers to cloud data centers, and security mechanisms for integration from communication from fog layer to cloud environment.

Cisco defines fog computing as a paradigm that extends cloud computing and services to the edge of the network. Fog computing will grow in helping the emerging network paradigms that require faster processing with less delay and delay jitter. Cloud computing would serve the business community, meeting their high-end computing demands lowering the cost based on a utility pricing model. By doing so, fog reduces service latency and improves quality of service, resulting in superior user experience. Fog computing supports emerging IoE applications that demand real-time/predictable latency (industrial automation, transportation, networks of sensors, and actuators). Fog supports densely distributed data collection points, hence adding a fourth axis to the often mentioned big data dimensions (volume, variety, and velocity).

Now there are different approaches to integration when connecting devices to the cloud. We could make integration happen on the data level, a point-to-point level where two applications are sharing chunks of data, or at a method level allowing them to share functionality apart from just data. Integration strategy plays a vital role in its success in the enterprise ventures. Also, the company needs to have a clear understanding of the requirements specifying what is to be achieved after integrating the applications and database from flog to cloud so that finite goals can be set. A very important and often neglected aspect of integration is the relevance of devices in the integration scheme.

In this architecture, we are suggesting to store the data into primarily fog servers, which is near-edge technology for IoT devices which are deployed in IoT applications. Edge computing plays a crucial role in IoT. Studies related to security, confidentiality, and system reliability in the fog computing platform is absolutely a topic for research and has to be discovered. Less demand for bandwidth, as every bit of data was aggregated at certain points instead of sending over cloud channels. Rather than presenting and working from a central cloud, fog operates on network edge. So, it takes less time. By putting small servers called edge servers in visibility of users, it is possible for a fog computing platform to avoid response time and scalability issues. Cloud computing would serve the business community, meeting their high-end computing demands lowering the cost based on a utility pricing

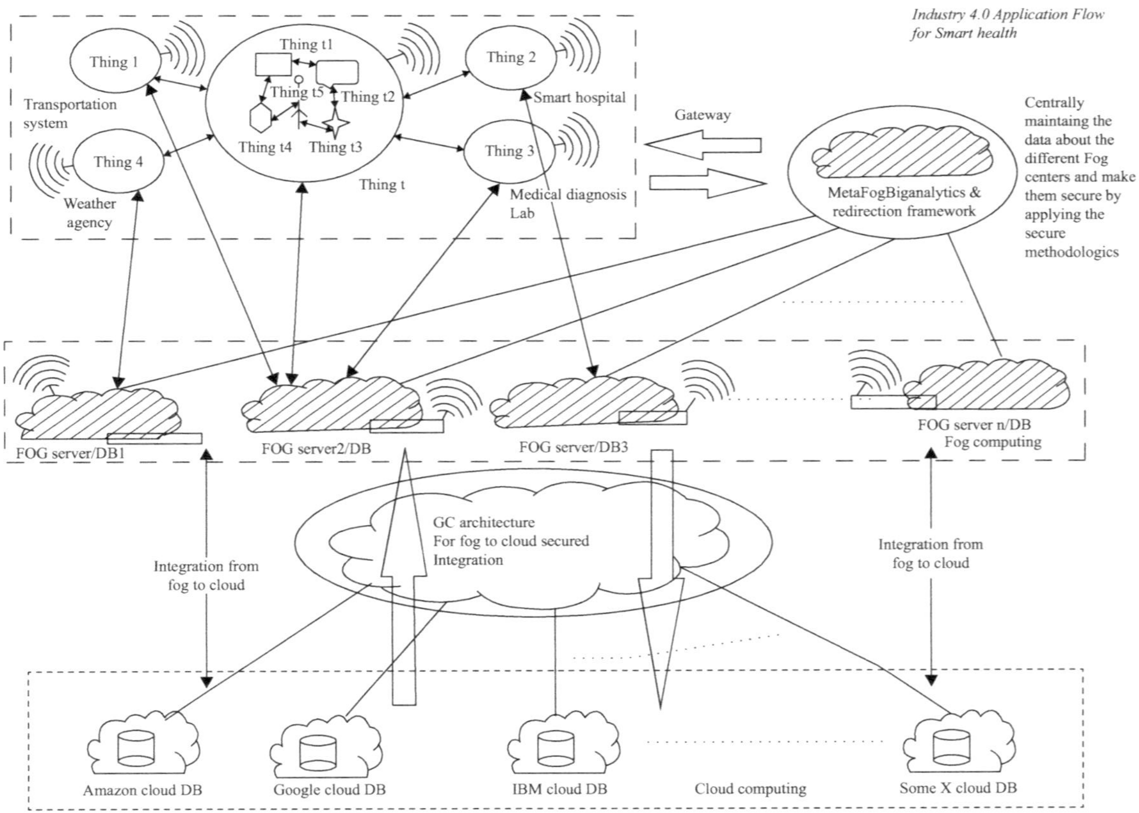

Figure 8.4 MetaFog-redirection (MF-R) architecture with grouping and choosing (GC) architecture

model. Processing the data of the applications of IoT in cloud data centers, we are using the cloud Big Data technologies and securing the data by storing the data into different cloud data centers provided by various cloud providers like Amazon, Google, Cisco, and Microsoft. We also categorize the data of the application and store the data into different data centers as per their categorization as critical, normal, sensitive, and personnel information of the end users of the applications.

GC architecture is embedded with MetaFog-redirection architecture for secure integration of fog to cloud computing and also protect Big Data against intruder. Sensor data is stored in multiple cloud data centers based on the importance and scope. Data categorization is classified into three levels such as sensitive, critical, and normal. Each categorized data is supposed to be stored in different data centers. Proposed architecture can efficiently redirect the user request to the appropriate data center in cloud provided by different vendors. Amazon web service (AWS) cloud trail is used in this proposed framework to process the log files. AWS key management service is integrated with AWS cloud trail that delivers log files to an Amazon S3 bucket. Cloud trail can easily integrate with any application using proper application programming interface (API). AWS cloud trail is capable of maintaining the time of the API call, IP address of the API caller, and the request and response parameters of the AWS service [29].

In this proposed framework, data centers will be separated into a sequence of *n* parts, where each part can be denoted by part *i* (*i*¬ (1, *n*)), and they will be stored at *m* different storage providers, where each provider is identified as provider *j* (*j*¬ (1, *m*)). In general, (parts of the data center) *n* is always far greater than (number of provide) *m*, these *m* storage providers belong to different organizations, such as Amazon, Google, and Sales force. Data parts stored on certain cloud storage providers will be allocated to some physical storage media that belongs to the storage provider. When Big Data is stored in the data center, it will form a unique storage path given as mapping Storage_Path={Data((P1(M1,M2 ... Mr))(P2(M1,M2 ... Ms)) ... (Pn (M1,M2 ... Mt)))}; where P denotes the storage provider and M denotes the physical storage media. Big data is always enormous and impossible to encrypt as a whole, so we propose a framework for encrypting the storage path of the Big Data and get a cryptographic value which can be called cryptographic virtual mapping of Big Data. So instead of protecting the Big Data itself, proposed framework protects mapping of the various data elements to each provider. The security for the MetaFog-redirection architecture is shown in Figure 8.4.

Although the proposed framework will distribute all data parts in different storage service providers, each provider holds some of the data parts. In order to provide high availability and robustness, the proposed framework will store multiple copies of same data on different cloud storage providers. Though Big Data is split and stored in different data centers, the administrator of the entire system will keep the storage index information for each data parts. When there is a problem in some data parts on the cloud storage, proposed framework can find another copy of the data parts according to their storage index information. The proposed security algorithm which is shown below protects the unauthorized access when trying to login into the application that has been deployed in cloud. Although this algorithm updates the following tables such as (1) Threat_updated table, (2) meta data storage cloud table, (3) Amazon cloud data storage table, (4) Google cloud data storage table, (5) Xcloud data storage table, and (6) Xncloud data storage table. Threat_updated table will store the entry related to malicious attempt, whereas meta data storage cloud table stores information regarding the data storage entry of different vendors. Critical, sensitive, and nonsensitive data are stored in other tables.

8.3 Applications of Internet of Everything (IoE) with 5G mobile network

8.3.1 Smart transportation applications

IoE mobile networks always use fiber optic and wireless network to connect with any vehicles all over the globe. The network capability of IoE is robust and scalable; hence, the connectivity problem will not arise in the IoE-based applications. For example, traffic-light system is one of the IoE-based application where the lights are switched on when the vehicle comes near to the street light. Sensor and actuators are used in these applications to switch on the lights. Traffic management system is another application of IoE, where Wi-Fi-based mobile devices are fixed with the vehicles; when a vehicle comes near to another one, the device sends an emergency notification to the car driver. In addition, the traffic system also works based on IoE application. For example, traffic lights are switched on based on the traffic available in the road. In addition to above-mentioned applications, smart parking system and water quality management system are also developed based on IoE. These types of applications are most often used in many countries. Though more advancement is achieved in IoE, there is a requirement to develop advance computing technologies to solve the speed issues and storage issues.

8.3.2 Smart healthcare applications

IoE also plays a vital role in healthcare applications by allowing the healthcare industry to expand the health services, clinical solutions, and reduce the cost for treatment and medicine. Nowadays, more number of sensors and medical devices are identified to observe the patients' health condition. This advancement is used to observe the glucose level, body temperature, blood pressure, and so on. The smart medical devices generally fixed with the human body and collect the patients' health consciously. The collected information sends to the doctor via cloud or Internet to take the better decision and clinical solution for the patient.

Internet-connected devices have been introduced to patients in various forms. Whether data come from fetal monitors, electrocardiograms, temperature monitors or blood glucose levels, tracking health information is vital for some patients. Many of these measures require follow-up interaction with a healthcare professional. This creates an opening for smarter devices to deliver more valuable data, lessening the need for direct patient-physician interaction.

IEEE has many standards in the eHealth technology area designed to help healthcare products, vendors and integrators create devices and systems for [30]

- Disease management
- Fitness tracking
- Health monitoring
- Independent living

8.3.3 Smart industrial applications

IoE is also used in the industries to improve the productivity and efficiency and reduce the overall cost, production time and efforts. IoE is used in the industries for monitoring the quality of service, performance meters of the machineries and consumer parameters. It will combine the global reach of the Internet with a new ability to directly control the physical world, including the machines, factories and infrastructure that define the modern landscape. It will change the basis of competition, redraw industry boundaries and create a new wave of disruptive companies, just as the current Internet has given rise to Amazon, Google, and Netflix. Companies will also use industrial Internet technologies to augment workers, making their jobs safer and more productive, flexible and engaging. As these trends take hold, and new skills are required, people will increasingly rely upon smart machines for job training and skills development. The convergence of physical industries and digital technologies will exacerbate the talent gap, especially among workers with both OT and IT skills. The industrial Internet requires analytical talent, including data scientists, yet most of our research participants agree that current education and training approaches are not up to the challenge.

The increased ability to make automated decisions and take actions in real time. The key business opportunities will be found in few major areas:

- The emergence of an outcome economy, fuelled by software-driven services; innovations in hardware; and the increased visibility into products, processes, customers and partners
- New connected ecosystems, coalescing around software platforms that blur traditional industry boundaries
- Collaboration between humans and machines, which will result in unprecedented levels of productivity and more engaging work experiences. As the industrial Internet gains broader adoption, businesses will shift from products to outcome-based services, where businesses compete on their ability to deliver measurable results to customers.

8.3.4 Smart cities

The idea is to embed the advances in technology and data collection which are making the IoT a reality into the infrastructures of the environments where we live. Already, large companies such as Cisco and IBM are working with universities and civic planning authorities to develop data-driven systems for transport, waste management, law enforcement, and energy use to make them more efficient and improve the lives of citizens. Climate, smart street lighting, transportation, smart parking, waste management and waste water management are the major services needed to provide the citizens of the smart cities. Smart city projects require expertise that spans many different fields including finance, planning, transport, energy safety telecommunications and more. They also require public–private partnerships that embrace all of these different dimensions. The IoT smart city concept is a holistic and layered framework that addresses the needs of multiple aspects of smart city projects and allows cities to use urban data to boost economic competitiveness and build more effective, workable solutions to many city challenges. Working with an ecosystem of partners, we offer products, tools, and services for public service providers, city network operators, application providers, and enterprises.

In [31], efficient IoE infrastructures for cities require two elements:

1. Smart, innovative solutions that break away from traditional, energy-intensive, waste-generating approaches
2. Solutions that eliminate silos of information within a city, allowing for more efficient and open sharing and utilization of information and resources

8.3.5 Smart cities in India

IoT service provide companies such as Sterlite Technologies Ltd. India and Aeris India are working towards building network infrastructure in smart cities to enable IoT technologies with the aim of linking intelligence and information with devices [32]. Sterlite is currently working in building Internet network capacities and system in two smart cities, Jaipur and Gandhinagar. For a smart city, a network is required that creates applications for e-governance, public safety, traffic and utilities, basically a high-level information and communication technology architecture. The opportunities here in India are immense, and India could potentially play a pivotal role in the development of global IoT ecosystem both as a market and as an innovation hub.

8.4 Tools and technologies

The tools and technologies for developing and deploying the powerful IoT applications are depicted in Table 8.1. It includes communications standard, encoding scheme, electronic product code, type of sensor, RFID type, and other network details.

Within mobile M2M infrastructure and IoE systems, smart phones play an unusual role. These are prepared with open environment and context sensors, and variety of cellular technologies like NFC, Bluetooth ZigBee, and Wi-Fi, etc. They will likely be used as sensors themselves, and as data relays for other nearby devices with additional restricted connectivity, for example, health sensors in a personal area network (PAN) or domotic sensors and actuators in a home automation environment: constrained application protocol (CoAP) and message queuing telemetry transport (MQTT). Pauls *et al.* [33] study the viability of using the general packet radio service for a low data rate long-lasting battery powered operation of M2M devices are common application-layer protocols.

Table 8.1 Technologies and standards

Technologies	Standards
Communication	IEEE 802.15.4 (ZigBee) IEEE 802.11 (wireless local area network, WLAN) IEEE 802.15.1 (Bluetooth, Low Energy Bluetooth) IEEE 802.15.6 (Wireless Body Area Networks) IEEE 1888 IPv6 3G/4G UWB
Data Content and Encoding	EPC Global Electronic Product Code, or EPCTM, EPC Global Physical Mark Up Language, EPC Global Object Naming Service (ONS)
Electronic Product Code	Auto ID: Global Trade Identification Number (GTIN), Serial Shipping Container Code (SSCC), Global Location Number (GLN)
Sensor	ISO/IEC JTC1 SC31, Sensor Interfaces: IEEE 1451.x, IEC SC 17B EPC Global, OSO TC 211 ISO TC 205
Network Management	ZigBee Alliance, IETF SNMP WG, ITU-T SG 2, ITU-SG 16, IEEE 1588
Middle	ISO TC 205, ITU-T SG 16
RFID	RFID air interface protocol: ISO 11785 RFID payment system and contactless smart card: ISO 14443/15693 Mobile RFID:, ISO/IEC 18092 ISO/IEC 29143 ISO 18000-2—for frequencies below 135 kHz ISO 18000-3—for 13.56 MHz ISO 18000-4—for 2.45 GHz ISO 18000-6—for 860–960 MHz ISO 18000-7—for 43 MHz

MQTT, according to IBM researchers, is a messaging protocol with following qualities: lightweight broker-based publish-subscribe intended to be open, simple, lightweight, and easy to implement with asynchronously [34]. It is an asynchronous protocol. A few MQTT messages enclose a variable header, present after the fixed header and before the payload that contains the protocol name, the protocol version, and flags. MQTT for sensor networks (MQTT-S) [35] is an extension of MQTT. MQTT-S optimizes the implementation on low-cost, battery-operated devices such as wireless sensor devices (mostly used in IoT systems) with more partial processing and storage resources.

CoAP is designed for constrained networks and nodes in M2M applications, and it is the representational state transfer (REST) paradigm based lightweight protocol [35]. CoAP also supports asynchronous communication. In the REST architecture, exchanges of client's operations on resources which stored at server is in the form of request and response, as in HTTP. CoAP easily translates to HTTP for integration with the Web, while accomplishing dedicated requirements such as multicast support, built-in resource discovery, block-wise transfer, observation, and simplicity for constrained environments. Like in HTTP, the clients do not need to maintain state, that is, clients can be stateless [35]. For the cases when the service side knows that it will take long time to answer a request, CoAP also supports asynchronous responses.

ZigBee is mainly designed to carry small amounts of data across medium distances. It is a mesh network protocol. It is based on a mesh topology network, which means that information from a single-sensor node travels across a group of modes until the broadcast reaches the gateway. ZigBee is a local area network (LAN); it is not designed like BLE to connect to device directly among users or devices. So, it connects to wide range devices. It is a best protocol for home automation and smart lighting.

ZigBee properties [36]:

- ZigBee is consistent and robust uses topology like multi-hop mesh networking to remove single points of failure and enlarge the reach of networks.
- ZigBee is low-power allowing battery-operated devices with the green power feature.
- It focuses on secure and uses different security mechanisms such as AES-128 encryption, device and network keys and frame counters.
- ZigBee is interoperable and standardizes network and application layers.

Bluetooth is now having two branches: traditional Bluetooth and BLE. Traditional Bluetooth will not be simply sufficient if any application needs to be battery operated for an extended period of time. Traditional Bluetooth design recommends 1 W of power consumption. But when we are using to wireless IoT applications, this is a lot.

BLE is a PAN, so the range is shorter than ZigBee, with a much higher data rate. The aim is to be able to connect to devices near a user. Traditional Bluetooth had a data rate between 1 and 3 Mb/s, and BLE data rate is 1 Mb/s for short bursts. Now, many operating systems (OSs), including Android, iOS, Windows 8/10, and OS X are supporting the BLE. On the other hand, Bluetooth isn't a great choice for high-density nodes or long-range applications.

However in Wi-Fi, well-configured access points inhibit the growth of the IoT over it. As long as Wi-Fi remains a uniform standard, security is implemented per-network. Its networks are discovered by their service set identifier (SSID). The sensor needs to know, out of the many SSIDs it scans, which one it should connect to. An IoT sensor must be configured to connect to the WLAN using three parameters: network discovery, authentication credentials and device identity. The promising IoT systems connect headless sensors over wireless connections to a cloud service that manages them and collects traffic. Connecting to random networks carries a risk that the sensor or its cloud service could be compromised. Credentials—usually passwords—are also specific to the network. They must be configured whether the network uses a pre-shared key or proper WPA2-enterprise authentication.

6LowPAN: A key IP-based technology is 6LowPAN (IPv6 low-power wireless PAN) network protocol. Moderately than being an IoT application protocols like Bluetooth or ZigBee, 6LowPAN is a network protocol. It defines encapsulation and header compression mechanisms. The standard has a freedom of frequency band and physical layer and can also be used across multiple communications platforms, including Ethernet, Wi-Fi, 802.15.4, and sub-1 GHz ISM. A key attribute is the IPv6 stack, which has been a very important introduction in recent years to enable the IoT. Particularly designed for smart home or building automation, IPv6 provides a basic transport mechanism to produce complex control systems and to correspond with devices in a cost-effective manner via a low-power wireless network.

8.4.1 IoT operating systems

Established OSs, such as Windows and iOS, were not designed for IoT applications. IoT applications need to save power but do not need great processors and memory storage. The desktop OSs can consume too much power, need fast processors, and in some cases, lack features such as guaranteed real-time response. They also have large memory footmark for small devices (things) and may not support the microchips, battery-based devices that IoT developers use. Therefore, a large variety of IoT-specific operating systems has been developed to suit many different hardware paths and feature needs.

8.4.2 IoT platforms

IoT platforms bundle many of the infrastructure components of an IoT system into a single product. These platforms provide different services, they can fall into three main types:

1. Low-level device control and operations such as communications, device monitoring and management, security, and firmware updates
2. IoT data collection, transformation, storing the data and data management and
3. IoT applications need development in the forms of event-driven logic, program the applications, visualization of the results, data analytics, and integration services to connect to enterprise systems.

Below are few IoT platforms currently playing the major role:

- ThingWorx is the technology platform that facilitates solutions to develop and deploy smart solutions for the IoT. It provides promptly a best platform to innovators who are putting the expectable profitable investments in smart, connected enterprises of the IoT.
- AWS IoT is Amazon's IoT platform is a best easier for developers to communicate among sensors for numerous applications from automobiles to turbines to smart home light bulbs. The vendor has associated with hardware manufacturers like Intel, Texas Instruments, Broadcom, and Qualcomm to provide utmost solutions to the clients.

8.5 Layered architecture of IoT

In order to achieve an efficient communication between the devices in the Internet, layered architecture (Figure 8.5) is identified with different layers as application, communication, security, embedded, hardware, integration, and data base (DB) layer (Table 8.2).

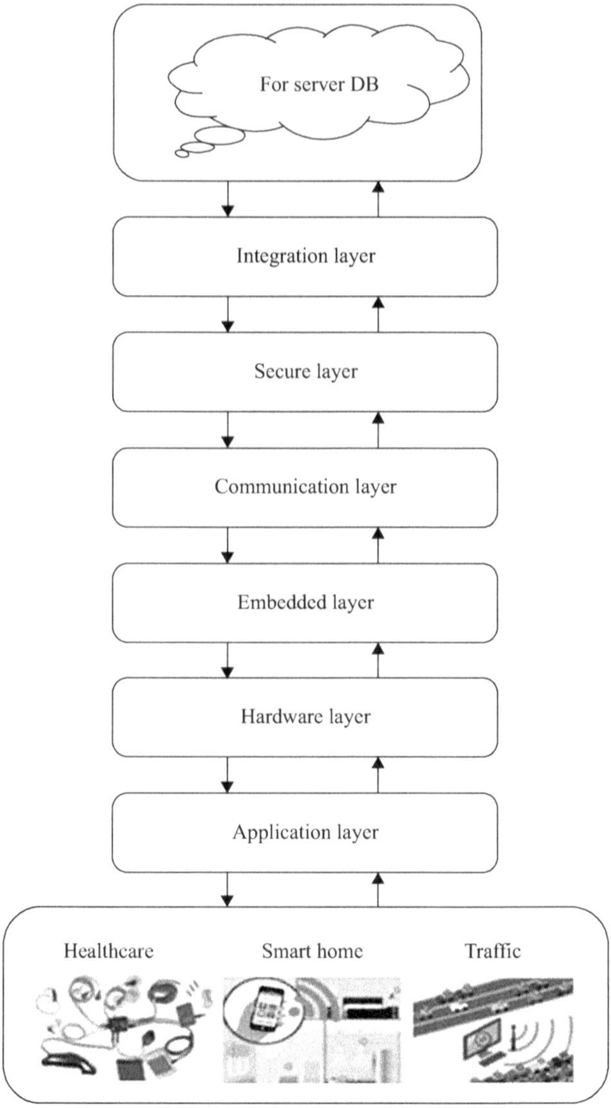

Figure 8.5 Layered architecture

Table 8.2 Layers and their tasks

IoT layers	IoT components	Tasks	Used technologies
Application layer	Applications	Provide the disabled with care and assistance, and enable the disabled to read/view their health information	Smart home technology, robotics, cloud computing, fog computing
Hardware layer	Device discovery, access control, data management	Enables communication between applications and things	CoAP, MQTT, REST, OMA lightweight, OMA DM, EPC, ONS
Embedded layer/sensing layer	Physical objects	Collect, monitor, identify, and provide data about disabled users in their environments	RFID, sensors, actuators
Communication layer/network layer	Communication technologies	Wireless WAN: transmit information over Internet from devices or gateway	Wireless WAN: 2G, 3G, long term evaluation (LTE), long term evaluation-advanced (LTE-A), 4G, 5G, satellite networks, etc.
		Wireless PAN/LAN: enables devices to share or exchange information themselves	Wireless PAN/LAN: RFID, Bluetooth, Wi-Fi, Li-Fi, ZigBee, 6LoWPAN
Secure layer	Embedded security, application security	Securing the things which are connected by Internet, applications deployed in IoT	PKI certificate, encryption and decryption technologies, cryptography tools
Integration layer	Hardware layer to fog to cloud integration, devices to fog server integration	Integration means communication from health devices to fog server and fog to cloud remote servers	Java web services, AWS
DB layer	Database technologies	Connecting the applications to data base in the cloud and fog	Oracle cloud, Microsoft Azure, AWS EBS, AWS EMR

1. The application layer uses the processed data from the smart layer to provide users with a variety of services without their intervention. IoT applications ranging from military to healthcare can be actuated by the functioning of this layer.
2. The sensing layer of the solution is a series of innovative wireless sensor devices; the data from the devices are collected through specific networking protocols. At the sensor layer, the sensor layer is made up of sensors and smart devices, real-time information to be collected and processed. Sensors use low power and low data rate connectivity. This is where we need our WSN formation to be made

such that this sensor information is connected and can be delivered to a targeted location for further processing. Sensors are grouped according to their purpose and data types such as environmental sensors, military sensors, body sensors, home sensors, surveillance sensors, and other things.

3. The DB layer deals with the accumulation and setting of data received from the Internet. It also takes into account, the storage and processing mechanisms that may deal with storing the data on cloud storage and data centers. The processing may include standardizing the data to be used for the smart layer.
4. The communication layer of the IoT is a very important layer that includes transmission of data and enables the exchange of information between different sectors. Thus, flow control assumes great importance in this layer. This layer also symbolizes the aggregation of different kinds of communication networks such as the mobile communication network, broadcast television network and thus will provide all types of address conversion, formatting techniques, etc. It helps in routing voluminous data to the data-processing layer. Batalla and Krawiec [37] propose a novel architecture of the ID (IDentifier) layer for IoT, which is embedded in the network level instead of traditional overlay solutions. Networking named content approach to specify rules for ID-based data transfer. The network nodes have capabilities of caching forwarded data for handling future requirements, what may decrease network overload and facilitate cooperation between applications and sensors that periodically move into sleep mode for saving energy. ID-based routing offers decoupling of identification of objects/services from their location.
5. Secure layer is mainly focus to secure the things connected to Internet communication channels. Security mechanisms here categorized as embedded security and application security.
6. Integration layer is concentrates on how the hardware layer objects are integrating to Fog layer and how communication is happening from Fog layer to Cloud layer?

8.6 Challenges of IoE

The primary challenges of IoE include the following:

8.6.1 Security

Nowadays, the number of devices connected to the Internet has increased. This would cause vulnerabilities in data transfer and communication between the devices. There is a need to develop an efficient security framework for IOE applications. As we increasingly connect devices to the Internet, new opportunities to exploit potential security vulnerabilities grow. Poorly secured IoT devices could serve as entry points for cyber-attack by allowing malicious individuals to re-program a device or cause it to malfunction. Poorly designed devices can expose user data to theft by leaving data streams inadequately protected. Failing or malfunctioning devices also can create security vulnerabilities.

8.6.2 Privacy

The recent advancement in IoE in location tracking, speech recognition and motion tracking are affecting the individuals' privacy. There is a need to protect the personal details of each user efficiently. IoT often refers to a large network of sensor-enabled devices designed to collect data about their environment, which frequently includes data related to people. These data presumably provide a benefit to the device's owner, but frequently to the device's manufacturer or supplier as well. IoT data collection and use becomes a privacy consideration when the individuals who are observed by IoT devices have different privacy expectations regarding the scope and use of those data than those of the data collector.

8.6.3 Standard

There is a need to develop the devices efficiently and accurately. If the devices not are designed properly then it may observe wrong values. In a fully interoperable environment, any IoT device would be able to connect to any other device or system and exchange information as desired. In practicality, interoperability is more complex. Interoperability among IoT devices and systems happens in varying degrees at different layers within the communications protocol stack between the devices. The standardization and adoption of protocols that specify these communication details, including where it is optimal to have standards, are at the heart of the interoperability discussion for IoT. Well-functioning and well-defined device interoperability can encourage innovation and provide efficiencies for IoT device manufacturers, increasing the overall economic value of the market.

8.6.4 Presence detection

It is important to show the details about the devices that are connected in the IoE. If any network problem arises then the device may drop the connection. Hence, all the devices should detect the presence details to the admin [38].

8.6.5 Power consumption

The IoE devices should run continuously without any break. Hence, the power consumption should be less. BLE has the potential for less power consumption than 802.15.4. Wi-Fi can be used in devices with less demand on low power consumption and as a wireless backbone in combination with other technologies.

8.7 Conclusion

The promising idea of the IoE is quickly finding its path throughout our present life, aiming to develop the superiority of life by connecting numerous smart technologies, devices, and applications. Overall, the IoE enable for the automation of everything around us. This chapter studies an overview of the principle of this concept, its enabling technologies, applications, protocols, and the recent research addressing various aspects of the IoE. This study is used to provide a good basis for practitioners and researchers who are involved to increase an insight into the IoE protocols and technologies to realize the general architecture and role of the diverse components and protocols that comprise the IoE. Here are discussed uses and advantages of IoE applications and marketing and business strategies for the investors. In addition, some of the issues and challenges that relate to the deployment and design of IoE implementations have been obtained. Furthermore, the interaction between the IoT, cloud, and Big Data analytics has been discussed. We are proposing a methodology, which can provide solutions to IoT applications with different architectures. At last, comprehensive application use cases were discussed to demonstrate typical protocol integration scenarios to carry desired IoT services.

References

[1] Gubbi J, Buyya R, Marusic S, Palaniswami M. Internet of Things (IoT): A vision, architectural elements, and future directions. *Future Generation Computer Systems*. 2013 Sep 30;29(7):1645–60.

[2] Mavromoustakis CX.On the impact of caching and a model for storage-capacity measurements for energy conservation in asymmetrical wireless devices. In Software, Telecommunications and Computer Networks, 2008. SoftCOM 2008. 16th International Conference on (pp. 243–247). IEEE, Piscataway, NJ; September 2008.

[3] Lopez D, Gunasekaran M.Assessment of vaccination strategies using fuzzy multi-criteria decision making. In Proceedings of the Fifth International Conference on Fuzzy and Neuro Computing (FANCCO-2015) (pp. 195–208). Springer International Publishing, Switzerland; 2015.

[4] Ron Sege, Echelon Corp. The hidden world of the IIoT: Between sensors and the cloud, Embedded Computing Design, Retrieved 29 December 2016 from http://embedded-computing.com/guest-blogs/the-hidden-world-of-the-iiot-between-sensors-and-the-cloud/.

[5] Joh H, Yang I, Ryoo I. The internet of everything based on energy efficient P2P transmission technology with Bluetooth low energy. *Peer-to-Peer Networking and Applications*. 2016 May 1;9(3):520–8.

[6] Mavromoustakis CX, Dimitriou CD, Mastorakis G.Using real-time backward traffic difference estimation for energy conservation in wireless devices. In Proceedings of the Fourth International Conference on Advances in P2P Systems (AP2PS 2012). September 2012 (pp. 18–23).

[7] Zieliński JS. Internet of Everything (IoE) in smart grid. *Przegląd Elektrotechniczny*. 2015;91(3):157–9.

[8] Walport M.The Internet of Things: Making the most of the second digital revolution. A report by the UK Government Chief Scientific Adviser. Retrieved 2014 from https://www.gov.uk/government/uploads/system/uploads/attachment_data/file/409774/14-1230-internet-of-things-review.pdf.

[9] Zhu Q, Wang R, Chen Q, Liu Y, Qin W. Iot gateway: Bridgingwireless sensor networks into internet of things. *In Embedded and Ubiquitous Computing (EUC), 2010 IEEE/IFIP Eighth International Conference* on (pp. 347–352). IEEE, Piscataway, NJ; December 2010.

[10] Atzori L, Iera A, Morabito G. The internet of things: A survey. *Computer Networks*. 2010 Oct 28;54(15):2787–805.

[11] Chui M, Löffler M, Roberts R.The Internet of Things, McKinsey & Company, Retrieved 29 December 2016 from http://www.mckinsey.com/industries/high-tech/our-insights/the-internet-of-things.

[12] Zotos N, Stergiopoulos C, Anastasopoulos K, *et al.* Unified platform for M2M virtual object interoperability. In *Resource Management in Mobile Computing Environments* (pp. 497–512). Springer International Publishing, Switzerland; 2014.

[13] Chellouche SA, Négru D, Arnaud J, Batalla JM. Context-aware multimedia services provisioning in future Internet using ontology and rules. In Network of the Future (NOF), 2014 International Conference and Workshop on the (pp. 1–5). IEEE, Piscataway, NJ; December 2014.

[14] Intel (2016). Retrieved 27 September 2016 from http://www.intel.com/content/www/us/en/internet-of-things/industry-solutions.html.

[15] 4G LTE D2D | Device to Device Communication | Radio-Electronics.Com (2016). Radio-electronics.com. Retrieved 7 July 2016 from http://www.radio-electronics.com/info/cellulartelecomms/lte-long-term-evolution/4g-lte-advanced-d2d-device-to-device.php.

[16] Booysen MJ, Gilmore JS, Zeadally S, Van Rooyen GJ. Machine-to-machine (M2M) communications in vehicular networks. *KSII Transactions on Internet and Information Systems*. 2017; 25(1):1–21.

[17] Mavromoustakis CX, Mousicou P, Papanikolaou K, Mastorakis G, Bourdena A, Pallis E. Dynamic cloud resource migration for efficient 3D video processing in mobile computing environments. *In Novel 3D Media Technologies* (pp. 119–134). Springer, New York; 2015.

[18] Ciobanu NV, Comaneci DG, Dobre C, Mavromoustakis CX, Mastorakis G. OpenMobs: Mobile broadband internet connection sharing. In *International Conference on Mobile Networks and Management* (pp. 244–58). Springer International Publishing, Switzerland; September 2014.

[19] Patil S, Patil V, Bhat P. A review on 5G technology. *International Journal of Engineering and Innovative Technology (IJEIT)*. 2012 Jan;1(1):26–30.

[20] Gupta A, Jha RK. A survey of 5G network: Architecture and emerging technologies. *IEEE Access*. 2015;3:1206–32.

[21] Mavromoustakis CX, Bourdena A, Mastorakis G, Pallis E, Kormentzas G. An energy-aware scheme for efficient spectrum utilization in a 5G mobile cognitive radio network architecture. *Telecommunication Systems*. 2015; 59(1):63–75.

[22] Lopez D, Sekaran G. Climate change and disease dynamics – A big data perspective. *International Journal of Infectious Diseases*. 2016(45): 23–4.

[23] Zheng K, Yang Z, Zhang K, Chatzimisios P, Yang K, Xiang W. Big data-driven optimization for mobile networks toward 5G. *IEEE Network*. 2016 Jan; 30(1):44–51.

[24] Saleem Y, Salim F, Rehmani MH. Resource management in mobile sink based wireless sensor networks through cloud computing. In *Resource Management in Mobile Computing Environments* (pp. 439–459). Springer International Publishing, Switzerland; 2014.

[25] Skourletopoulos G, Mavromoustakis CX, Mastorakis G, Batalla JM, Sahalos JN. An evaluation of cloud-based mobile services with limited capacity: A linear approach. *Soft Computing*. 2016;20(3): 1–8.

[26] Skourletopoulos G, Mavromoustakis CX, Mastorakis G, Rodrigues JJ, Chatzimisios P, Batalla JM. *A fluctuation-based modelling approach to quantification of the technical debt on mobile cloud-based service level*. In 2015 IEEE Globecom Workshops (GC Wkshps) (pp. 1–6). IEEE, Piscataway, NJ; December 2015.

[27] Papanikolaou K, Mavromoustakis C. Resource and scheduling management in cloud computing application paradigm. In *Cloud Computing* (pp. 107–132). Springer, London; 2013.

[28] Lopez D, Gunasekaran M, Murugan BS, Kaur H, Abbas KM.Spatial big data analytics of influenza epidemic in Vellore, India. InBig Data (Big Data), 2014 IEEE International Conference on (pp. 19–24). IEEE, Piscataway, NJ; 27 October 2014.

[29] Manogaran G, Thota C, Kumar MV. MetaCloudDataStorage architecture for big data security in cloud computing. *Procedia Computer Science*. 2016 Dec 31;87:128–33.

[30] Chen KC. Machine-to-machine communications for healthcare. *Journal of Computing Science and Engineering*. 2012;6(2):119–26.

[31] Lopez D, Sekaran G. Climate change and disease dynamics – A big data perspective. *International Journal of Infectious Diseases*. 2016(45):23–4.

[32] Zhang Y, Yu R, Xie S, Yao W, Xiao Y, Guizani M. Home M2M networks: Architectures, standards, and QoS improvement. *IEEE Communications Magazine*. 2011 Apr;49(4):44–52.

[33] Pauls F, Krone S, Nitzold W, Fettweis G, Flores C. Evaluation of efficient modes of operation of GSM/GPRS modules for M2M communications. In *Vehicular Technology Conference (VTC Fall)*, 2013 IEEE 78th (pp. 1–6). IEEE, Piscataway, NJ; 2 September 2013.

[34] MQTT. MQ Telemetry Transport (MQTT) V3.1 Protocol Specification. Retrieved 20 July 2014 from http://www.ibm.com/developerworks/webservices/library/ws-mqtt/index.html.

[35] CoRE Working Group. Constrained Application Protocol (CoAP) RFC 7252. Retrieved 20 July 2014 from https://datatracker.ietf.org/doc/draft-ietf-core-coap/.

[36] Coetzee L, Eksteen J. The Internet of Things – Promise for the future? An introduction. *In IST-Africa Conference Proceedings*, 2011 (pp. 1–9). IEEE, Piscataway, NJ; 11 May 2011.

[37] Batalla JM, Krawiec P. Conception of ID layer performance at the network level for Internet of Things. *Personal and Ubiquitous Computing*. 2014; 18(2):465–80.

[38] Weber RH. Internet of Things – New security and privacy challenges. *Computer Law & Security Review*. 2010 Jan 31;26(1):23–30.

Chapter 9

Combining FIWARE and IoT technologies for smart, small-scale farming: the case of QUHOMA platform architecture

Harris Moysiadis[1], Nikolaos Zotos[1], Marinos Kardaris[1], George Bogdos[1], Charalampos Stergiopoulos[1], Kostas Anastasopoulos[1] and Kostas Mavropoulos[2]

Abstract

This work aims to integrate the technical designs of Future Internet (FI) Architecture of the European Community (FIWARE) with state-of-the-art Internet of Things (IoT) technologies and the platform's business requirements and specifications for realizing efficient, small-scale qualitative, farming. An innovative business model is introduced through a set of offered services that are based on networking 'things' and passing contextual data (information) to business entities to further process, distribute and monetise the derived knowledge to their organisations (farmers, agronomists/mentors, Quality Certification Bodies). All these follow, technologically, the IoT concept. A FIWARE-enabled platform exploits Future Intelligence's end-to-end standardised modern wireless sensor network (Future Intelligence's Internet of Things (FINoT)) that performs tedious tasks and makes field-data available anytime and 'everywhere'. Everywhere currently means within the under-development community or for public use only under farmers' (on a per demand case) permissions. The platform enables access to the sensor data and facilitates process automation, resource management and data handling. The main target of the solution is to establish an ecosystem of technology services that lead to very specific business opportunities: a data consolidation mechanism acquiring data from different sensor controllers bought from various vendors. In that sense, the platform aims to continue the integration of FI-enabled software tools with emergent technologies, architectures and business concepts. Creativity and quality of usage of the Generic Enablers and FIWARE's Technology chapters is profound: the proposed solution takes advantage of the already built-in application programming interfaces and tools provided by the FIWARE platform, like the IoT/context chapter smoothly

[1]Research and Innovation, Future Intelligence, Patr. Gr. and Neapolews Aghia Paraskevi, Athens, Greece
[2]TÜV AUSTRIA Hellas, Headquarters 429, Mesogeion Avenue, 15343, Agia Paraskevi, Athens, Greece

integrated with FINoT platform and co-developing outstanding B2B Marketplace opportunities (Business Framework) consumed through an ideal User Experience Web Environment. Overall, the agriculture marketplace and community's (QUalitative HOrticulture Marketplace) vision is to promote and reward quality and sustainable farming in fruit and vegetables' production by bringing together Internet of People with the IoT; a bridge realised by a proper business model.

List of acronyms

Acronym	Meaning	Acronym	Meaning
IoT	Internet of Things	QUHOMA	QUalitative HOrticulture Marketplace
FINT	Future intelligence	API	Application programming interface
CB	Certification body	GE	Generic enabler
WS	Weather station	CAPEX	Capital expenditure
OPEX	Operational expenditure	FI	Future Internet
EU	European Union	member states	MS
DAQ	Data acquisition	6LoWPAN	IPv6 low-power wireless personal area network
TCP/IP	Transmission control protocol/internet protocol	WSN	Wireless sensor network
PHY	Physical layer	MAC	Media access control
TIM	Transducer interface module	NCAP	Network capable application processor
TEDS	Transducer electronic data sheets	REST	Representational state transfer
AES	Advanced encryption standard	PSU	Power supply unit
USDL	Unified service description language	IP66	International protection marking 66

9.1 Introduction – the WHAT

This document aims to integrate the technical designs of Future Internet (FI) Architecture of the European Community (FIWARE) [6] with Future Intelligence's (FINT's) in-house state-of-the-art Internet of Things (IoT) technologies and the platform's (QUalitative HOrticulture Marketplace (QUHOMA [7])) business requirements and specifications for realising efficient, value-added services' exchange for small-scale qualitative, farming.

The system's architecture is introduced along with a brief functional description of the modules in use. Details on layered FIWARE components to match business needs are also displayed.

The report focuses on realising real-life (farmers', mentors' and Certification Body's (CB)) service offerings and purchases that facilitate non-siloed collaboration among partners that has never in the past attempt such disclosures in their business cases and activities.

FINT aims to support farmers' transition to a modern business model based on advanced technological tools that facilitate the sharing of knowledge gained through applied research to currently unforeseen recipients and beneficiaries. However, along with FIWARE's competencies, the company does comprehend, takes into account and aims to mitigate all the foreseen drawbacks (lack of privacy and security, information overwhelming, free-riding, self-value extraction) that are associated with the extensive usage of online platforms and communities. Summing up, this report is the first attempt to coherently draw the architectural lines between FIWARE-envisioned trade of agronomical services and agriculture certifications under a validated multi-users' perspective that in fact corresponds to business users' tangible needs.

9.2 The business project/use case – the WHY

This section aims to document the exact process of the steps' to be followed by the business users in order to make a service available to multiple customers. These processes must be aligned to the purpose and syntax of each one involved generic enablers (GEs). Ultimately, these services must enact certain IoT resources from the field/farm/plot which are actually available through FINT's enabled hardware (following FINoT platforms' design principles such as communication nodes and sensor controllers). The hardware will be provided for free to our test users and to our later customers as long as they become tactical or strategic QUHOMA subscribers.

After the completion of the project, FINT will launch to the market an agriculture-specific, low-cost FINT-FIWARE Weather Station that will utilise certain, value-added precision agriculture metrics (soil and microclimate temperature/humidity) while getting additional data from proximate, open-sourced and well equipped, public weather stations. Furthermore, the company along with FIWARE will provide scalable application programming interfaces (APIs) and agriculture Services-As-A-Service for attracting future providers and customers to jump in the marketplace reducing their capital expenditure to zero, while maintaining a profitable margin for their operational expenditure (Figure 9.1).

In the following sections, the author introduces the ideas on the platform development and elaborates on the business and technical endeavours. The first section describes a business model that includes the scheme of sales of services, definition of players, costs, etc. The second section introduces the implementation model which involves the FIWARE chapters (Business Framework, Context

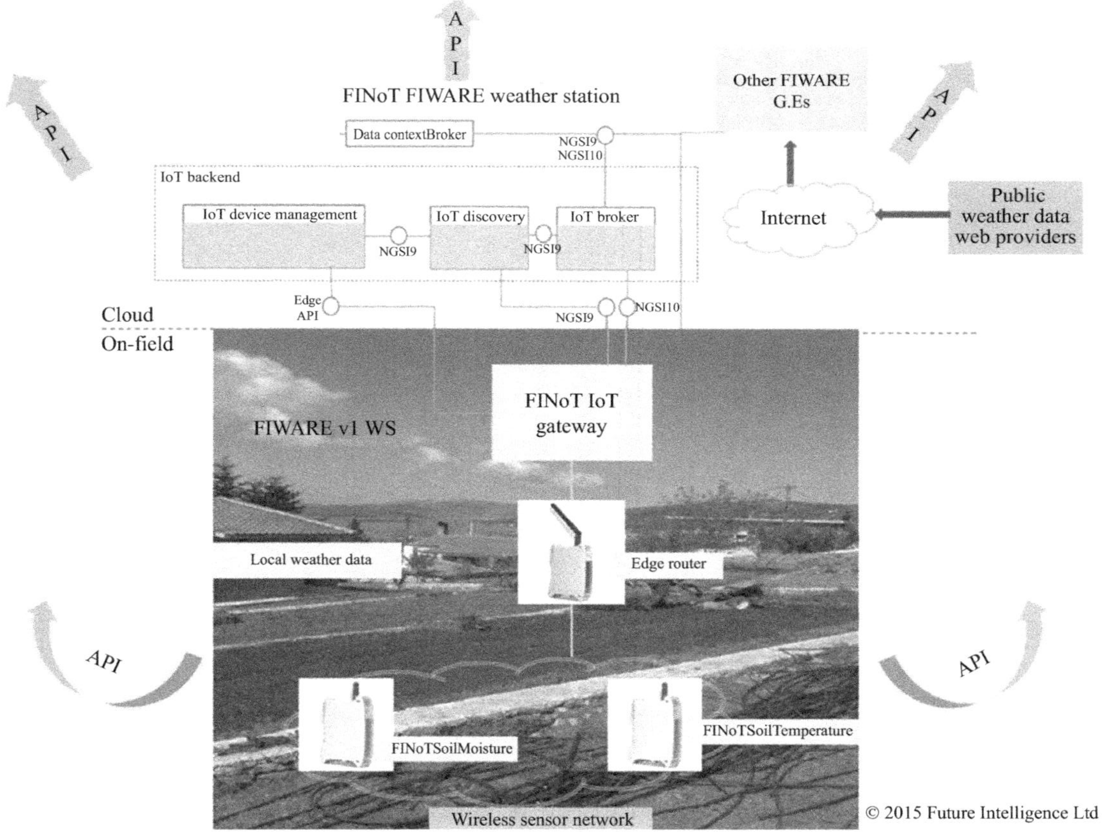

Figure 9.1 The general concept of QUHOMA platform

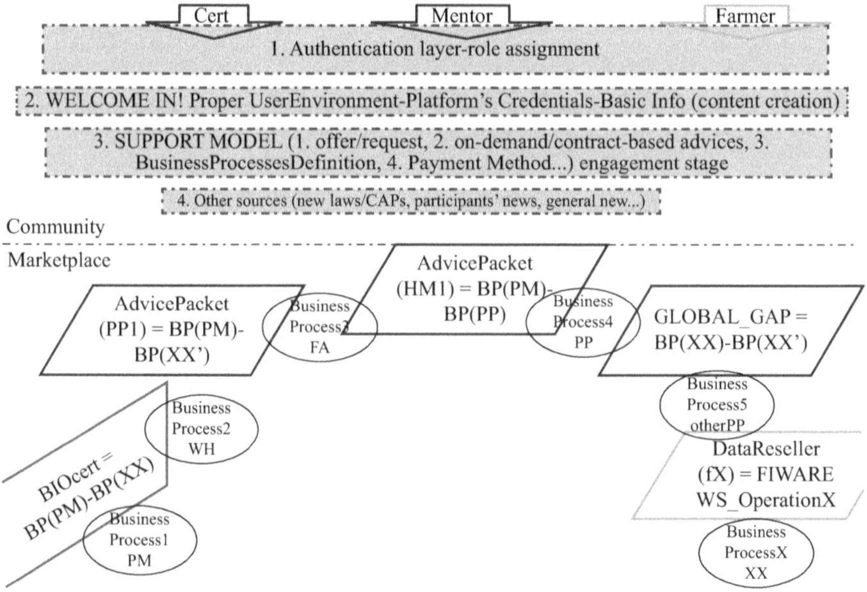

Figure 9.2 Content sources

Framework, IoT Chapter and User Interface Framework), extensively analysed in the relevant technical section.

9.2.1 The marketplace creation

QUHOMA Business Approach is based on two principles: content and its commercialised exchange, the marketplace.

Content is being constantly increased and updated. The sources are the FINT-FIWARE devices, external databases, the farmers (personal and farm's info) and the mentors-agronomists that provide advice and the CB inspection's directives and guidelines (Figure 9.2).

Content providers are among the most important members of every electronic community. For keeping such contributors active, QUHOMA aims to adopt a model of rewarding the active over the lurkers (free riders). This concept was included in the initial design in order to distinguish users among the ones that they consider their participation as part of a strategic investment and long-term collaboration from the ones that make short-term or occasional usage of the platform and its services.

The data that are stored and available in a fully exploitable format can contribute to the production of quality agro-food products; thus, they constitute the core of the QUHOMA platform. The data sources may be sensors or/and various other connected devices as well as users that will manually insert them through user-friendly processes and interfaces. Having this concept in mind, several data-related services are envisioned and are analysed in the paragraphs below. All current stakeholders are able to offer and make use of services which are all associated with the products' quality increases fully exploiting Europe's small-scale, family farming competitive advantage.

9.2.1.1 Supply side

As already indicated, all participants can create and consume services. However, the main reason for QUHOMA creation is the support of farmers for producing qualitative products that globally set a premium price to all markets. They will be the main users of QUHOMA receiving consultancy services and customised advice by agronomists-experts. What's more, they can sign contracts for specific quality certification schemes (BIO and GLOBALG.A.P.) – negotiating a better price rather than using the mainstream market channels (i.e. physically contacting CBs). FINT does understand the critical part the farmers play in QUHOMA as their buying engagement and the cultural adoption of technology into their business model will critically influence QUHOMA success. That is why along with the support of FIWARE technologies and FINISH Acceleration open-source scheme, the piloted so-called FINoT [8] Agri Nodes will be given for free at least to innovators (the very first members of QUHOMA community). This locally deployed electronic equipment initially enables farmers' convenience, precise 24/7 field conditions' awareness, forthcoming remote intervention (e.g. irrigation system at a future stage) and of course services' prosume and optimised farming performance.

Available mentoring services

The consultants'/agronomists' primary purpose for jumping in the platform is to increase their customer base. The first stage is to create a system account and create a profile including personal data, working experience, education and skills. The experience will be related to specific crops and species along with their method of dealing with potential issues. The profile will also include the type of service (advice) packets that are able to provide to the farmer on-site or online. The services, provided by the QUHOMA platform, may be part of service packages, or independent, operational services (multiple operational services constitute a service packet). It is mandatory for the mentor/agronomist to provide at least one service category (advice packet) through the platform in order to become active member and have access. By browsing and selecting available services, the farmers could interact and communicate with the corresponding agronomist/consultant either online or in person when the consultant visits the field.

Many user and business options regarding services will be available (e.g. modular, on-demand, per payment user schemes, etc.) by the mentors. This way the farmers will be able to choose tailor-made solutions. Specifically a farmer who is registered in QUHOMA can choose either independent services combining available packages or the purchase of a single suitable service package, in a very competitive price.

Each service package is designed to provide help for every stage related to the farming process. This division allows the farmer to make strategic decisions regarding his/her business in a more efficient way. This may be related to a set of services about crop planning, cultivation practices, the implementation of quality systems (certification of the products) and the commercial outlet of all farm products (strategic level). Alternatively, a farmer may need a vertically integrated service scheme for the farming practices of a particular crop (tactical level). Finally, the farmer may purchase certain independent services, from the total advices packages (operational level).

The cooperation between mentor and farmer, via QUHOMA platform is not only restricted to services about the management of farming practices (e.g. harvest), but it is also a tool of engagement by promoting and supporting goodwill and mutual trust in the produced products (additional marketing channel).

Available quality certification services

For the time being, FINT engaged TÜV AUSTRIA HELLAS [9] in QUHOMA as the single entry on behalf of legislatively-approved (accredited) CBs. However, in terms of business scalability and growth potential, the success of the platform will be critically defined by its horizontal diffusion in each one of the actor's business positions (farmer, agronomist).

TÜV AUSTRIA Hellas motivation to be part of the QUHOMA is encapsulated in two very important business drivers: *efficiency and effectiveness*. The first one describes the process of lowering intermediate costs when conducting core activities anticipating the exactly same results (doing the thing right). This is profoundly the case, as long as QUHOMA promises optimisation of audit resources (audit time and cost effectiveness) such as reduction of the on-field documentation inspection from the CB's auditors and inspectors, since the locally deployed FINoT AGRI nodes and QUHOMA's online environment and user-completed forms (images' upload, etc.) promise to provide adequate and reliable information about the critical factors under CB's assessment (e.g. fertiliser receipt/description form). Effectiveness in business is translated into accomplishing the right results, while keeping resources unaltered (doing the right thing). As such, CB can now contact additional customers that are already informed, educated and ready to collaborate for responsible farming in QUHOMA's all-in-one community for qualitative, agricultural products, methods and practices.

A CB can create a profile and then can upload a profile, a portfolio of certification services based on internationally approved and recognised standards (e.g. Organic Products-BIO, Products of integrated farm assurance against the GLOBALG.A.P requirements), which promote the sustainable and responsible production of certain crops and food species, with main concern on the quality and safety of the products for consumers, as well as the prevention of environment and

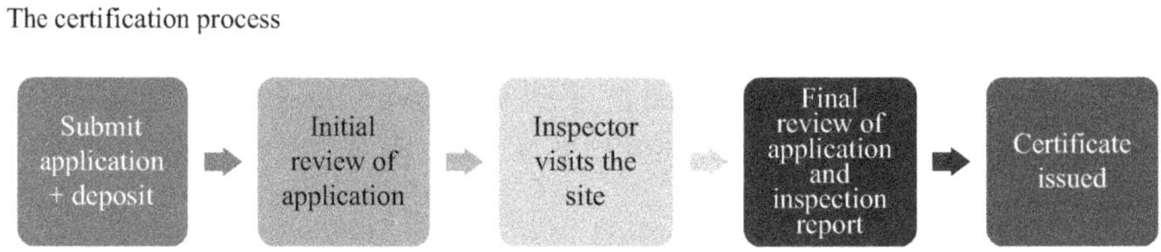

Figure 9.3 The certification process

the producers' health. Any CB that cannot provide accredited services at least on one of these quality certification services, it will not be able to get a QUHOMA account.

The roadmap for initialising and finally delivering the quality certification to a candidate is largely imposed by the relevant standard's requirements and Member's State and European Union (EU)'s legislation (specifically for organic farming) and is presented in Figure 9.3.

Any operator who places products in the market as organic or in conversion to organic has to

1. register his enterprise to the organic product inspection and certification system,
2. notify his activity to the competent authorities of the Member State where the activity is carried out,
3. conform with the production requirements of the relevant standard (Reg. EU 834/2007),
4. hold a product certification granted by an accredited CB.

Before a decision on certification is made, the certifying body must conduct an on-site inspection. The farm should be in some stage of production at the time of the inspection so that compliance can be demonstrated. The farmer, or any other person in position of being aware about the farm operation, needs to be on hand to answer any questions the inspector may have. All aspects of the organic enterprise will be examined. If the inspector considers that it is necessary, some product test samples may be taken at that time, for thorough chemical analysis, such as for pesticide residue test on plant tissues, fruits, etc. The inspector's job is to observe and gather information and assess farmer's compliance towards the regulation requirements; not to make any decision regarding the status or issuance of the farm's certification.

According to GLOBALG.A.P, there are five steps that a farmer has generally to follow in order to get certified.

- Get or download the relevant GLOBALG.A.P. standard documents and checklists from the website document centre (http://www.globalgap.org/) or follow the link on the relevant standard page.
- Compare offers from the certification bodies in the country, register with the one you choose and get your GLOBALG.A.P. Number (GGN). There is a

full list of GLOBALG.A.P.-approved certification bodies in http://www.globalgap.org/.

- The farmer should carry out a self-assessment using the checklist and correct all the points that he does not comply with. Farm assurers/mentors can provide the farmer with valuable assistance during his audit preparations. Farm assurers are independent, on-site advisors and consultants who help producers navigate the steps necessary to implement Good Agricultural Practices and to obtain GLOBALG.A.P. Certification. With first-hand knowledge of the GLOBALG.A.P. System and the latest industry developments, farm assurers use their expertise to make the standard easier to understand and simplify audit preparations.
- Then he must arrange an appointment with a GLOBALG.A.P, an approved CB. An inspector will then conduct the first on-site inspection.
- Once the farm is successfully complied with the standard's requirements, he will receive a GLOBALG.A.P. Integrated Farm Assurance Standard certificate for the relevant scope.

Farmers participating in QUHOMA platform and through the selection of a specific accredited CB could gain access to new markets for their certified products. Some CBs have worldwide representatives, and there certificates are more recognised and reliable in many countries. For example TÜV AUSTRIA HELLAS, utilises the expertise and scientific support of the international network of TÜV AUSTRIA GROUP – is active abroad, with its own subsidiaries, branches and local representatives in Cyprus, Albania, Turkey, Egypt, Israel, Yemen, Jordan, Pakistan, Korea as well as in Doha of Qatar.

Available farmers' services

An innovative concept of QUHOMA is that apart from limiting the 'resistant-to-change' attitude of the farmers to install electronic equipment at their farms and collaborate with relevant players through the offer of a – really hard to refuse – holistic services' packet is the moderator's intention to tackle the 'value-extraction' factor often encountered in such online platforms. By value-extraction the writer describes the process of free disclosure of (non) personal information that others actually over-exploit, making money out of it.

This phenomenon is very common in social media sites (Facebook, LinkedIn) where content created by active communities' participants (status' news, images, check-ins in public places) is further analysed, commodified and shadowy exchanged with third parties (online marketers) in order to deliver fully customised ads back to the content creators.

In QUHOMA, we have predicted that access to on-field data and farmers' input practices and methods will be available on-demand to mentors and especially CBs rather than permanently in order to protect farmers' privacy, security and their resources' valuable insights. Business partners should get farmers' on-click permission to get access to their resources' bind. On-field conditions might be of particular interest for mentors/agronomists in order to statistically estimate a specific intervention instrument success under certain environmental conditions.

In addition, farm(er) data might serve CB to actually confirm that his GLOBALG. A.P. certification applicant applied the appropriate quantity of an approved pesticide for the crop, in order to confront weeds that was indeed crucial to be tackled at a specific production stage.

These are just some examples of how a farmer can monetise his data and get revenue back from them. As the concept advances, some other cases may as well appear so it is very important to book an open place for famers in order to offer their services in QUHOMA marketplace.

9.2.1.2 Demand side

As already underlined, the money flow will be largely triggered by the farmer. However, the other business users will also consume services and of course there might be scenarios for business synergies currently not taken into account (a potential cooperation among mentors–CBs). All these buying procedures must be in accordance with FIWARE Services' Delivery Framework processes and principle GEs functions have to serve such a framework. In short, the buying activity for farmers and mentors/CBs is presented in the schemas below.

There are many opportunities for new market transactions for all the involved parties of the vague, quality-driven farm industry actors. The example of CBs or agronomists-researchers purchasing data available from a farm, in order to use them in other projects at the same area or for their literature review has already been defined and clearly described. Input suppliers could join QUHOMA, in order to reach new clients, thus directly promote new products (fertilisers, pesticides, organic inputs, etc.) and information about their application. In general, additional third parties (e.g. labs) might reserve a specific area for posting their offerings or buy already used farm by-products. This might take the form of standardised APIs to facilitate each of those postings, and it would be very interesting to be realised within QUHOMA, even if these concepts are way beyond platform's initial setting.

9.3 The technical approach – the HOW

This section aims to precisely describe the integration of FINT innovative hardware and software platform for heterogeneous objects' interconnection (FINoT) with FIWARE. In the later paragraphs, the FIWARE-based business services' exchanges are precisely presented.

9.3.1 Interconnecting the generic enablers

Here, the integration of FINT's FINoT Platform is customised for use in smart agriculture applications, with FIWARE and the QUHOMA project. It provides detailed information about the FINoT platform architecture and the protocol adapter of the FINoT Gateway which enables the communication with FIWARE. Finally, it presents the deployment plan of the QUHOMA AGRI Nodes including the minimum installation parameters.

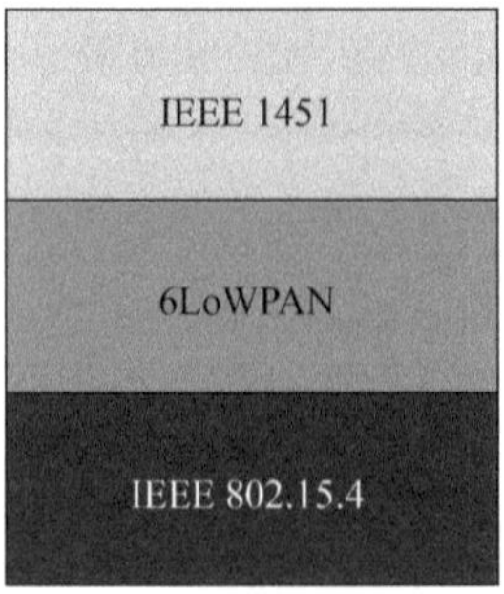

Figure 9.4 The main standards the FINoT platform is built upon

9.3.1.1 FINoT platform introduction

The FINoT platform is based on the innovative fusion of a sensor data acquisition (DAQ) framework on top of a IPv6 low-power wireless personal area network (6LoWPAN)-enabled network implementation. The key concept of the platform is to enable the interface of a vast variety of sensors with a minimum setup effort which can communicate using proven transmission control protocol/internet protocol (TCP/IP) technologies over a wireless mesh network. What truly differentiates the FINoT platform from similar wireless solutions is the level of abstraction that offers; it allows the seamless configuration and DAQ of virtually any type of sensor or actuator, no matter its complexity under a common operator interface without sacrificing almost any of the low-level access capabilities.

Rather than reinventing the wheel, the platform's architecture was based on a set of open standards selected specifically on par with the core platform's main concept; a flexible generic wireless sensor network (WSN). The use of open standards enhances its design flexibility since we are able to optimise their implementation from the ground up, facilitating the platform's customisation according to the vertical's specifications.

In Figure 9.4, we can see the main open standard families that the FINoT platform architecture relies on. Starting from the bottom level, we use the widely spread IEEE 802.15.4 standard for the physical layer (PHY) and part of the media access control (MAC) communication layers, the common reference between the vast majority of WSNs currently IEEE 802.15.4 allows the creation of a low power multi-channel and multi-topology secure wireless network which can be configured according to needs. On top of the IEEE 802.15.4 PHY resides the 6LoWPAN stack which allows the network's abstraction to standard TCP/IP technologies. Finally, the application layer is based on IEEE 1451 which enables a powerful common transducer interface.

9.3.1.2 Network architecture

The network layer is based on the implementation of a low power wireless mesh network. All the standard mesh features are here; multi-hopping, self-healing routing. On top of thc MAC layer resides a 6LoWPAN-compatible stack which

Table 9.1 Network specification list

MAC	IEEE 802.15.4
Frequency band	2.4 GHz
Transmission power	<20 dB m
Security	AES 128
Modulation	O-QPSK
Network stack	RFC4944, RFC6282, etc. (6LoWPAN)
Routing protocol	RFC6550

allows the communication with the gateway and between nodes to be made using the standard TCP/UDP protocols. Each node is configured with a unique IPv6 address.

An Advanced Encryption Standard-128 encryption engine is utilised in both the edge router and the nodes to allow the secure communication between each network entity.

Table 9.1 shows a brief list of the network specifications of the FINoT platform.

9.3.1.3 Sensor DAQ framework

On the heart of the FINoT platform lays the DAQ framework which is responsible for the interface and management of the various sensors connected to the system. The framework is based on the IEEE 1451 set of standards which allow the access of transducer data through a common set of interfaces. The highly efficient binary data format used by the standard is suitable for the low data rate, low power network topology used, whereas the use of embedded transducer datasheets in conjunction with the powerful command set that is exposed by the standard allow the implementation of complex transducer functions independently of the sensor type connected.

9.3.1.4 IEEE 1451.0

This standard provides a common basis for members of the IEEE 1451 family of standards to be interoperable. It defines the functions that are to be performed by a transducer interface module (TIM) and the common characteristics for all devices that implement the TIM. It specifies the formats for Transducer Electronic Data Sheets (TEDSs). It defines a set of commands to facilitate the setup and control of the TIM as well as reading and writing the data used by the system. APIs are defined to facilitate communications with the TIM and with applications. The relationships between the IEEE 1451.0 standard and the other members of the family are shown in the following diagram (Figure 9.5).

The underlying purpose of this family of standards is to allow manufacturers to build elements of a system that are interoperable. To accomplish this goal, the

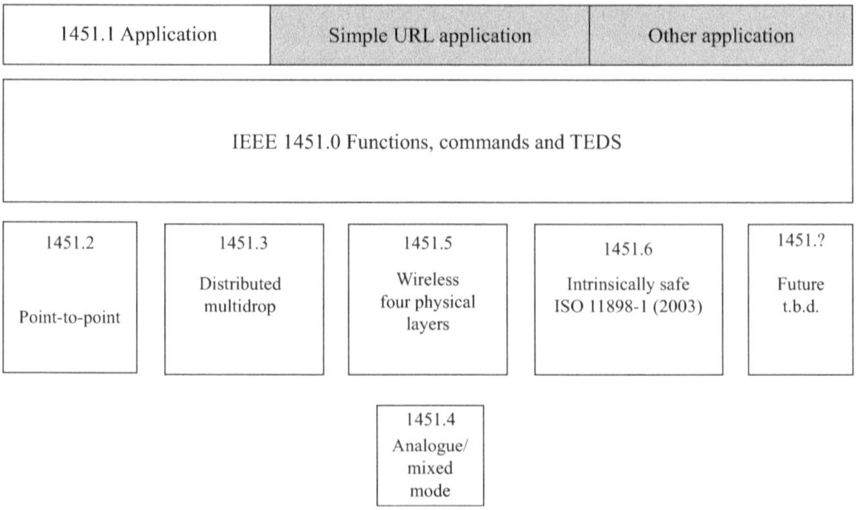

Figure 9.5 IEEE 1451 standards relation [1]

IEEE 1451 family of standards divides the parts of a system into two general categories of devices. One is the network capable application processor (NCAP) that functions as a gateway between the users' network and the TIMs. The NCAP is a processor-based device that has two interfaces. The physical interface to the users' network is not specified in any of this family of standards. IEEE Std 1451.1 provides a logical object model for this interface between the NCAP and the TIMs is defined in the remaining members of the family of standards. Different manufacturers may build the NCAPs and TIMs, and if both comply with this standard, they should be interoperable. This standard provides a description of the functions that are to be performed by a TIM or TIM. Provisions are made for a high level of addressing that is independent of the physical medium-level and low-level protocols that are used to implement the communications. It defines the common characteristics for all devices that implement the transducer modules. The timing of the acquiring or processing of the data samples is described. Methods of grouping the outputs from multiple transducers within one TIM are defined. Common status words are also defined. A standard set of commands are defined to facilitate the setup and control of the transducer modules as well as to read and write the data used by the system. Commands are also provided for reading and writing the TEDS that supply the system with the operating characteristics that are needed to use the transducer modules. A method of adding manufacturer unique commands is included. In addition, this standard provides formats for the TEDS. Several TEDS are defined in the standard. Four of these TEDS are required, and the remaining TEDS are optional. Some TEDS are provided to allow the user to define information and to store it in the TEDS. This standard provides areas that are 'open to manufacturers'.

Table 9.2 Enumeration of Phy ID

Value	Meaning
0	IEEE 802.11
1	Bluetooth
2	ZigBee
3	LoWPAN
4-254	Reserved for future expansion
255	Manufacturer specific

9.3.1.5 IEEE 1451.5

This standard introduces the concept of a Wireless Transducer Interface Module (WTIM), connected wirelessly over an approved radio Communication Module to a NCAP Service Module. The IEEE 1451.5 approved radios (Dot5AR) are IEEE 802.11™, IEEE 802.15.4™, IEEE Bluetooth™ and IEEE ZigBee™ technologies. In essence, it standardises the supported wireless interfaces that an IEEE 1451 – compliant device may use. In Table 9.2, we can see the list of Phy IDs that the IEEE 1451.5 PHY TEDS support.

We notice that the third enumeration value stands for Low-Power Wireless Personal Area Networks and as noted in the body of the IEEE 1451.5 standard, for 6LoWPAN in specific.

As shown in Figure 9.6, from left to right, the NCAP IEEE 1451.0 Services interface with the NCAP IEEE 1451.5 Communication Module through the IEEE 1451.0/5 Communications API. The NCAP IEEE 1451.5 Communication Module communicates with the WTIM IEEE 1451.5 Communication Module through the IEEE 1451.5 Radio wireless PHY. On the WTIM, the WTIM IEEE 1451.0 Services interfaces with the WTIM IEEE 1451.5 Communication Module through the IEEE 1451.0/5 Communications API. What is shown shaded in Figure 9.1 includes the logical and physical partitioning that is covered by the radio sub-specifications for IEEE 1451.5 services.

9.3.2 FINoT devices

9.3.2.1 FINoT Node

As the name suggests, the FINoT Node is the device which plays the role of the communication entity within the WSN. The nodes form the mesh network and transmit the sensor data to the gateway. Nodes do not support the direct connection of sensors to themselves; however, they use an RS485 bus to connect to S/AP (sensor/actuator peripheral) modules. This feature adds great flexibility to the actual deployment of the sensors which can now be placed at a substantial distance from the node and thus optimizing both the sensor installation topology and the wireless coverage of the network, especially in indoor and constrained places (Figure 9.7).

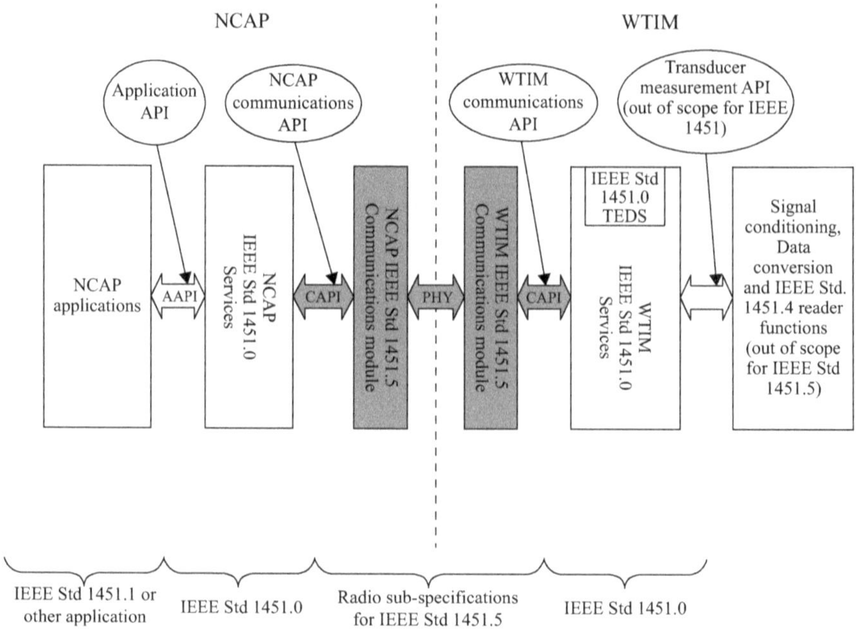

Figure 9.6 Functional context for radio sub-specifications for IEEE 1451.5 services

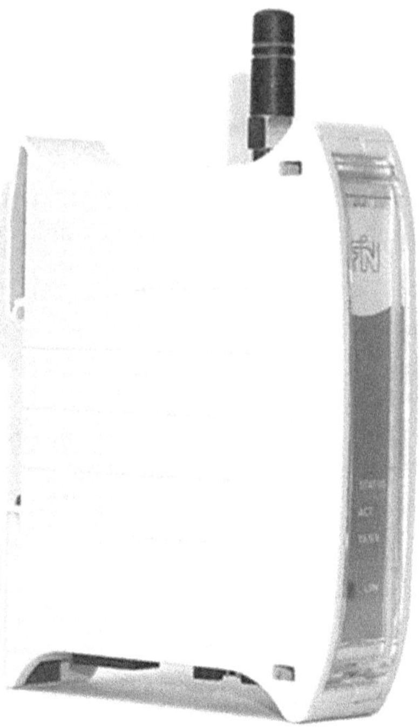

Figure 9.7 The FINoT Node

Figure 9.8 A typical example of a FINoT S/AP

9.3.2.2 FINoT S/AP

The sensor/actuator peripherals are the devices which are responsible for the direct connection and DAQ of sensors. Each S/AP is designed according to the type of transducers that the deployment will use, including interfaces for analogue and digital sensors as well as actuators. They do not possess any wireless connectivity capability themselves; instead they are connected using the RS485 bus to the FINoT Node. Up to 16 S/APs of the same or different functionality can be simultaneously connected to each node, allowing the creation of a diverse wireless DAQ point (Figure 9.8).

9.3.2.3 FINoT Gateway

The FINoT Gateway acts as the edge node of the WSN network. All sensor data are forwarded to the application framework through it, while it maintains optional features like data logging and initialisation sequencing. The gateway is responsible for the coordination of the WSN network, whereas the access to Internet is possible by means of an Ethernet port, Wi-Fi or mobile broadband connection.

9.3.3 FIWARE interoperability

According to the FIWARE IoT specifications, the FINoT Gateway must be adapted to an IoT gateway which will represent an aggregation point for all sensors/actuators inside a farm. The IoT gateway will support all the IoT backend features, taking into consideration the local constraints of devices such as the available computing, power, storage and energy consumption. The level of functional split between the IoT backend and the IoT gateway will also depend on the available resources on the IoT gateway, the cost and quality of connectivity and the desired level for the distribution of intelligence and service abstraction (Figure 9.9).

In order to the FINoT data concentrator fit the FIWARE architecture, a number of software-plug-ins have been developed as part of the project. The new software add-ons convert the FINoT Platform to a FINoT IoT Gateway. It features a dedicated REST management API and a partial implementation of the standardised NGSI API.

9.3.4 FINoT deployment

9.3.4.1 Introduction

In order to acquire data about the qualitative indicators of a farm's soil and microclimate, a set of agricultural sensors is required to be deployed. For this purpose, a special FINoT sensor/actuator peripheral is used, which is capable of interfacing a multiple of these sensors, acquire accurate measurements and through the FINoT Node, send them upstream to the QUHOMA application. A minimum set of sensor types have been identified as crucial for the aforementioned process, and a deployment plan has been carried out in order to successfully install them preserving the platform's expandability options and at the same time ensure a strong resistance to the elements.

9.3.4.2 Minimum setup

The architecture of an on-field QUHOMA measurement station is described in Figure 9.10.

Each FINoT AGRI Node consists of a FINoT Node, an analogue DAQ sensor/actuator peripheral, the sensors, an antenna, a PSU unit, an electrical safety sub-circuit and an IP66 steel cabin. This setup allows the expansion of the AGRI Node with a multitude of optional sensors which can be interfaced either by the installed S/AP or by adding another one with minimum effort. The AGRI Node is power supplied by the mains grid or optionally using a solar panel and a battery.

In order to complete the deployment, a FINoT Gateway is installed either to the farm's premises or if not available, to a dedicated cabin. The data are forwarded to the QUHOMA platform using an available Internet connection or usually by means of a mobile broadband connection (GPRS/3G).

9.3.4.3 Sensor types

As shown in Figure 9.10, the minimum setup consists of a soil temperature sensor, a soil humidity sensor and an ambient temperature sensor. In order to have a better understanding of the farm's microclimate, a set of additional weather sensors can also be installed like air humidity, solar radiation, etc. (Table 9.3 and Figure 9.11).

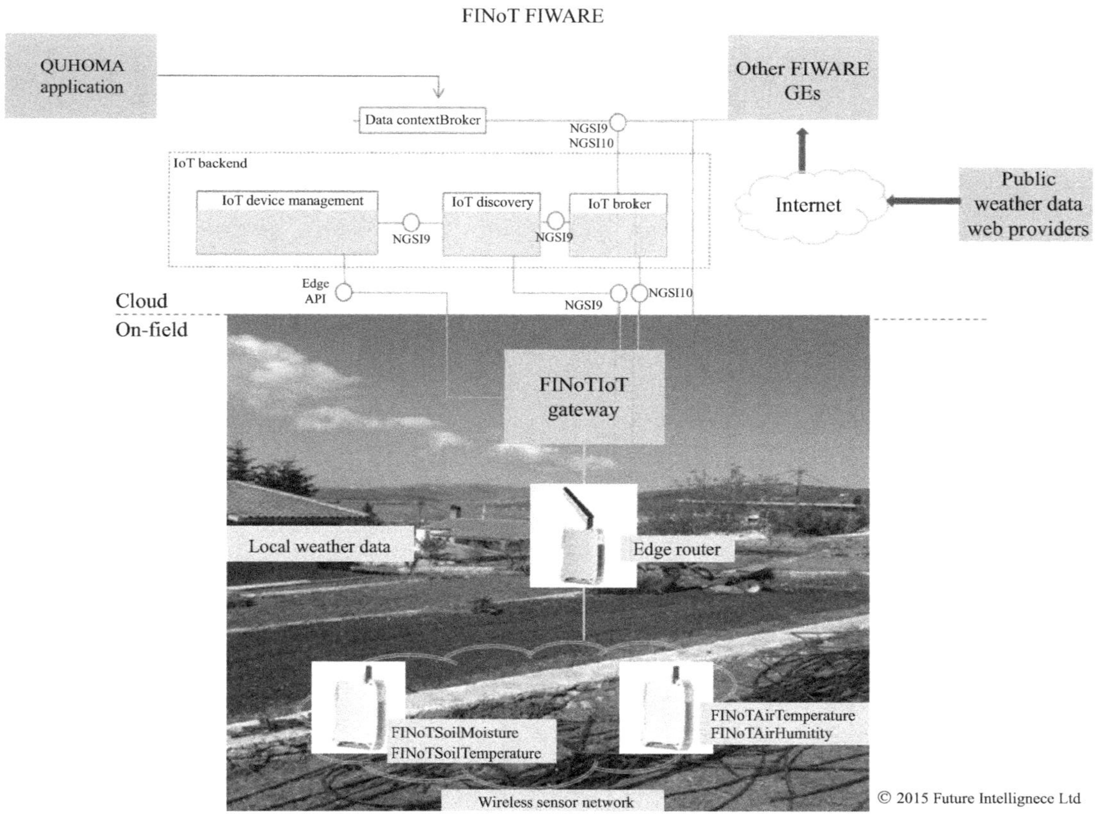

Figure 9.9 FINoT-FIWARE WS

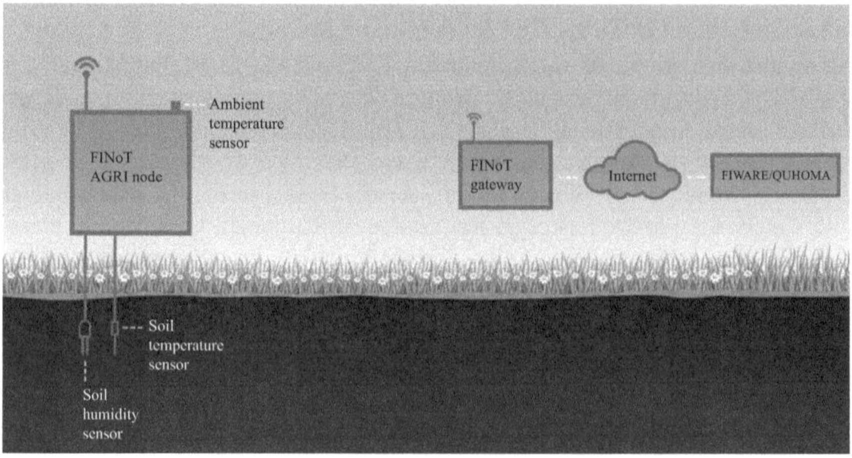

Figure 9.10 QUHOMA measurement station

Table 9.3 Sensor types

Sensor	Type
Soil temperature	RTD type, 3-wire
Soil humidity	Capacitive, 4–20 mA output
Air temperature	Thermistor or thermocouple type

Figure 9.11 An installed QUHOMA AGRI node

9.3.5 Service offerings in FIWARE

During the first phase of implementation QUHOMA will be based totally on FIWARE GEs and the so-called Business Chapter as described in FIWARE Architecture of Applications, Services and Data Delivery manual [2].

The GEs of the Applications/Services and Data Delivery together support the creation of an ecosystem of applications, services and data that is sustainable and fosters innovation as well as cross-fertilisation. In particular, GEs that are part of this reference architecture can be grouped in three mayor architectural blocks as described below.

9.3.5.1 Business framework

This framework includes the following:

A *Store*, which enables selling digital assets (i.e. applications, services and data) for consumers as well as developers and is responsible for managing offerings and sales. The store supports (1) registration and publication of new offerings by application/service and data providers (e.g. Mentors, Farmers), (2) contracting of applications/services and data, (3) gathering application/services (including data services) usage accounting info, and (4) charging for the acquisition and usage of application/services, on the basis of the predefined price model.

A *Marketplace*, which allows consumers to find and compare offerings published on different stores and provides further functionality to foster the market for FI applications, services and data in a specific domain.

A *Revenue Sharing System* (RSS Engine), which allows the calculation and distribution of revenues according to the agreed business models. The RSS GE requires a revenue sharing model to calculate the incomes and revenue shares to be distributed among parties (service providers). The RSS GE also offers expenditure limits functionality for limiting the amount of money spent by a customer. Moreover, the RSS GE offers reporting functionality for administrators.

A *Repository*, which provides a consistent uniform API to service descriptions (models) and associated media files for applications of the business framework (Figure 9.12).

- *FI Application Mashup Framework:* The FI Application Mashup Framework aims at offering support for application mashup, with a focus on the creation of visualisation dashboards and operation cockpits from the underlying services and data. The framework leverages the notion of app and data mashup to allow domain experts and other knowledge workers without programming skills to easily develop application mashups as highly configurable cockpits and dashboards based on data- and event-based wiring of widgets and operator chaining, being these widgets and operators linked to backend services and data.

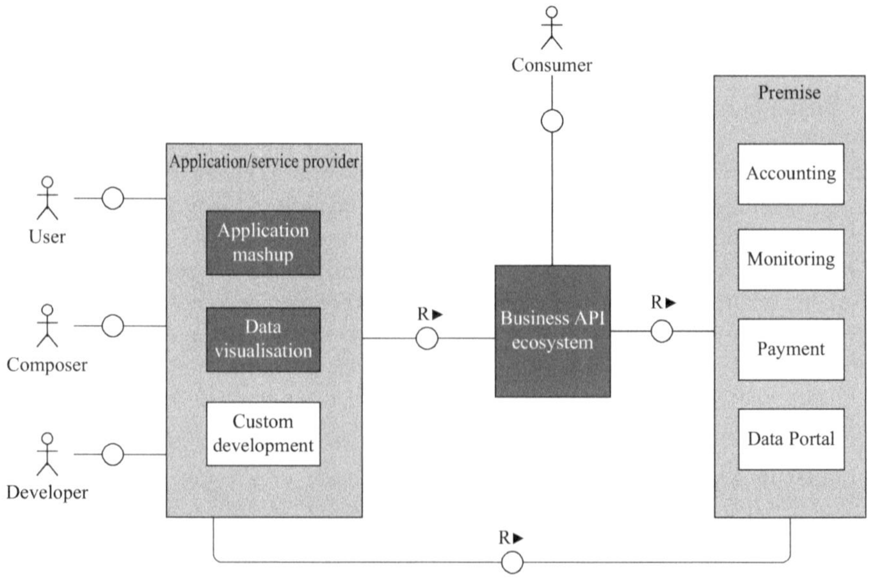

Figure 9.12 The business framework [2]

- *Data Visualisation Framework:* The Data Visualisation Framework aims at creating agile, beautiful visualisations and meaningful reports useful to present the large variety of datasets data stakeholders will bring in the play as well as providing customisable data analytics.

FIWARE will not build an ecosystem, FIWARE will rather provide Generic Enablers for a core business platform, which will offer tradability, monetization, revenue sharing, payment … important ingredients for a business ecosystem, after customization and domain-specific adaptation for USDL and the GEs as well as some complementary components.

The *Orion Context Broker* runs on top of the *IoT Broker*. This is a module introduced to handle the complexity of a large setup with 1,000s of devices and IoT agents connecting to them. In QUHOMA, it is obvious to many existing sensors in agriculture over large patches of lands over an entire country. In that case, it will need a unit which handles and aggregates data for user and a unit to discover sensors in a farm. A rule of thumb apparently says: with more than 1,000 sensors, you should add an IoT broker to your application.

The Orion Context Broker itself only holds the *last value of an entity*. To have the historical view on sensors values (e.g. temperature), the Orion Context Broker is connected via a FIWARE component called *Cygnus* with *MySQL*, where the data will be stored and can be analysed. Cygnus uses the subscription/notification feature of the Context Broker that provides notification for updates regarding the attributes of the selected entities. Cygnus is based on apache flume and will allow one to persist data from the Context Broker not only to MySQL but also Apache hadoop or CKAN.

The complex event processing (CEP) GE, also known as Proton, is a scalable integrated platform to support the development, deployment and maintenance of event-driven and CEP applications. Although standard reactive applications are based on reactions to single events, the Proton engine component reacts to situations rather than to single events. A situation is a condition that is based on a series of events that have occurred within a dynamic time window called a context. Situations include composite events (e.g. sequence), counting operators on events (e.g. aggregation) and absence operators. The Proton engine is a runtime tool that receives information on the occurrence of events from event producers, detects situations and reports the detected situations to external consumers.

The *PEP Proxy* GE interacts with two components in order to check authentication and authorisation:

- Identity Management GE: When PEP Proxy receives a request, it retrieves the authentication token from the specific header and validates it with the Identity Management GE (FIWARE Account).
- Authorisation PDP GE: If the component is configured to check not only the authentication but also the authorisation, PEP Proxy will check with Authorisation PDP if the user (from the token) has the correct permissions to access the resource specified in the request.

Rush notification relayer is used together with Orion Context Broker in 'standalone' mode. The advantage of Rush as notification relayer is the following: Instead of managing the notifications itself (including waiting for the HTTP timeout while the notification receives responses), Orion passes the notification to Rush, which in turn deals with it. Thus, Orion can implement a 'fire and forget' policy for notification sending, relaying in Rush for that.

Each device with sensors installed in a QUHOMA field will be mapped as an Entity associated with a Context Provider by FINoT IoT Gateway collector. The type and name of the entity will be created also by the FINoT IoT Gateway.

Each of the measures obtained from the device should be mapped to a different attribute. The type and name of the attribute will be configured by the FINoT IoT Gateway automatically or optionally by the administrator.

9.4 QUHOMA's Road Ahead for sustainability – the WHO and WHEN

The main issue to be discussed in this report is the governance of QUHOMA project. This role is currently undertaken by FINT but it is believed that as the platform attracts wider audience (both from farmers' and agronomists' side) the harder it would be for a technology provider to play that role. This is easily explained by the fact that FINT can never get (and actually does not want to have) full understanding of power issues and dominance positions related to a community that is interested in qualitative farming. For the time being, the main approach is to create something like a communal Code of Conduct which will underline the basic rules, obligations and rights that QUHOMA participants will engage to as soon as they become QUHOMA members.

Last, quality certification bodies of other EU countries must be contacted in order to unveil their interest in joining in and ways to overcome scepticism over cross-countries horizontal competition (meaning same quality accreditations in different countries).

A preliminary exploration of related issues had already begun but more details will be available in the following weeks.

9.4.1 A brief exploration of market dynamics, interests and power potential over smart technologies in smart farming [3]

Each group of stakeholders in the agro-food chain has its own business issues. Introduction of smart technologies can impact differently those business models. On the production side, the potential linked to smart technologies is high. Some of the expected benefits of smart farming are as follows:

- increase productivity: increase yields by optimizing growth and harvesting processes for example,
- reduce cost: cost of resources (water, energy), lower fertiliser and pesticide usage for examples,
- enhance environmental issues: water and energy consumption, animal feed, health and welfare, plant health, soil pollution, etc.,
- help predict the property value of farms and have insight into the commodities market,
- move closer to consumer demands,
- improve communication with consumers and food-processing companies,
- strengthen position in the value chain,
- reinforce governance support of farmers' local communities and improve decision processes.

The needs and benefits between large farmers and small farmers are different.

For food manufacturers, food safety has become a critical concern. Smart technologies can help them to enhance product labelling and traceability in order to improve supply chain transparency. IoT could also reinforce their positioning compared to retailers with more access to consumers data.

On the distribution side, smart technologies can mainly contribute to optimise and improve freight, transport and storage. IoT brings two main elements: information instantaneity and increase of the number of available data. It could allow checking some constraints (temperature, humidity, package opening, etc.) and having information on trucks filling ratio or driver tiredness.

For retailers, smart technologies help meet the changing needs of consumers who expect to have full pricing and product transparency before making their decisions. Active packaging and smart tagging can offer value-added functionality. For example, smart tags using temperature and/or quality sensors can monitor freshness of a product through the entire value chain. Indicators of product status can be available to both sellers and consumers.

However, IoT could challenge the positioning of retailers in the value chain with the risk to be disintermediated by food manufacturers or producers which will have also access to consumers' data.

Finally, for consumers, smart technologies answer to the demand of more quality and transparency such as food components, breeding conditions, cultural practices, etc. IoT could also facilitate new ways of consumption such as periodic unfixed fresh products, or cooperatives of organic food consumption.

Regarding costs, farmers have very low margins. Investments in innovation are difficult and farmers usually count on public support. Cost for smart farming is still high, especially for small-field farming. Some technologies such as RFID or NFC are still problematic due to the cost associated with this technology compared with the cost of the product.

Exceptions are largest farms with stronger financial capabilities, such as in the United States.

9.4.1.1 Similar market approaches from the IoT providers' side

Several business models could be considered on how ICT providers can sell IoT in agricultural and farming sector:

- Sale of hardware (sensors, etc.) by manufacturers directly or through service providers, with free basic applications,
- Premium subscription for value added applications,
- Advertising-based model: free value added applications with advertising,
- Data value based model: free value added applications in order to retrieve many data in platforms and reuse or re-sell data in specific ecosystems.
- Some options are to be considered in successful IoT business models:
- Open innovation and collaboration which imply the development of strong ecosystems able to share data, know-how and experiences across the overall ICT value chain,
- Supplies of end-to-end solutions (conception, integration, maintenance, etc.),
- Strong knowledge of the agro-food sector,
- Promotion of solutions through associations related to each specific agricultural and industrial food sectors.

Costs of IoT solutions include hardware, development but also deployment (installation and equipment), future updates, replacements, scalability and maintenance. The quantity of sensor nodes and deployed systems is a key cost element. Moreover, costs will be higher with a fragmented market compared to generic solutions using standard interfaces, ensuring interoperability between different providers.

Finally, open source solutions can be promoted as they are usually cheaper than proprietary systems. Also they can be much more flexible and customised for the application purposes. But the main problem in open models is related to support, maintenance and after-sale. Indeed API can change and old versions cannot be available anymore. In addition, it can be more difficult in rural area to find open source experts.

9.4.2 Lessons learned from the above and additional sources

In all the above cases the focus has been moved from small-scale, independent, family farmers to big associations the interests of those might surpass the interests of the Qualitative Horticulture Community/Marketplace. Here, the project team aims to build a sustainable organisation to promote sustainable farming and vice versa: sustainable farming only applies to sustainable members of a distributed, non-hierarchical, self-autonomous and thus sustainable community.

However, it should be highlighted that all the above concepts, approaches and technologies limit themselves to what QUHOMA identifies as its Minimum Viable Product. More precisely, QUHOMA's scope is not simply to construct an IoT solution in order to facilitate farming activities just for the sake of farmers' easiness. The main innovation of QUHOMA is that there is already a very specific plan on how such IoT technology utilisation will facilitate a community's development and then how to engage community members in services' buys and sells. Such an approach is profoundly innovative, and it is also used to persuade farmers to collaborate technically.

The proposed business model as published in the best practices to guide EU's Large Scale Pilots' implementation on Smart Farming and Food Security [3] assumes that the following: QUHOMA is an example of FIWARE-FINT's farm services. The QUHOMA platform is a data-centred community and marketplace for promoting qualitative horticulture. Hardware (FINoT equipment) is provided for free to farmers and access to relevant data is provided upon subscription to agronomists/mentors and Quality Certification bodies:

- Basic (operational) service packet: farmers who have subscribed to QUHOMA can remotely manage their farms through a WebApp. Then, they can purchase operational (WeedHandling, PlantProtection, etc.) advice packets from mentors on a pay-per-use model,
- Tactical service packet: additional to the basic service, farmers can now enjoy training and holistic farming management advices with a discount,
- Strategic service packet: farmers can now buy business intelligence advices and discounted certification products.

9.4.2.1 Champion approach

Merging QUHOMA with national and EU initiatives like LEADER and/or Smart Specialisation Strategies (RIS3) would enable vast diffusion of smart farming tools like FINoT_AGRI devices used in QUHOMA and support side challenges that the project aims to address like actual increase in young professionals' employability, digital divide reduction, technology for the masses and a knowledge-intensive economy creation for EU.

Moreover, innovation schemes will actually be facilitated by state policies which can also deploy a fair regulation framework for similar marketplaces and communities that their existence largely influences public good (in terms of food safety).

9.4.3 What is then proposed?

To build a business community around IoT technology:

- This community will favour access to data instead of data ownership within the community. For publishing/re-using data outside the community farmers have to give their permission to do so.
- FINoT equipment will be given to land owners with a set of clearly defined terms and obligations for FINT (as technology enabler the company does retain the right to exploit anonymous/pseudonymous data), for the farmer and what are his obligations within the community and for the moderator of the community (farmers' union, state, else). He is also in charge of describing the community's own self-regulation instruments (e.g. General Assembly, voting rights, ways to deploy decisions, ways to control decision-making).
- On-site data release will be possible but the releasers should be securely facilitated by technology to do so and get a reward for any data set given further (e.g. to mentors/agronomists in order to scientifically run correlation tests among pest usage and micro/macroclimate conditions or to quality CB to run probability/risk assessments tests for specific certification schemes for specific crops in specific areas).
- When (land) ownership changes the terms and conditions of the community might also change.
- Overall, the business model will be used to eliminate reluctant to change behaviours from farmers' side and a realisation of circular economy in accordance with EU official roadmap that is to be communicated later this year [10].
- IoT data at the centre of a circular community: massive lakes of structured and unstructured data gathered by sensors allow everyone who needs that data to improve their work and decision making and instantly share that data.
- A distributed community instead of a hierarchical top-down governing body presented in Section 9.3. When data about operations and products were fragmentary and little and only available for limited use this dictated both company organisation and management styles. Companies were hierarchical, because managers controlled distribution of what data was available, on a top-down and need-to-know basis. Typically, one department would analyse data based on its area of expertise and decision-making, then pass it along to the next department, whose purview would be similarly limited. Production mode was linear while similarly, the supply chain and distribution network were also linear. But this no more the case so novel management styles have also to emerge from technological innovations. And this is what QUHOMA needs to accomplish (Figure 9.13).

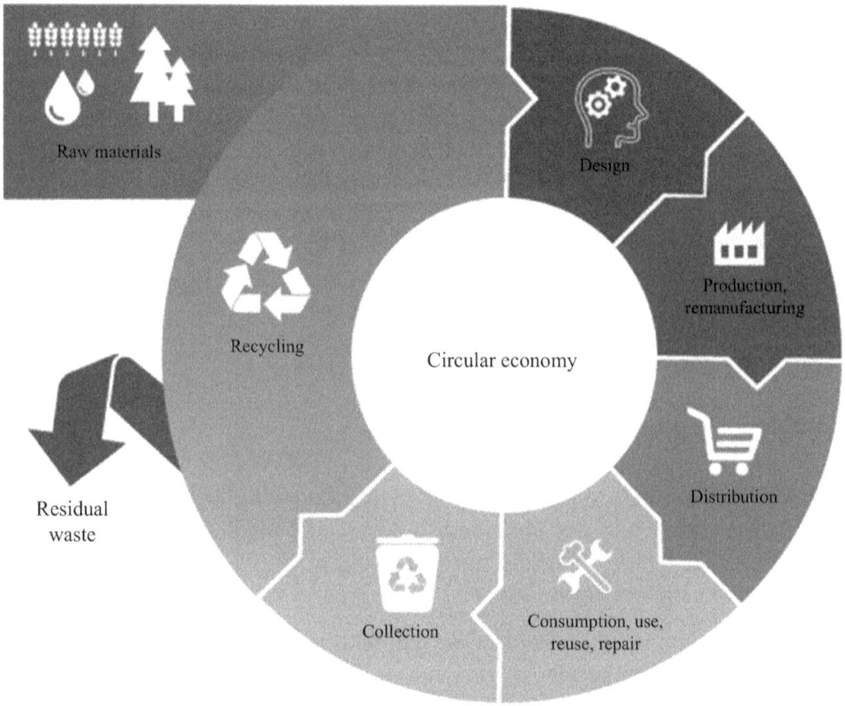

Figure 9.13 The circular economy [4]

Such a community-based, horizontal approach as a management style of an organisation/association gets more and more momentum as numerous publications and real-life example constantly unveil to us [11].

9.4.3.1 Who else can be engaged?

- The European Food Safety Authority [12] is an independent European agency funded by the EU budget that operates separately from the European Commission, European Parliament and EU Member States.
- Drivers of top-down approaches for holistic rural developments, for example LEADER [13], RIS3 [14] and more under national or European initiatives.
- Top-down approaches in international context. For example close collaboration with United Nations and its branch on Food Safety and Security, FAO [15]. 'FAO recognizes that the private sector is a key stakeholder in the fight against food insecurity, malnutrition and rural poverty, and acknowledges the potential that better coordination and collaboration between the public and private sectors can offer in the delivery of the Organization's Strategic Objectives. The Organization, therefore, takes an open and pro-active approach to optimizing the benefits of closer collaboration. In this regard, FAO will consider engaging with all private sector entities, including small and medium enterprises (SMEs), cooperatives and producers' organizations, local companies through to MNCs'.

- IFOAM (International Federation of Organic Agriculture Movements) [16] is a global organisation that leads, unites and assists stakeholders from every facet of the organic movement. In addition to support a diverse global membership scheme IFOAM organises high profile events where organic stakeholders can share their knowledge and expertise and establish valuable partnerships; implements projects, with global and regional partners, which facilitate conversion to organic agriculture, empower local stakeholders and strengthen supply chains as well as help raise consumer awareness; guides through the increasing complexity of organic standards and regulations, and promotes alternatives to certification that are adapted to the diverse needs of organic farmers.

9.4.3.2 Key message for all involving parties

It should be noted that the American Farm Bureau Federation published a potential risks outline relating to the data mining in the agricultural industry and on farm tools [17]. Farmers especially fears that price discrimination may appear if big input suppliers use data to charge them a different amount for the same product or service. In addition, farmers identified three specific challenges on the usage of their farms' data.

Top three concerns from farmers

1. *Liability* – In the case of a data breach, who is liable for my farm data? Can misuse of my data be used against me if not obtained legally?
2. *Usage* – How is my data being used by each company and who is it being shared with?
3. *Privacy* – Is my data anonymous so it cannot be traced back to my site specific operation?

Research's background: Companies have used farm level data for years, but the level of real-time information gained at a micro-level unit is a concern. If a large agribusiness firm had access to real-time information from 1,000 combines randomly spread across the Corn Belt, that information would be extremely valuable to traders dealing in agricultural futures. Traders have traditionally relied on private surveys and USDA yield data. These yield estimates are neither timely nor necessarily accurate. But now, real-time yield data is available to whoever controls those databases. Virtually every company says it will never share, sell or use the data in a market distorting way – but we would rather verify than trust.

One of the most important issues around 'big data' goes directly to property rights and 'who owns and controls the data'. The risks to privacy that the farmer faces, such as his pesticide or GMO usage that may be an accepted practice but politically unpopular, are great.

In addition, farmers' information is valuable to the companies, so farmers should have a say in and be compensated when their data is sold. Farmers need to protect their data and make sure that they bargain wisely as they share their data with suppliers and companies who desire access to their information.

Farmers are rightly concerned about data privacy. Even if an individual operator does everything to the best of his ability, following all the applicable rules, regulations and best management practices, there is still concern that the EPA or one of the numerous environmental organisations that bedevil agriculture might gain access to individual farm data through subpoenas or an overall-clad Edward Snowden.

9.5 Encouraging local adoption and use – the FOG case

As the World Economic Forum [5] pointed out in the 2014 report – *Delivering Digital Infrastructure: Advancing the Internet Economy* – digital ecosystems that produce local content and apps are vital for building digital literacy, attracting local users and serving local needs. Digital services can be a big step towards addressing local problems and boosting competition in an increasingly international digital services market. In addition, using the Internet can have a large impact on local businesses, especially SMEs. Internet awareness and relevant digital content obviously have a symbiotic relationship; an increase or improvement in one will help drive an increase or improvement in the other. In developed markets, where factors such as infrastructure and cost are minimal constraints, content and usage have become a double-barrelled growth engine. To reach the goal of global connectivity, the problem of relevance as it relates to awareness and language must be addressed. The public sector, private sector and civil society can encourage adoption and use by facilitating local content development and putting policies in place that make it easier for businesses, especially SMEs, to benefit from digital technology, as shown in Figure 9.14.

9.6 Conclusion – the WHERE

QUHOMA is now in its MVP commercial phase which targets farmers and mentors as its primary revenues' channels. Continuous technical developments are being undertaken as well as additional business cases are included. The project has raised much attention from academic and market area while it is considered among the most promising and successful FI BUSINESS ideas. It has also been presented in prestigious Greek business events for identifying IoT business opportunities and modern tools for agriculture and it has raised awareness during some of the most profound European conferences of the agriculture domain (e.g. Berlin's Fruit Logistica 2016) as well as participated in FAO and EIP-AGRI closed meetings (more info at quhoma.com/news). Strategic partnerships with respective players of both demand and supply side are being finalised, whereas respective envisioned versions are already been validated so that QUHOMA keeps its leading role in the cutting edge Smart Farming wave of innovation. Interesting partners can always subscribe to http://quhoma.com/login/ to join QUHOMA community [10] and be continuously updated on the project's advancements. A project that was grasped, developed, implemented and disseminated in the middle of Greece's debt crisis since 2015 by FINT, a Greek SME with expertise in advanced Telecom Engineering services.

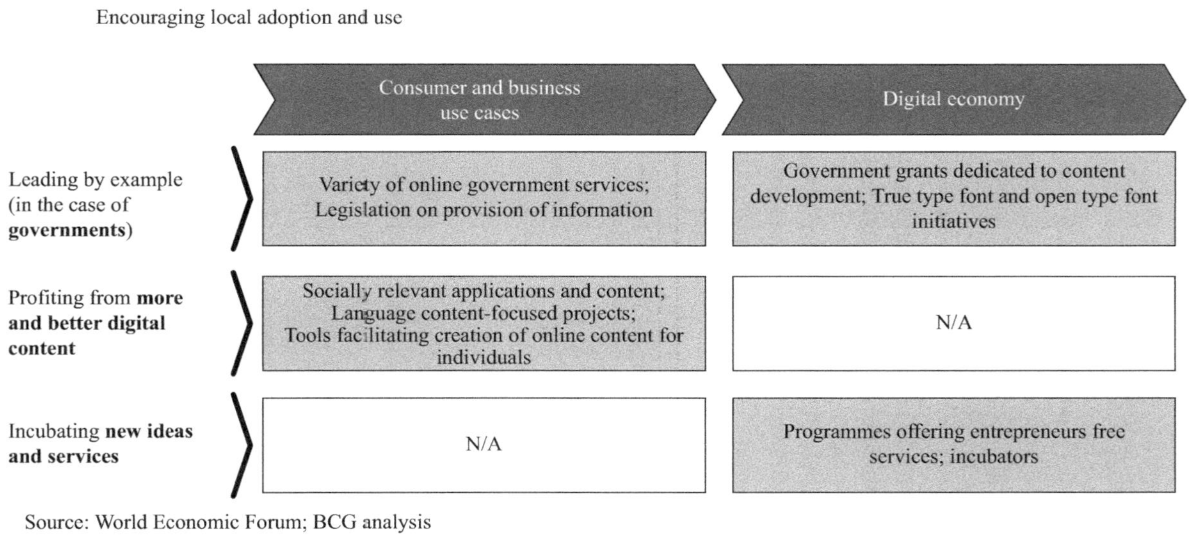

Figure 9.14 Local adoption and use (5)

Acknowledgement

This material is based upon work partially supported by Finish Accelerator, Future Internet Architecture of the European Community (FIWARE) and has received a sub-grant from the European Union Seventh Framework Programme (FP7/2007-2013) under grant agreement no. 632 857.

References

[1] IEEE (2007) IEEE Std 1451.0 TM-2007, IEEE Standard for a Smart Transducer Interface for Sensors and Actuators—Common Functions, Communication Protocols, and Transducer Electronic Data Sheet (TEDS) Formats

[2] FIWARE (2016) Architecture of Applications, Services and Data Delivery Manual, available at https://forge.fiware.org/plugins/mediawiki/wiki/fiware/index.php/Architecture_of_Applications,_Services_and_Data_Delivery, accessed 29/5/2016

[3] AIOTI (2015) WG06: Report on Smart Farming and Food Safety Internet of Things Applications, available at https://ec.europa.eu/digital-agenda/en/news/aioti-recommendations-future-collaborative-work-context-internet-things-focus-area-horizon-2020

[4] Circular Economy (2015), https://s3-eu-west-1.amazonaws.com/europarl/circular_economy/circular_economy_en.svg in http://www.europarl.europa.eu/news/en/news-room/20150701STO72956/circular-economy-the-importance-of-re-using-products-and-materials, accessed 29/5/2016

[5] WEF (2016) Internet for All Framework Accelerating Internet Access Adoption, available at http://www3.weforum.org/docs/WEF_Internet_for_All_Framework_Accelerating_Internet_Access_Adoption_report_2016.pdf, accessed 21/5/2016

Online sources

[6] https://www.fiware.org/ accessed 24/5/2016

[7] http://quhoma.com/ accessed 24/5/2016

[8] FINoT Platform, *Future Intelligence*, available at: http://www.f-in.gr/products/finot-platform accessed 24/4/2016

[9] http://www.tuvaustriahellas.gr/ accessed 24/4/2016

[10] http://ec.europa.eu/environment/circular-economy/index_en.htm accessed 9/4/2016

[11] https://microsite.sonnenbatterie.de/de/home/ accessed 30/1/2016

[12] http://www.efsa.europa.eu/ accessed 22/5/2016

[13] https://enrd.ec.europa.eu/en/leader accessed 30/3/2016

[14] http://s3platform.jrc.ec.europa.eu/, accessed 25/3/2016
[15] http://www.fao.org/home/en/ accessed 9/4/2016
[16] http://www.ifoam.bio/en accessed 22/5/2016
[17] Privacy and Security Principles for Farm Data, *Farm Bureau*, available at: http://www.fb.org/issues/technology/data-privacy/privacy-and-security-principles-for-farm-data accessed 24/5/2016

Chapter 10

Stable real-time video distribution by exploiting cloud and peer-to-peer interaction

Maria Efthymiopoulou[1] *and Nikolaos Efthymiopoulos*[1]

10.1 Introduction

Real-time video distribution over internet has already become enormously popular. Real-time video distribution is a continuous evolving and growing application because users of it increase their presence and because of the extraordinary growth of network technologies. In the future, similar application will have as its requirement to distribute video content with high playback rate with a way that will be able to cope up with dynamic and heterogeneous network environments. The rapid, reliable, and efficient transmission of the video content consist of the core of the problem.

The first approach with which took place the distribution of video through the network was the client-server model. In this architectural approach, a client (user) interacts with a server. Video is sent to the participating clients from the former. The next approach was content delivery networks where a server initially forwards video to a set of servers that have the role to deliver the content (cloud). Users acquire the video by contacting one of the aforementioned servers. Despite the fact the applications that provide live streaming (LS) have become popular; they challenge video servers, and they highly stress on internet traffic. Scalability problems make the video streaming solution that is based in client-server architecture expensive. This is the reason that constituted peer to peer (P2P) networks attractive and motivate research community and industry to do research on them. They provide an alternative solution towards video streaming because they have low cost, and they are scalable. A major advantage in P2P is that each peer, which takes part in the system, puts its bandwidth (BW) and processing resources to the distribution process. In more detail in a P2P, LS system peers not only acquire data from the network, but they also forward (by exploiting their upload capabilities) data to other peers in the system. In this way, the upload BW of end peers is used in an efficient way and minimizes the BW load, which would be put on the servers in the other case.

[1]*Department of Electrical and Computer Engineering, University of Patras, Patras, Greece*

In more detail, the video content could have a stable or dynamic *playback bit rate* measured in bits per second (bps). The sum of the upload BW that servers and peers contribute divided by the number of the participating peers is the average available distribution bit rate and introduces a physical constraint that represents the maximum bit rate that the system is able to provide to its users irrelevant of its architecture. Every video has a fixed size noted as the *video size*. Every second of each video is divided into a set of data objects that noted as *video blocks*. Video blocks are the fundamental entity that is exchanged among participating users. Their size, measured in bits, noted as video block size, and the frequency in which the system has to deliver those represents the *distribution block frequency*. Each user maintains a data structure named video block buffer, which holds the state of each block. Two states are of interest: received blocks and missing blocks (blocks that have not been delivered yet to it). Every video block is correlated with a playback deadline that is defined as the time instant which the video player will consume this block. A *buffer bitmap* is the buffer status of a user at any given time instant. Usually received blocks noted with 1 and missing with 0. It is said that any two user have overlapped buffers only if they have one or more common blocks into their buffers. As *content bottleneck* defined the state in which a user does not have sufficient number of useful blocks (as useful defined a block that a user has, whereas other users have not) to exchange.

According to the fragmentation of distributed video into video blocks and its distribution policy, systems can be categorized into two categories. The first, which due to its simplicity is more widespread today, is the distributed *video encoding agnostic* category, in which is not exploited the media encoding format. The second is distributed *video encoding aware* category in which distribution architectures and transmission policies are able to exploit the encoding mechanism of the distributed videos. Another important architectural decision is if the video blocks will be sent encoded or unencoded. The unencoded approach prevents and reconstructs from errors, although it induces more overhead compared to the first approach.

Latency is the time interval between the generation of a video block and its playback. Startup latency or *startup delay* is the time period starting from the instant that a user joins the system to the instant that the user starts playing out the video after buffering a few seconds of it. Startup delay is strictly correlated with user's buffering level; the number of consecutive blocks should be received and cached in user's buffer. The *prebuffering* time is a trade-off between having a short waiting period and having low block loss during playback. This makes the prebuffering period an important metric.

Every user, from system's perspective, can be described from its upload and its download BW. Download BW expresses the rate in which a user (peer) can download blocks from other users or the server, whereas the upload BW expresses the rate in which the user (peer) uploads blocks to other users. User's capacity represents the total number of blocks that a user can upload to other users in a certain time interval and correlated with user's upload BW and the video playback rate.

Users exchange video blocks each other or with the server in discrete time called round. So, as throughput is defined the total number of blocks exchanges per round. It is for a metric that relates to the utilization of the system resources. As a consequence, high throughput implies high utilization of system's resources in terms of upload BW that used for block diffusion. While, user's throughput is the amount of information that downloaded by a user in a unit of time. As *goodput* expressed the bit rate in which a user is able to fill with video blocks its buffer. Goodput determines the maximum bit rate that the system is able to achieve full playback continuity. As continuity index, is defined the ratio of video blocks that arrive before their playback deadline over the total number of arrived video blocks. High continuity index ensures an uninterrupted video playback.

Users' different arrival patterns and their duration into the system constitute user behavior. A *flash crowd* is a situation in which a large set of users join the system suddenly for a brief time. As *peer churn* is defined the phenomenon in which users dynamically and suddenly leave the system. Peer churn introduces dynamics and uncertainty into the P2P network and maybe degrades the viewing quality of the remaining users. Although it is expected that a user leaves the system as soon as the video finishes, early departures occur. As early departure is defined the phenomenon in which a user can leave at any time without advance notice.

Finally, the server stress is the amount of BW that allocated from the server. It is an indicator of how congested (on average) the video server is. While the server capacity is the maximum BW provisioned from the server to all peers and consequently expresses the number or requests served at each round. Especially, server's service rate could be defined as server's upload BW divided by video playback rate. Server stress is strictly correlated with users' upload BW utilization and with its buffer size (larger buffer implies higher probability of exchanges among users). In scalable P2P LS systems, server stress is very important as it correlates the architecture of the system with its business model and the financial requirements that LS service has.

In a nutshell, this chapter presents a P2P live video streaming system that is scalable and stable. The proposed system is able to guarantee the complete and on time video distribution to every participating peer based on the three aforementioned strategies. The contribution of this chapter is summarized to the development of these strategies with respect to the aforementioned P2P LS requirements. The rest of this chapter is structured as follows. Section 10.2 analyzes our system's architecture. Section 10.3 presents the playback rate adaptation strategy. Section 10.4 analyzes the provision of quality of service (QoS) through cloud assistance, whereas Section 10.5 analyzes the provision of QoS through other peer's assistance. Finally, in Section 10.6 we conclude and we give some hints on our future work.

10.2 System's requirements and architecture

As analyzed earlier, video distribution over internet has already become enormously popular. It is a "killer" application due to users' growing demand and extraordinary growth of network technologies. In the future, internet will be able to deliver video of high quality in a way that will be efficient and personalized through very dynamic and heterogeneous network conditions.

With P2P networks service providers avoid and these systems are also very promising in terms of scalability. In this way, industry and researchers consider these architectures a promising approach that could be adopted in the future. The main advantage in the exploitation of P2P architectures is that the peers that take part in the delivery of the video put their own BW, storage, and processing resources in order to self-organize the distribution of the multimedia content. A P2P LS allows peers not only to acquire media blocks from the system by using their download BW, but also use their upload BW in order to send media blocks to other peers that participate in the system. Thus, the upload BW of participating peers is utilized in an optimal way and in this way is minimized the BW that is needed from the media servers towards the complete and on time distribution.

As analyzed in the aforementioned paragraph, LS requires that the participating peers will use each piece of the video in the same time. In order to achieve this, they have to download the video with a rare that approximates its playback rate. Thus, a vital objective in these systems is the distribution of the audiovisual content in a way in which all participating peers acquire all the media blocks before their playback deadlines. In addition, each peer has to acquire each media block only one time in order to avoid the waste of resources. By trying to annotate all the above, and without harm of generality, the main requirements from a live video streaming system are as follows:

- *Efficiency* of the scheduling of the distribution of the media blocks in a way that maximizes the upload BW that is used from the peers. This is the way to minimize the BW that will be needed from the media servers towards the complete distribution and/or the distribution with the maximum possible playback rate. In other words, efficiency determines the trade-off between BW, the BW that media servers put and the playback rate that the P2P architecture is able to deliver.
- *Stability* is determined as the robustness of the distribution architecture to continue to distribute the video in a time effective and complete way in the case of disturbances. As disturbances can be considered: (i) the changes in the congestion of the underlying network, (ii) the entrance of new peers in the system, (iii) the case in which a set of peers leave the system, (iv) the change in the BW of peers, and others. These disturbances have a temporal but significant impact in the stability and robustness of the system. As a consequence, it is triggered degradation in the quality of LS service (QoS).
- *Scalability* is the correlation of the BW, processing, and storage amount that media servers have to contribute with the number of participating peers. In order to design a system that is scalable, there is the need of distributed algorithms in case that it is possible and the design of low overhead algorithms in case that functionality has to be centralized. In order to have a system that is scalable, there is a need for low overhead especially in cases when the number of the users (peers) in the system is high.

More analytically in P2P real-time distribution systems, all the users download the video with a rare one that approximates its playback rate. Thus, one of the most important requirements from such systems is the timely distribution of the media block to every user in a way that each user will acquire each media block only one time in order to avoid the waste of BW resources.

Every system should be efficient, reliable, and secure. The efficiency, as analyzed, implies the trade-off between the fulfillment of the needs of service consumers and resource costs of the service and network provider. As costs assumed the upload BW that the service provider contributes and the traffic introduced to the network during the provision of the service. The exploitation of the peers upload BW and the ability of P2P systems to adapt their distribution paths to the topology of the underlying network act as motivations towards the selection of P2P architectures for the provision of streaming services. As reliability we define the ability of the system to provide stable and uninterrupted services. The factors that could affect the reliable functionality of the system are peers arrivals and departures, flash crowds, insufficient total upload BW, sudden faults in network paths, and faults in components of the service provider. A successful streaming system could be assumed reliable if it is able to prevent all these scenarios and handle them efficiently in case. Finally, security is a well-known research problem out of the scope of our work, thus will not be analyzed further.

Service provider desires the system to be scalable despite the large number of users and the unpredicted arrival pattern. In more detail, it will have low cost in terms of management overhead and so distributed algorithms towards a self-managed system constitute one of the system requirements. In addition, it requires the minimization of the storage and BW resources that it has to contribute as the system grows in terms of participating peers. These act as a motivation behind the selection of P2P architecture towards the provision of a streaming service.

Moreover, the whole implementation should be fault tolerant in different scenarios could be caused either from the system or from the user perspective (such as peers' behavior, server failures, etc.). The system should be robust to peers' dynamics and links failures by offering uninterrupted video playback. Also another issue that needs to be regulated is to avoid free riding. Peers require a minimal download speed to sustain playback, and so free-riding is especially harmful as the altruistic peers alone may not be able to provide all the peers with sufficient download speeds. So, the system should give incentives to participating peers to maximize the upload BW that they contribute.

User satisfaction is correlated with four factors. The quality of the video that the service is able to deliver, the latency that the service introduces, the uninterrupted video playback rate that often referred to as playback continuity and the functionalities that the service offers (e.g. Video storage, Video seeking, etc.). Average available distribution bit rate is strictly correlated with video quality that users enjoy. Users require as high as possible average available distribution bit rate and a sophisticated video encoding protocol that will maximize the quality of experience (QoE) or streaming quality that the service provides. Thus, QoE implies an uninterrupted video playback in high definition quality of video. In technical terms, this determined as a peer should have a buffering level of more than 80% of the total size of its playback buffer (by this way we can also calculate the percentage of high quality peers).

Participating users desire to start consuming the distribution objects as soon as possible after their request. Thus, the startup delay should be as small as possible. In some systems, each peer has different startup delay and in some others it is a system parameter. As mentioned earlier in system requirements, reliability is a crucial factor. From user's perspective reliability could be translated as playback continuity. This requirement is satisfied if video blocks received before their playback deadlines and if all peers can receive the streamed video at the desirable distribution bit rate. Playback continuity and startup delay can also be assumed as performance metrics.

Major goal from network provider perspective is to minimize the traffic arises from peers and server contributions. That implies, the lower the traffic the lower the cost. Traffic within the network has to be minimized. Obviously, the network should have as little overhead traffic as possible. Control messages have to be as few as possible.

According to the recent progress of the researchers, we have three ways towards the harmonization between the playback rate and the dynamic conditions that occur in the BW of the participating users. The first [1] can be expressed as the dynamic adjustment of the playback rate by taking dynamically as input the available upload BW of the users that are in the system each time instant. The next one is the dynamic control of upload BW from media servers (e.g. clouds). The final (third one) [2] is relevant with the dynamic provision of upload BW form other users that they do not need temporarily. This solution does not require the existence of media servers from a BW provider or a cloud. In order to choose the appropriate way to stabilize the distribution, we have to examine the desired QoE of the users that participate in the system and the use case that the system will be used. In an example use case in which we desire a system that does not cost the first way is more appropriate. Alternatively in an example use case in which we desire a system with high playback rate and very high fault tolerance, the second way is more appropriate. Furthermore, in a use case in which the cost and the video quality are important, the third way is more appropriate [3].

Our P2P architecture, as most of the existing architectures, has a set of media server(s) in a cloud (noted by *S*) and a set of peers (noted by *N*). The servers, *S*, are handling (i) the initial diffusion of the video to a small subset of nodes among participating peers, (ii) the tracking of the network addresses of participating peers in order to assist the construction and the management of the P2P overlay, (iii) the dynamic and scalable monitor of resources of participating peers, and (iv) the dynamic allocation and release of auxiliary BW. The rest of the functionalities, which we will describe below, take place in each peer in *N*. Figure 10.1 illustrates our proposed P2P LS system's architecture.

The video stream, which the system disseminates, is divided into video blocks. In order to allow peers to exchange video blocks, each peer maintains network connections with a small subset of other peers that will be noted as its neighbors. The sets of these connections change dynamically and form a dynamic graph called the P2P overlay that is self-organized (no centralized management from *S*). In our previous work [4], we present a graph topology and P2P overlay management

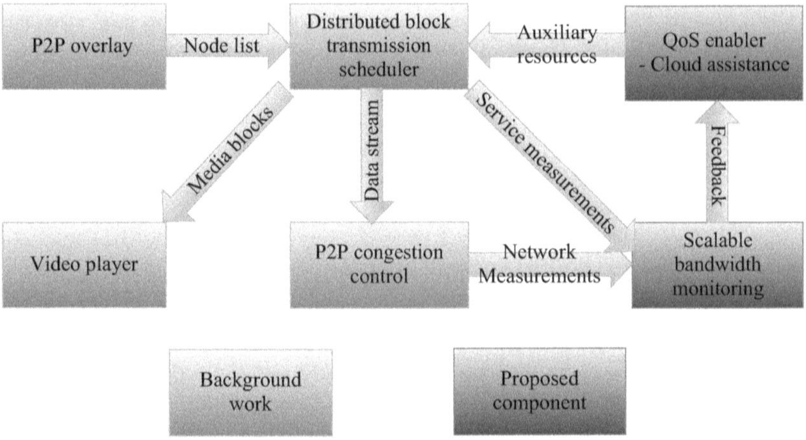

Figure 10.1 Proposed P2P live streaming system's architecture

(dynamic and distributed optimization) algorithms, which peers in N periodically execute, which result in the dynamic reconfiguration of the P2P overlay. We use distributed optimization theory in order to dynamically ensure in a distributed (scalable) and dynamic fashion that (i) peers have connections proportional with their upload BW, (ii) peers have connections with other peers close to the underlying network, and (iii) our P2P overlay is adaptable to underlying network changes and peer arrivals and departures. As the reader may observe in [4–7], this allows us to efficiently exploit all the available BW resources even if they are highly heterogeneous.

The dynamic output of the P2P overlay management algorithms that run in each participating peer is a neighbor list that is passed to the distributed block transmission scheduler (DBTS).

The DBTS coordinates block exchanges. In order to achieve this it has a set of algorithms executed by every peer in N, which dynamically communicates with its neighbors. The major objective of DBTS is to ensure the timely delivery of each video block to every peer in N by exploiting the upload BW of participating peers and the additional BW resources that S may contribute. In more detail each peer periodically sends to its neighbors control messages that encapsulate information about video blocks that it owns. Thus, periodically each peer (through a matching algorithm) is able to request from each one of its neighbors a different video block or nothing if there is no video block to request. In order to perform the requests, a matching algorithm is executed periodically by each peer and its objective is to request as many unique blocks as possible. These requests are served sequentially by each peer who prioritizes them by selecting each time its most deprived neighbor to serve its block request. As most deprived is defined the neighbor that has the smallest total number of blocks compared to the video blocks that the sender peer owns. Our proposed DBTS is analyzed in detail in our previous works [4,5].

The DBTS sends the video blocks, which have to be sent, in the P2P congestion control component and the ordered stream with the blocks, that it receives, to the video player, that each peer in *N* has.

Our proposed P2P overlay and our DBTS enhance our P2P LS system with two properties that we will exploit in order to create our proposed components. The first property (Property 1) is that if idle BW exists it is derived from BW surplus in the system and not from the inefficiency of the system to exploit it. In other words, we guarantee that the presence of idle BW implies (testifies) the complete stream delivery. The second property is that the percentages of the idle resources among participating peers are almost equal (Property 2). We highlight here that in the case of heterogeneous peer upload BW, various peers send with various bitrates (analog with their upload BW capacity), but the percentage of their BW utilization, and consequently the percentage of their idle time, is very similar.

Our P2P congestion control [8] is totally distributed and executed in each peer in *N* and is able to manage sequential transmissions of video blocks to multiple locations that DBTS sends to it and to provide to the scalable bandwidth monitoring (SBM) the dynamic estimation of: (i) the upload BW capacity, and (ii) the idle BW resources of each participating peer with the way that will be requested from the latter.

In the rest of this chapter, by exploiting the features of the aforementioned components (background work), we develop two new components. We note the first as SBM, in which a scalable gossip protocol is executed in each peer in *N* and is connected with a centralized component in *S*. In more detail, it (i) aggregate the monitoring information from DBTS and P2P congestion control, and (ii) forms all the required metrics that the bandwidth allocation control (BAC) needs.

The BAC, which is noted as QoS Enabler in Figure 10.1, is the second proposed component that is executed exclusively in *S*, and its purpose is to calculate dynamically the amount of total system's upload BW surplus or deficit towards the control of the idle BW resources or to adapt the playback rate towards this goal.

10.3 Quality of service through playback rate adaptation

In order to summarize this section, we analyze here a way that enriches real-time P2P media distribution with stability by adjusting in a dynamic fashion the playback rate, and in this way it ensures that every user in the system will acquire every media block even in cases where the dynamic BW of the system varies in a very "strong" manner. Towards the achievement of this goal, we created a scalable system that monitors dynamically the BW of a very small subset of users in the system and a functionality that dynamically adjusts playback rate according to the available BW of the users. By using discrete time differential equations is expressed analytically the correlation among the playback rate and the dynamic average upload BW of the participated users. Thus, it is feasible to exploit control theory in order to solve this problem. In addition, inaccuracies in the process of BW monitoring are taken into account, and in this way we are in a position to model analytically, and in a dynamic way, the aggregate BW that remains idle.

More analytically we created (i) a monitoring functionality of the existing BW, and (ii) a playback rate adjustment functionality that

- Offers a scalable monitoring, that achieves to monitor the average BW of the participating users by introducing very small BW monitoring towards this process. In order to achieve this, we exploit the properties the P2P overlay and the DBTS that we have developed.
- Guarantees the timely distribution of all the media blocks and even in cases where the conditions are very dynamic. In more detail, it is grounded on a theoretical model that we developed in order to use control theory towards guaranteed results.
- It exploits the BW of the users that are in the system. In more detail, we find dynamically the largest playback rate that we are capable of distributing, and we correlate it with the accuracy of the BW monitoring algorithm and the maximum magnitude of BW disturbances.

10.3.1 Problem statement

A set of peers, in which we refer to as N, acquire the same video. The objective is to design a system that is able to allow the requests of the peers that participate in N to be fulfilled from a tiny portion of nodes in N which we call as their neighbors. The bit rate of these requests is p_k, and it is equal with the playback rate of the video. k is a positive number and denotes the time instant. The requests of the peers that the system addresses are the incoming flows. The requests that peers announce are taken care by their selves though the use of their upload BW. These are the out-going flows of the system. P2P congestion control uses these flows and in this way generates a passive monitoring that estimates in a dynamic fashion: (i) upload BW, $u_{(i)k}$, of users in the system, and (ii) idle percentage of the upload BW, $id_{(i)k}$, of the users that participate in the system and described earlier with N. In the rest of this subsection, we analyze the issues that we addressed towards dynamic playback rate adaptation which are as follows:

1. The creation of an analytical model that correlates, in a dynamic fashion the playback rate with the idle BW of the users in N.
2. The use of the features that our P2P overlay and DBTS have towards the creation of a monitoring system. It works in a scalable way and is able to calculate in an online and accurate fashion the total idle resources. It introduces very low network aggregated overhead by communicating only with a small number of users which is negligible when compared to the total number N.
3. The development of a control strategy that uses the aforementioned analytical model and controls $id_{(i)k}$ of users in N to a point id_{REF} that the system administrator desires. In order to achieve this, it changes dynamically p_k. In this way, it is guaranteed that the system works below its limits by delivering every block to every user and furthermore it does this on time.
4. The analytic determination of the lower value of id_{REF} that is able to ensure the uninterrupted delivery of the video even in cases that (i) there are inaccuracies

in the system model and (ii) there are dynamic disturbances of in the BW conditions (fluctuations in $u_{(i)k}$ of users that are in N). This is how the proposed architecture works in the largest feasible playback rate and ensures the delivery of the video in it.

5. The determination of the largest $id_{(i)k}$ towards the prediction of the largest possible amount of BW resources that the system wastes. These are correlated with the inaccuracies that our equations introduce and the changes that occur in the BW conditions of the system.

The next section analyzes how we model the aforementioned phenomena with equation of differences and the design methodology of the system, more details are explained in [1] that have to do mainly with evaluation issues.

10.3.2 Modeling and controller design

The core control unit of the system noted as playback rate control (PRC). PRC is executed in a periodic fashion (period T). A media server (S) acquires all the data that we need for its execution and PRC instantiated with a centralized architecture. Preferably S could be the same server that encodes the video and prepares it for the distribution. The goal of PRC is to control idle, $id_{(i)k}$ of users in N to the value id_{REF}. In order to have this, it controls its input p_k. In the following lines is explained this activity which has been modeled with equations of differences and the control strategy that selects with a period T p_k. There are several symbols that we use in order to describe our model. These have been gathered together and shown in Table 10.1. The index i represents the user id (brackets), and k is the index that represents an instant of time.

Table 10.1 Notation

Symbol	Definition
S	Generator (source) of the media object
N	Set of participating peers (in the equations below N is used as the number of participating peers)
p_k	Media playback rate at time instant k
$u_{(i)k}$	Upload capacity (upper limit) of peer i at time instant k
$id_{(i)k}$	Idle time percentage of peer i that at time instant k between 0 and 1
id_k	Average estimated idle time percentage of N peers at time instant k between 0 and 1
id_{REF}	Average idle time percentage reference value that is between 0 and 1
T	Period of execution of PRC
w_k	System input that represents the change in the playback rate that is determined from PRC
w_{REF}	System input in the equilibrium point
$\delta 1$	Percentage of modeling and monitoring inaccuracies. It belongs to $(-\delta 1_M, \delta 1_M)$
$\delta 2$	Percentage of average upload bandwidth change. It belongs to $(-\delta 2_M, \delta 2_M)$

Towards a smooth presentation of the model, we have two assumptions. These two assumptions could be broken as we do in [1]. In more detail, we have the following:

- *Assumption 1*: According to *Property 2*, as analyzed in the previous section, we have $id_{(i)k} = id_k$ for each user i that is in N. With id_k we express the average $id_{(i)k}$ of users N. Thus, according to this, we can estimate id_k if we probe only one user in N, and we acquire only one $id_{(i)k}$.
- *Assumption 2*: T is the time interval between two consecutive executions of PRC, and it has to be less than the time that usually takes to have important changes in the BW conditions of users in N. According to this, we can assume that the aggregated BW of the system is approximately identical during this period T.

If the system (users in N) has enough BW, our proposed overlay and scheduler (see Section 10.2) are able to ensure that the stream will be given to every user in N and thus the incoming flow of every user i is p_k. Thus, the aggregated incoming flow of users in N is Np_k. In addition, the total amount of incoming flows that users have is the same with the aggregated outgoing flows that users send. The sum of outgoing flows is the sum of their nonidle upload capacity $u_{(i)k}$. By examining the aforementioned and by keeping *Property 1* as analyzed in the previous section, we have the following:

$$Np_k = \sum_{i \in N} \left(1 - id_{(i)k}\right) u_{(i)k} \tag{10.1}$$

Assumption 1 leads us to rewrite (10.1) as

$$Np_k = (1 - id_k) \sum_{i \in N} u_{(i)k} \tag{10.2}$$

Rewriting (10.2) for time instant $k+1$, it exists:

$$Np_{k+1} = (1 - id_{k+1}) \sum_{i \in N} u_{(i)k+1} \tag{10.3}$$

Now, by dividing (10.2), (10.3), under *Assumption 2* holds that

$$(1 - id_{k+1})p_k = p_{k+1}(1 - id_k) \tag{10.4}$$

By definition, at time instant $k+1$, the dynamic playback rate, p_k, is expressed as the sum of the playback rate at time instant p_k and w_k. Thus, holds that

$$p_{k+1} = p_k + w_k \tag{10.5}$$

By combining (10.4) and (10.5), it raised that

$$id_{k+1}p_k = (p_k + w_k)id_k - w_k \tag{10.6}$$

Setting $id_k = id_{k+1} = id_{REF}$ in (10.6) is obtained w_{REF}, which can be defined as the input in the equilibrium point and is equal to 0. Thus, the equilibrium point is (id_{REF},0). In order to have a system which has as its equilibrium point (0,0) let's set the following:

$$x_k = id_k - id_{REF} \tag{10.7}$$

The idle time percentage, id_k, belongs to the interval (0,1) by definition. Thus, x_k ranges between $(-id_{REF}, 1 - id_{REF})$. By substituting (10.7) for $k=k$ and for $k=k+1$ in (10.6), it holds that

$$x_{k+1} = x_k + \left(\frac{x_k + (id_{REF} - 1)}{p_k}\right) w_k \tag{10.8}$$

Equation (10.8) is nonlinear. In order to have a linear closed loop system is selected a feedback linearization control strategy [9]. Feedback linearization is a strategy that introduces a state feedback such that the closed loop system becomes linear. To this end is selected a control strategy $w_k(x_k, p_k)$ of the form:

$$w_k = \frac{p_k}{x_k + (id_{REF} - 1)}(k_c - 1)x_k \tag{10.9}$$

In (10.9), k_c is a parameter that can be chosen. By combining now (10.8) and (10.9) is arising a linear system with eigenvalue k_c, which is

$$x_{k+1} = k_c x_k \tag{10.10}$$

In this way is easy to see from (10.10) that the series $\{x_k\}$ converges to 0, and so id_k to id_{REF}, for any value of k_c that belongs to $(-1,1)$.

Since k_c is a designer's choice, the eigenvalue of the system can be explicitly defined by just setting k_c. The implementation of the proposed control strategy is

$$w_k = \frac{p_k}{id_k - 1}(k_c - 1)(id_k - id_{REF}) \tag{10.11}$$

10.4 Quality of service through cloud assistance

As we explain in the first section, an alternative way to harmonize the relationship between playback rate and the aggregated upload BW is to provide upload BW in a dynamic fashion through auxiliary sources. Media servers that could be in clouds could be considered as an auxiliary source towards QoS. In the rest of this section, we see how we can achieve this in an effective way.

More specifically, it is analyzed here a P2P live video streaming system that is reinforced with BW from media servers (cloud). In this way, we have system with high degrees of scalability and stability. In order to achieve this, we created (i) a gossip protocol that has minimum overhead in terms of BW that it consumes while it simultaneously achieves to monitor the aggregated upload BW that is available and (ii) a control strategy which (i) in case that there is lack of BW allocates in a dynamic fashion the exact quantity that is needed in order to ensure the distribution of the video in a time effective and complete way and (ii) in the case of the existence of more BW than the necessary monitors the surplus and holds it towards its exploitation in the distribution of other videos.

Towards these goals, we model with equations of differences, and by embedding the system dynamics in our model, the correlation among the aggregated BW that peers have (deficit–surplus) and the video playback rate. Thus, we are able to exploit the tools that control theory features and the aggregated BW that is required is controlled. More analytically, it is designed a BW monitoring allocation and control architecture that

- Features an aggregated BW monitoring architecture that has important properties as scalability with respect to the BW overhead that it introduces and fault tolerance with respect to the behavior of the system users (arrivals/departures). Towards this goal we exploit (i) the attributes (balance of the idle percentage) of our proposed scheduler as analyzed in Section 10.2 and (ii) a P2P LS aware gossip protocol that is analyzed in detail in Section 10.4.2.
- Guarantees the delivery of the video in situations in which the BW changes dynamically. In these cases, the analytical model that we propose considers these changes (underlying network and peer behavior) and adapts to them. In this way, P2P live video streaming uses control theory and ensures QoS.
- It utilizes efficiently the upload BW of participating peers without sacrificing the uninterrupted distribution of the stream. In specific, is calculated analytically (by proposing an innovative nonlinear model) the minimum amount of BW overprovision that ensures the successful distribution of the stream as a function of the accuracy of the measurements and the maximum possible disturbance in total available upload BW.

10.4.1 Problem statement

A set of peers, in which we refer to as N, acquire the same video. The objective is to design a system that is able to allow the requests of the peers that participate in N to be fulfilled from a tiny portion of nodes in N that we call as their neighbors. The bit rate of these requests is p_k, and it is equal to the playback rate of the video. k is a positive number and denotes the time instant. The requests of the peers that the system addresses are the incoming flows. The requests that peers announce are taken care by their selves though the use of their upload BW. These are the outgoing flows of the system. P2P congestion control uses these flows and in this way generates a passive monitoring that estimates in a dynamic fashion: (i) upload BW, $u_{(i)k}$, of users in the system and (ii) idle percentage of the upload BW, $id_{(i)k}$, of the users that participate in the system and described earlier with N. In the rest of this subsection, we analyze the issues that we addressed towards dynamic BW allocation.

The first problem that tackled with is the exploitation of the properties of the proposed system towards the creation of SBM, through a distributed gossip protocol. This will allow us to calculate in a dynamic fashion and accurately the idle resources of the whole system under a scalable and fault tolerant way. The second problem is the creation of the equation of differences (BAC model) that correlates dynamically the BW that should be dynamically allocated or released with the idle BW of the users that take place in the distribution of the video. The third problem that solved is the creation of a BAC strategy through which is exploited the proposed analytical model in order to control $id_{(i)k}$ of each participating peer in N to a reference value id_{REF}. This can be achieved by adapting dynamically, through the use of auxiliary resources (cloud), system's total upload BW. By this way, is allowed to the proposed system to ensure the on time distribution of every video block to every participating peer by using exactly as upload BW as needed for the distribution. Thus, if the total uploads BW of participating peers is greater than the required, then is dynamically estimated this surplus in order to be allocated for other purposes (e.g., distribution of another stream). Otherwise, if total system's upload BW is less than the required, then is dynamically estimated the amount of the deficit and is demanded from the cloud S in order to ensure the stability of the distribution. The fourth problem that solved is the analytical calculation of the minimum id_{REF} that guarantees the stable distribution of the stream as a function of the inaccuracies that the proposed model introduces and the disturbances of the system (dynamic changes in $u_{(i)k}$ of peers in N). By this way, is created a robust system that minimizes the overprovision of upload BW, whereas simultaneously it guarantees the distribution of the stream. Finally, is calculated analytically the upper bound of average $id_{(i)k}$ among participating peers, and thus is feasible to predict in advance, the maximum percentage of BW resources that possibly remain idle, quantified again as a function of the inaccuracies that the proposed model introduces and the disturbances of the system. In the rest of this subsection is presented a brief analysis in the modeling and the design of the controller.

10.4.2 Scalable bandwidth monitoring

In this section, is described the SBM that is a gossip protocol whose architecture is depicted in Figure 10.2. The first goal of SBM is to determine in a dynamic and totally distributed (scalable) and fault-tolerant fashion a set L of controller peers (where L is a small subset of N), and its second goal is to aggregate to peers in L all the required information that it must be sent in the cloud S that will execute BAC.

Each controller peer L_j has a double role. First, is responsible for gathering through control messages information that needed in order to allow S to execute BAC. Second, in case that there is surplus of upload BW in the system it is instructed dynamically from BAC to release a fraction of its upload BW in order to be used for other purposes.

In more detail each peer i in N (Figure 10.2 nodes with strong gray color) selects periodically among its neighbors in the P2P overlay the peer (Figure 10.2 nodes with medium gray color) with the highest upload BW, and it considers it as its controller peer L_j (Figure 10.2 nodes with medium gray color). The highest upload BW criterion serves the highest exploitation ratio of the surplus of the upload BW, in case that it will be exploited to facilitate the distribution of other streams.

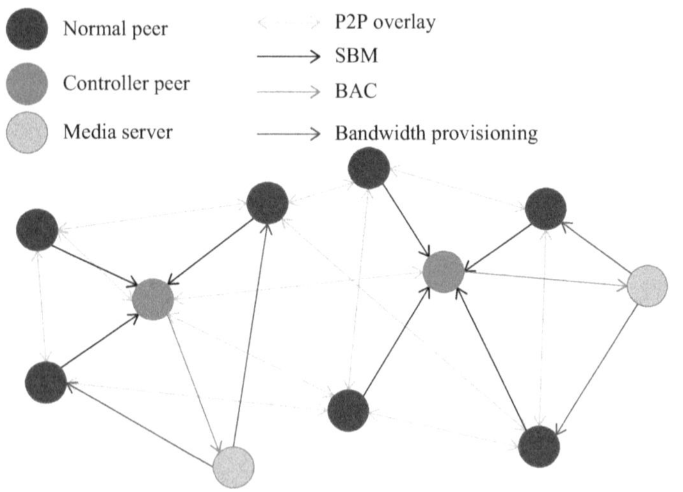

Figure 10.2 BAC monitoring

As a second step each peer i in N forwards periodically, with a period T, to its controller peer two types of control information:

1. Its estimated upload BW capacity, $u_{(i)k}$, for the last T seconds
2. Its idle percentage of its upload BW capacity, $id_{(i)k}$, which is defined as the ratio between the time interval that peer i remains idle during the last T seconds divided by this period T.

For the dynamic and accurate measurement of the upload BW and the idle time of each participating peer is exploited the P2P congestion control algorithm that developed in [8].

Consequently, each controller peer j in L (L_j) acquires periodically from all the peers that selected j as their controller peer (noted from now on as N_{L_j}) the two aforementioned types of information. Then, each controller peer in L forwards the average $id_{(i)k}$ in N_{L_j} and the sum of $u_{(i)k}$ in N_{L_j} to S (cloud—media server with light gray color in Figure 10.2). In this way, S, as will be described in the next section, is able to execute BAC, by exploiting the information that receive from peers in L.

In order to describe the scalability properties of SBM is highlighted that according to the P2P overlay architecture, L is approximately 16 times less than N, and S receives a control message (with two floats) from each peer in L every T seconds (which has a typical value around 5 in the proposed system). Under this analysis, the BW overhead that SBM introduces to S is around 3.2 bps (including UDP-IP headers) multiplied by the size of N (or 3.2 bps/participating peer) and is considered as very low.

Finally, the SBM has satisfactory fault tolerance properties because is executed periodically and in case that a peer in L, L_j leaves the system, the peers in N_{L_j} will select a new controller peer the next period that SBM will be executed.

10.4.3 Bandwidth allocation control

With BAC, we note a process that the system executes in a periodic fashion, (period *T*). The architecture of BAC is centralized. In order to achieve this, we use a media server (*S*) potentially, which creates the video that is distributed. The purpose of it is the adjustment of the idle (%), $id_{(i)k}$, of each user *i* (that belongs to *N*) to a reference value id_{REF}. This is done through the periodic adjustment of $U_{(S)k}$. With $U_{(S)k}$ we define the quantity of the aggregated upload BW that BAC has to add or remove from the P2P LS system at a time *k* in which BAC takes place. This is correlated with the available BW in *N* and the required BW towards the complete and on time distribution of the video. Later in this section, we model this functionality with equations of difference and in this way we quantify $U_{(S)k}$. Towards the calculation of $U_{(S)k}$, we use a variety of variables that we depict/summarize in Table 10.2. In Section 10.3, we gave more information on the formalization of the notation.

Towards a clear, progressive and step-by-step presentation of the system, we did two assumptions that we break, as a next step, and thus we have a more accurate and robust system model. In more detail our assumptions are as follows:

- *Assumption 1*: According to *Property 2*, as analyzed in the previous section, we have $id_{(i)k} = id_k$ for each user *i* that is in *N*. With id_k we express the average $id_{(i)k}$ of users *N*. Thus according to this, we can estimate id_k if we probe only one user in *N*, and we acquire only one $id_{(i)k}$.
- *Assumption 2*: Period *T*, with which BAC is executed, is lower than the time interval that is needed for significant changes in the total upload BW of participating peers. So, it exists the approximation that total upload BW remains the same between two consecutive executions of BAC.

Table 10.2 Notation

Symbol	Definition
S	Generator (source) of the media object
N	Set of participating peers (in the equations below is used N as the number of participating peers)
P	Media playback rate
L	Set of controller peers
L_j	Controller peer j
$U_{(S)k}$	Amount of upload bandwidth that should be added/removed from the P2P overlay at time instant k as it determined from BAC
$u_{(i)k}$	Upload capacity (upper limit) of peer i at time instant k
$id_{(i)k}$	Idle time percentage of peer i at time instant k between 0 and 1
id_k	Average estimated idle time percentage of N peers at time instant k between 0 and 1
id_{REF}	Average idle time percentage reference value that is between 0 and 1
T	Period of execution of BAC
$\delta 1$	Percentage of modeling and monitoring inaccuracies. It belongs to $(-\delta 1_M, \delta 1_M)$
$\delta 2$	Percentage of average upload bandwidth change. It belongs to $(-\delta 2_M, \delta 2_M)$

In a given time, instant k by assuming the case that the aggregated upload BW is enough our proposed graph and scheduler (as they analyzed in Section 10.2) ensure the distribution of the video to all the users in N in case that the incoming flow to each one of them is not less than playback rate p. Consequently, the sum of the incoming flows of N peers is Np. It is self-proven that the aggregated incoming flows that peers receive equal the sum of the outgoing flows of users N. These are derived from the contribution of their upload BW. In addition, the sum of these flows is the sum of their upload BW $u_{(i)k}$ that is not idle. By examining all these and by taking into account *Property 1* as it is analyzed in Section 10.2, we have (10.12) that we demonstrate below:

$$Np = \sum_{i \in N} \left(1 - id_{(i)k}\right) u_{(i)k} \tag{10.12}$$

Under *Assumption 1*, (10.12) can be written as

$$Np = (1 - id_k) \sum_{i \in N} u_{(i)k} \tag{10.13}$$

Rewriting (10.13) for time instant $k + 1$, it holds that

$$Np = (1 - id_{k+1}) \sum_{i \in N} u_{(i)k+1} \tag{10.14}$$

By definition, at time instant $k + 1$, total system's upload BW resources, can be expressed as the sum of total system's upload BW resources at time instant k plus $U_{(S)k}$. Thus, holds that:

$$\sum_{i \in N} u_{(i)k+1} = \sum_{i \in N} u_{(i)k} + U_{(S)k} \tag{10.15}$$

By combining (10.14), (10.15) it arises that

$$Np = (1 - id_{k+1}) \left(\sum_{i \in N} u_{(i)k} + U_{(S)k} \right) \tag{10.16}$$

Now by dividing (10.13), (10.16) under *Assumption 2* it holds that

$$id_{k+1} = 1 + \frac{(id_k - 1) \sum_{i \in N} u_{(i)k}}{\sum_{i \in N} u_{(i)k} + U_{(S)k}} \tag{10.17}$$

By setting,

$$q_k = \frac{\sum_{i \in N} u_{(i)k}}{\sum_{i \in N} u_{(i)k} + U_{(S)k}} \tag{10.18}$$

From (10.17) by the use of (10.18) it arises that

$$id_{k+1} = 1 + (id_k - 1) q_k \tag{10.19}$$

Setting $id_k = id_{k+1} = id_{REF}$ in (10.19) is obtained q_{REF} which is defined as the input in the equilibrium point is equal to 0. Thus, in this case arise that q_{REF} is equal to 1. In order to have a system which has as its equilibrium point (0,0), we set

$$x_k = id_k - id_{REF} \tag{10.20}$$

$$u_k = q_k - q_{REF} \tag{10.21}$$

The idle time percentage, id_k, belongs to the interval (0,1) by definition. Thus, x_k ranges between $(-id_{REF}, 1 - id_{REF})$. By substituting (10.20), (10.21) in (10.19) it holds:

$$x_{k+1} = 1 - id_{REF} + (x_k + id_{REF} - 1)(u_k + q_{REF}) \tag{10.22}$$

By observing (10.22) it results that is nonlinear. In order to have a linear closed loop system is selected a feedback linearization control strategy [9]. Feedback linearization is a strategy that introduces a state feedback such that the closed loop system becomes linear.

To this end, is selected a control strategy $U_{(S)k}$ of the form:

$$U_{(S)k} = \frac{(1 - k_c)x_k}{k_c x_k + id_{REF} - 1} \sum_{i \in N} u_{(i)k} \tag{10.23}$$

In (10.23), k_c is a parameter that will be chosen. By combining now (10.21)–(10.23), it arises a linear system with eigenvalue k_c which is

$$x_{k+1} = k_c x_k \tag{10.24}$$

In this way, it is easy to see from (10.24) that the series $\{x_k\}$ converges to 0, and so id_k to id_{REF} for any value k_c that belongs to $(-1,1)$. Since k_c is a designer's choice, the eigenvalue of the system can be explicitly set by just setting k_c. So the implementation of the proposed control strategy is

$$U_{(S)k} = \frac{(1 - k_c)(id_k - id_{REF})}{k_c(id_k - id_{REF}) + id_{REF} - 1} \sum_{i \in N} u_{(i)k} \tag{10.25}$$

As it analyzed in Section 10.4.2, each controller peer L_j in L forwards the average $id_{(i)k}$ in N_{L_j} and the sum of $u_{(i)k}$ in N_{L_j} to S. In a second step, the server S (i) calculates the total upload BW in N by adding the sums of $u_{(i)k}$ in all sets N_{L_j} that all controller peers in L send to it, and (ii) produces id_k by calculating the average of the averages of $id_{(i)k}$ in all sets N_{L_j} from all controller peers in L. In this way, the S is able to calculate $U_{(S)k}$ according to (10.25) in order to send id_k to a specific value id_{REF}.

After the calculation of $U_{(S)k}$ and by considering also the upload BW that S already contributes at time instant k, two cases are possible. In the first case, the total BW of participating peers is less than required, and S is responsible to add the exact missing amount. In the other case, where the total BW of participating peers is greater than the required, S orders the set of leaders, L, to allocate only a fraction of their upload BW (or the upload BW among N_{L_j}) and save the rest for other purposes.

10.5 Quality of service through auxiliary peers assistance

In the two previous sections (Sections 10.3 and 10.4) are analyzed two strategies, whose purpose is to dynamically control playback rate or total upload BW respectively towards the effective and stable distribution of the video. These two strategies require the existence of a centralized management component that will aggregate the required monitoring information and will apply the appropriate control strategy.

In this section, is presented not only a scalable but also a totally distributed mechanism, which monitors dynamically the total system's available BW and a totally distributed control strategy that dynamically allocates the required BW by exploiting the resources of other (auxiliary and/or idle) participating peers.

More analytically we present a novel system that monitors and controls the aggregated BW of the system that

- Is scalable and does not need any centralized components as it executes a totally distributed architecture. Towards this goal it implements a gossip protocol and control strategy.
- It ensures stability as its dynamical model factorizes the changes in upload BW.
- It uses in an efficient way the BW resources of the users in N.

10.5.1 Problem statement

We have a set N of users that consume the same video (media object) and towards this goal they create a media distribution graph (P2P overlay). These users in N ask for content (video blocks) from their neighbors in the P2P overlay with a rate p equal with the video playback rate. Thus a time instant k, network nodes (peers) have (i) upload BW $u_i[k]$ as derived from the P2P congestion control, (ii) a set of neighbors $neigh_i[k]$, as calculated dynamically from media distribution graph, and (iii) the idle percentage of the upload BW of each user (node), $id_i[k]$, as P2P congestion control derives it [8].

We use these metrics, and in this way we control the aggregated idle resources of participating peers in a way that our proposed scheduler is able to guarantee the distribution of the video to all the users in the system. We do this in a dynamic fashion by putting in real time, in case that is needed, upload BW from a set of users (peers) that have idle BW. If we compare the architecture that we propose in this section with the architecture of the previous section, we realize that the role of the servers is taken from peers (that are idle at a specific time instant) and the monitoring and control algorithms become from scalable and centralized distributed and very scalable. According to these in order to analyze our architecture here as media servers we note the "helper" peers. Furthermore, when we have upload BW more than needed, it is taken from the P2P overlay and it is used in other distributions of videos.

Towards the aforementioned objectives and in order to guarantee stability with no centralized components, we have to

1. Find the subset of nodes (peers) L that will act as controllers that of course belong to set N that are used in order to gather monitoring data from the rest of users in N. This is done through an innovative way that we developed and noted as scalable monitoring protocol (SBM). The second step is the execution of distributed bandwidth control algorithm (DBCA).
2. Find a vector $\mathbf{u_D}[k] \in_{R^{1\times L}}$ that is dynamic and each element in it is created periodically at a time instant k. This is done through the execution of DBCA that takes place in peers that belong to L. The sum of the elements of this vector is the sum of the upload BW that DBCA will put dynamically (if $u_{Dj}[k]$ is positive) or remove (if $u_{Dj}[k]$ is negative).
3. Categorize effectively between two occasions. The first is when the upload BW of users must be extracted from the P2P overlay. The second is when media servers must put upload BW. In this case, the system must determine the set of peers that will put it and the quantity that they will put.

10.5.2 Scalable bandwidth monitoring

In this subsection, it is analyzed SBM with the use of Figure 10.3 (that depicts the way in which the set L of nodes that act as controllers). In more detail is presented how L is dynamically formed and the way that information from N is mined in L towards the execution of DBCA.

More analytically users j in N (with strong gray color), select every T seconds among their neighbors ($neigh_i[k]$) (with medium gray color in Figure 10.3) the one that has the largest upload BW. This user j becomes its controller peer L_j (medium

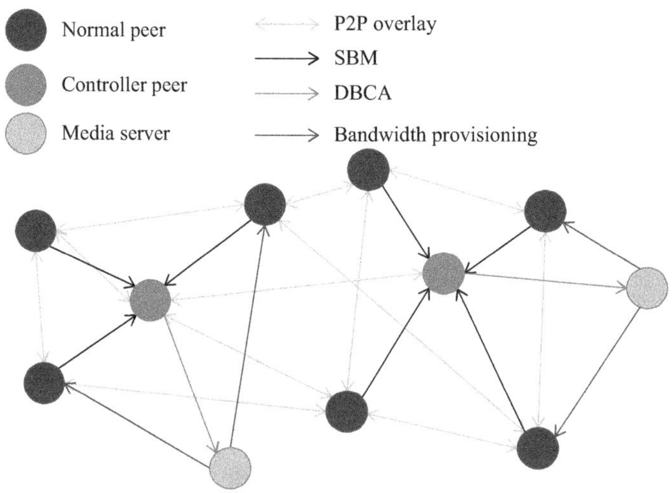

Figure 10.3 DBCA monitoring

gray nodes). The highest upload BW criterion serves the highest exploitation ratio of the surplus of the upload BW in case that it will be exploited for other purposes.

In this way users that form L are able to know in a collective fashion and dynamically the surplus or deficit of aggregated BW. In order to achieve this each user propagates every T seconds the three important metrics for the functionality of DBCA that are

- The size of the set of the neighbors that it has $|neigh_i[k]|$.
- The degree of the utilization of its upload BW, $id_i[k]$, that is the time interval that user i is idle at T divided by the time interval T.
- A calculation from P2P congestion control of the upload BW $u_i[k]$.

For the dynamic and accurate measurement of the upload BW and the idle time of each participating peer is exploited the P2P congestion control algorithm that developed in [8]. The next action that takes place from each peer is to forward all these three metrics that acquired from its neighbors to its controller peer. The set of peers that selected L_j as their controller peer is noted as N_{L_j}. Each controller peer L_j acquires dynamically control information about all the neighbors of the peers that belong to N_{L_j}. This set of peers is noted as $neigh(N_{L_j})$ and for simplicity it will be referred from now on in the text as "clique of L_j".

The next step is to execute DBCA, by exploiting the information that $neigh(N_{L_j})$ sent to L_j. In case of deficit of upload BW media servers (nodes with light gray color – auxiliary peers) take over and contribute upload BW, whereas in case of surplus of upload BW controller peers (nodes with medium gray color) save an amount of their upload BW in order to be used for other purposes.

10.5.3 Distributed bandwidth control algorithm

DBCA is executed periodically, with a period T, by each controller peer that belongs to L. The major objective of the proposed algorithm is to control $id_i[k]$ of each peer i in N by dynamically adjusting the upload BW of controller peers and the BW of media servers (auxiliary peers).

In order to achieve this, it should be expressed the amount of the total upload BW resources, which is contributed from each peer i in the set N that consumes the video object with rate p, as a function of $u_i[k]$, $id_i[k]$ and $|neigh_i[k]|$.

At an arbitrary time instant k and in case that there are idle BW resources according to the first property of the P2P overlay and the DBTS (Section 10.2) every peer among the set N consumes BW with a rate p. Thus, the total BW that is consumed is Np. Without loss of generality, this is equal with the total BW that is contributed, which is the sum of nonidle BW that peers in N contribute (Table 10.3):

$$Np = \sum_{i \in N} (1 - id_i[k])u_i[k] \tag{10.26}$$

The same happens at time instant $k + 1$:

$$Np = \sum_{i \in N} (1 - id_i[k+1])u_i[k+1] \tag{10.27}$$

Table 10.3 Notation

Symbol	Definition
S	Media server
N	Set of participating peers
p	Video playback rate
$neigh_i[k]$	Peer's i neighbor set at time instant k
$u_i[k]$	Peer's i upload bandwidth at time instant k
L	Set of controller peers
L_j	Controller peer j
$neigh(N_{L_j})$	Set of peers belong in clique of L_j
$id_i[k]$	Peer's i idle time percentage at time instant k
$id[k]$	Average estimated idle time percentage of N between time instances $k-1$ and k
id_{REF}	Idle time percentage reference value
$U_{Dj}[k]$	Upload bandwidth difference that j has to allocate or release at time instant k
$\delta_{j,i}[k]$	Number of $neigh_i[k]$ that belong in clique of L_j
T	Period of execution of DBCA
k_{C2}	Eigenvalue of the controlled system

In (10.26) and (10.27), the first parts are equal and consequently the second parts are equal. In this way is derived (10.28) where is also exploited *Property 2* of the proposed system (Section 10.2). By this way is approximated the idle percentage of each participating peer with the average idle in the P2P overlay:

$$(1 - id[k+1]) \sum_{i \in N} u_i[k+1] = (1 - id[k]) \sum_{i \in N} u_i[k] \tag{10.28}$$

In more detail, $id[k]$ is the average estimated idle time percentage between time instances $k-1$ and k. The same holds for $id[k+1]$, which is the average estimated idle time percentage between time instances $k+1$ and k. Alternatively, the total upload BW in the P2P overlay can be rewritten as:

$$\sum_{i \in N} u_i[k] = \sum_{j \in L} \sum_{i \in N} \delta_{j,i}[k] \frac{u_i[k]}{neigh_i[k]} \tag{10.29}$$

where $\delta_{j,i}[k]$ is the number of peers among $neigh_i[k]$ that have j as their controller peer at time instant k. Each controller peer j among the set L will adjust at time instant k the upload BW of the system according to the output of DBCA by a value $u_{Dj}[k]$. According to this, the sum of upload BW at time instant $k+1$ will be described from:

$$\sum_{i \in N} u_i[k+1] = \sum_{j \in L} \sum_{i \in N} \delta_{j,i}[k] \frac{u_i[k]}{neigh_i[k]} + \sum_{j \in L} u_{Dj}[k] \tag{10.30}$$

By now using (10.28) and exploiting (10.30) it arises:

$$\begin{aligned}&(1-id[k+1])\left(\sum_{j\in L}\sum_{i\in N}\delta_{j,i}[k]\frac{u_i[k]}{neigh_i[k]}+\sum_{j\in L}u_{Dj}[k]\right)\\&=(1-id[k])\left(\sum_{j\in L}\sum_{i\in N}\delta_{j,i}[k]\frac{u_i[k]}{neigh_i[k]}\right)\end{aligned} \tag{10.31}$$

Equation (10.31) can be reformed as:

$$\begin{aligned}&\sum_{j\in L}\left[(1-id[k+1])\left(\sum_{i\in N}\delta_{j,i}[k]\frac{u_i[k]}{neigh_i[k]}+u_{Dj}[k]\right)\right]\\&=\sum_{j\in L}\left[(1-id[k])\sum_{i\in N}\delta_{j,i}[k]\frac{u_i[k]}{neigh_i[k]}\right]\end{aligned} \tag{10.32}$$

By observing (10.32) and by considering N_{L_j} as the set of peers that have as their controller peer L_j and $neigh(N_{L_j})$ the set of neighbor peers that the set N_{L_j} has, can be created L desired equalities, which each one of them concerns a controller peer j. This arises by assuming that if each two terms of (10.32) are equal, then (10.32) holds and can be derived from (10.32) $|L|$ equations each one for each controller peer j as follows:

$$\begin{aligned}&(1-id[k+1])\left(\sum_{i\in neigh(N_{Lj})}\frac{u_i[k]}{neigh_i[k]}+u_{Dj}[k]\right)\\&=(1-id[k])\sum_{i\in neigh(N_{Lj})}\frac{u_i[k]}{neigh_i[k]}\end{aligned} \tag{10.33}$$

By setting now:

$$q[k]=\frac{\sum_{i\in neigh(N_{Lj})}(u_i[k]/(neigh_i[k]))}{\left(\sum_{i\in neigh(N_{Lj})}(u_i[k]/(neigh_i[k]))+u_{Dj}[k]\right)} \tag{10.34}$$

From (10.33) and by the use of (10.34) it arises:

$$id[k+1]=1+(id[k]-1)q[k] \tag{10.35}$$

By setting as id_{REF} the value in which is desired to set the average idle percentage of the participating peers, then from (10.35) is obtained q_{REF} which is defined as the input in the equilibrium point and is equal to 1. Then, let set

$$w[k] = id[k] - id_{REF} \tag{10.36}$$

and

$$y[k] = q[k] - q_{REF} \tag{10.37}$$

So, from (10.35) and by the use of (10.36) and (10.37) it arises

$$w[k+1] = 1 - id_{REF} + (w[k] + id_{REF} - 1)(y[k] + q_{REF}) \tag{10.38}$$

By observing (10.38), it arises that this is a bilinear single input single output system. These systems can be controlled and stabilized by using feedback linearization [9].

In more detail, is selected a feedback $y[k]$ which is described from the equation below:

$$y[k] = \frac{(k_{C2} - 1)w[k]}{w[k] - 1 + id_{REF}} \tag{10.39}$$

The denominator in (10.39) is from (10.36) equal to $id[k] - 1$ and is not zero unless $id[k]$ is equal to 1, which is a case that never occurs if the P2P overlay delivers a stream. So form (10.38) with the use of (10.39) the system becomes

$$w[k+1] = k_{C2} w[k] \tag{10.40}$$

In this way, it is proven from (10.40) and control theory [9] that the proposed system is stable for any value of k_{C2} that belongs to (0,1). Now, from (10.39) and by the use of (10.36) and (10.37) it arises

$$q[k] = \frac{k_{C2} id[k] - k_{C2} id_{REF} + id_{REF} - 1}{id[k] - 1} \tag{10.41}$$

By the use of (10.34) in (10.41) it arises

$$u_{Dj}[k] = \sum_{i \in neigh(N_{Lj})} \frac{u_i[k]}{neigh_i[k]} \left(\frac{(1 - k_{C2})(id[k] - id_{REF})}{k_{C2} id[k] - k_{C2} id_{REF} + id_{REF} - 1} \right) \tag{10.42}$$

Finally, is defined $totalu_{Dj}[k]$ according to (10.43):

$$totalu_{Dj}[k] = \sum_{l=0}^{k} u_{D_j}[l] \tag{10.43}$$

There are two possible cases. In the first case, where $totalu_{Dj}[k]$ is less or equal to 0, controller peer j releases any media server (auxiliary peers) that may facilitate it and sets the upload BW that it contributes to $u_j[k] + totalu_{Dj}[k]$. In the second case, where $totalu_{Dj}[k]$ is greater than 0, it sets its contributed BW to $u_j[k]$, it allocates from its media server (auxiliary peers) $totalu_{Dj}[k]$ and gives its neighbors to it in order to allow the media server (auxiliary peers) to provide to them video blocks.

10.6 Conclusions and future work

In this chapter, is presented a peer-to-peer live video streaming system that is scalable and stable. The proposed system is able to guarantee the complete and the on-time video distribution to every participating peer by adapting the peer-to-peer LS service to the dynamic total upload BW of participating peers. Towards this goal developed two different strategies. The selection of the strategy is correlated with the QoS that participating peer's desire and the business model of the service provider.

The first strategy is the video playback rate adaptation according to the existing upload BW of participating peers. This strategy can be used in cases, where peers and the service provider desire a costless LS service. In Section 10.3, presented a system that is able to monitor dynamically, in a scalable way, the upload BW resources of participating peers in a P2P live video streaming system and to adapt dynamically the playback rate of the video stream according to the aforementioned resources.

The second strategy is to dynamically allocate upload BW from auxiliary sources (e.g. clouds). This strategy can be used in cases, where peers and the service provider desire a high QoS LS service. In Section 10.4, presented a system that is able to monitor dynamically, in a scalable way, the upload BW resources of participating peers in a P2P LS system and to allocate dynamically the extra amount of BW resources that required for the stable and high-quality stream distribution.

The two aforementioned strategies require the existence of a centralized management component that will aggregate the required monitoring information and will apply the appropriate control strategy. Motivated by this fact, proposed a P2P LS architecture that with not only a scalable but also totally distributed way determine the required BW (hence the equivalent in surplus/deficit) for the video distribution and in the case of deficit dynamically allocate it by exploiting the resources of other (auxiliary and/or idle) participating peers. In Section 10.5, presented a system that is able to monitor and control, in a totally distributed and scalable way the upload BW and the idle resources of participating peers in order to guarantee the smooth peer-to-peer LS service.

10.6.1 Future work and system exploitation

The future work could be focused on four major areas. The first area could be the evolution of the P2P overlay and the DBTS in order to be more balanced and enhance the monitoring systems with higher level of accuracy. In more detail, as far as it concerns the P2P overlay various topologies could be tested towards this goal. In addition, DBTS could be improved by examining the handshakes between peers in order to achieve higher levels of fairness.

The second area could be the development and the evaluation of a hybrid and more evolved control framework that controls playback rate and BW simultaneously. Special attention could be given in correlating this control framework with QoE and cost parameters and in the study of the impact of various physical constraints that playback rate and BW introduce.

The third area could be the combination of these controllers with more advanced video coding techniques in order to further enhance QoE. For instance, MDC is a very promising and widespread technology, and its correlation with the proposed controllers could further reveal its uses and the uses of these controllers simultaneously.

Finally, the fourth and the most interesting area, could be the exploitation of the proposed modeling and control methodologies in other areas. In example, networks as Software Defined Networks are able to offer dynamically to the network edges and to internal network points information about the network capacity and the network resource utilization. By using as a basis of the proposed architecture, it could be developed routing and congestion control algorithms that will deliver networks with lower delays, higher reliability, and higher stability.

Beyond the technical future work that remains in order to finalize our P2P LS system another important aspect is its commercial exploitation (Figure 10.4). As Internet users continue to grow their networks online, social media become an essential channel for information dissemination, consensus seeking, collective action, and decision making.

Human resource departments use social media to recruit personnel and, based on a better contextualization of individual environments, create better corporate resource allocation policies and employee benefit schemes. Governments use social

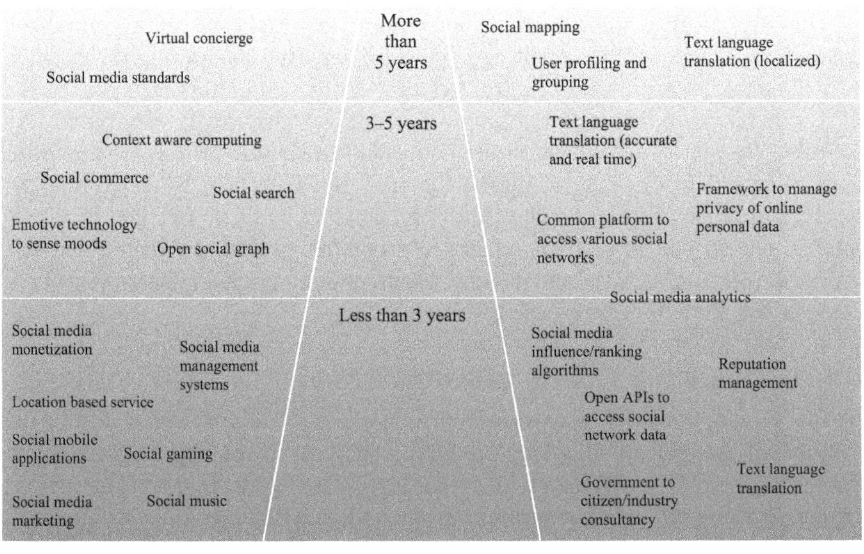

Figure 10.4 Social media exploitation landscape

media to engage the citizens, solicit feedback on policy proposals, and communicate new policies and political agendas. Social media are used by individuals for producing, retrieving, and sharing information that might not be reported by national media channels.

In the crowded landscape determined by a large number of ever-changing alternatives, understanding the main drivers of social media is imperative to derive appropriate strategies for their effective socioeconomic exploitation. The development of deeper and effective ways (e.g. LS) that will allow communities in social media to interact deeper is one of the key objectives.

In the rest of this section, we will provide a short overview of some of the possible ways to exploit our system in social media, an extended list of which is shown in Figure 10.4.

10.6.1.1 Social casting

"Social casting" refers to the use of lightweight tools for the creation of scheduled and extemporary live broadcasts on mobile terminals, particularly smartphones, for both professional and amateurish use. Professional news organizations can use social casting to cut costs or create live broadcasts in areas that are difficult to reach. Social casting has also been used as a means for personal promotion in "distributed" scouting by TV agents and producers, a method perceived as simpler, more manageable and more effective than traditional competitive auditions in studios.

Amateurs or occasional users can record live video content from their mobile phones while the action is happening and multicast it directly to the mobile phones, PCs or televisions of other individuals or communities (family members, friends, and fans), involving them in viewing and commenting. Alternatively, contents can be posted at an Internet site to allow for delayed view or re-casting to larger audiences through other channels (e.g. news agencies, online newspaper, etc.) or mainstream media (TV news, radio, etc.).

Social casting, just like citizen journalism, described below, emphasizes the immediacy of the experience and the engagement of the audience as both producer and consumer of the generated content.

10.6.1.2 Citizen journalism

Similar to the layman use of social casting, "citizen journalism" concerns the practice of producing multimedia news "reports" of an event directly by the public, in the form of videos and comments taken by mobile phones. In this way, people become observers and commentators that generate real-time, on-the-ground perspectives of facts as they unfold valuable information that can spread rapidly over social networking and through user generated LS. There have been already many episodes of great relevance when this happened on a large scale, such as the 9/11 attacks, the big earthquakes in Haiti (2010) and Christchurch (2011), Japan earthquake and tsunami (2011), and Boston Marathon bombings (2013).

Besides providing different and original point of views about events, collaborative reporting may become the principal source of information at sites where it is not possible to deliver professional reporters in time to get the news. It is as well a precious resource as a spontaneous fallback for traditional news media in situations where the dedicated communications infrastructure are damaged by disruptive events like earthquakes, incidents, terrorist attacks, and others.

In countries that practice censorship and restrictions to press freedom, citizen journalism has already played an important role in informing the world about facts. A known example is the Arab Spring and Occupy movements, where social media were used to leverage mass support, organizing, communicating, and raising people awareness to overturn governments, in the end effecting enormous levels of political and societal change.

If the objectivity, quality and accuracy of citizen reports can be arguable, they undoubtedly possess appealing advantages such as variety, timeliness, wide reach and scope, and the possibility of exploiting crowd sourcing, for example to locate key witnesses in huge events. Rather than constituting an alternative to mainstream news sources, which retain their superior authority and reliability, citizen journalism can therefore efficiently complement traditional journalism. As such, citizen journalism is even becoming adopted by media channels in their main news feeds.

10.6.1.3 Use during emergencies

The characteristics highlighted above when describing Social Casting and Citizen Journalism are the same that makes social media and LS powerful tools to collect and spread a huge amount of information in crisis situations.

Survivors of natural disasters resort to social media applications at pervasive rates for a number of scopes such as contacting friends to ensure safety, asking online friends to contact responders, and using information to find shelter and supplies. Approximately 80% of Americans expect emergency response agencies to monitor and respond to social media platforms. This happened for example during Hurricane Sandy, when the Federal Emergency Management Agency tweeted to advice about expected phone line congestion and suggest use of social networks to reassure relatives and friends. In that occasion, Red Cross monitored 2.5 million related postings, of which 4.5k were official requests for aid.

10.6.1.4 Social media and politics

The increasing usage of social media by policy makers to establish and reinforce communication networks and move toward their objectives is an important example of the recognition of free and bottom up generated LS as providing a great potentiality for influencing society.

Several studies of citizen voting habits have shown that voting decisions are not usually based on one-step communication but are rather determined through conversations with opinion leaders, colleagues, friends, and acquaintances who can either consolidate or weaken the voter's opinion. Social networks expand the opportunities for such interactions, allowing each single individual to share his knowledge, wisdom, and personal experiences with his peers. Even if the effectiveness of social media as a means to influence politics cannot be taken as granted, there are many examples of cases where this has happened, to a greater or lesser extent:

- President Barack Obama in 2008 campaign and his 2012 State of the Union Address at a Google+ Hangout, the first virtual interview from the White House.
- Republican Darrell Issa, Chairman of the House Oversight and Government Reform Committee, website (KeeptheWebOpen.com) which encouraged US citizens to comment, and add to the conversation, on the Online Protection and Enforcement of Digital Trade Act, an example of crowd sourcing effort in US policy crafting.
- The 2012 posting of the first draft of the Icelandic Constitution on the Icelandic Constitution Council website, which encouraged citizens to comment on a Facebook page.
- The creation of a committee by King Mohamad VI to revise the Moroccan Constitution, by the creation of a crowd sourcing website to gather opinions from 150,000 Moroccans on the constitutional amendments.
- The setup of a crowd sourcing platform by a presidential candidate in Egypt in July 2011 to engage policy administrators and citizens in drafting a new constitution.
- Nicolas Sarkozy's victory over the opposing socialist candidate Ségolène Royal for the French presidency, when 40% of Internet users reported that conversations and other activities on the Internet had an effect on their voting decisions.
- The collection by the German Pirate party in 2011 Berlin state election of 8.9% of the vote and 15 seats in state parliament, achieved with just a €50,000 budget: votes came from many different sources, including those who had just reached voting age, past silent voters, the Greens, Social Democrats, the left-wing, liberals, and Christian Democrats, 20% voters being aged 18–34.
- The effect of the introduction of the "Living Platform" open wiki project on Canadian Green Party federal parliament electoral scores.

10.6.1.5 Social media marketing

Social media marketing is a key contributor to social media revenue. According to Gartner's analysis, social media advertising revenue will increase from the US $11.83 billion of 2011 to US $33.5 billion in 2016, whereas an eMarketer's analysis showed that social media advertising revenue will hit US $9.99 billion in 2013, up from US $5.54 billion in 2011.

Private companies, nonprofit organizations and government agencies are all increasingly exploiting various social media marketing channels for their campaigns and causes. Various reports indicate that senior management is increasingly involved in company-wide social media strategies.

According to the 2013 Social Media Marketing Industry Report, 86% of interviewed marketers say that social media is important for their business, a slight increase with respect to the 83% reported in 2012 and the 78% resulting from a

2011 analysis among business-to-business (B2B) organizations. The reported top two benefits of social media marketing are increasing exposure (89%) and increasing traffic (75%), whereas most marketers are using social media to gain marketplace intelligence (69%) and develop loyal fans (65%).

In general, the greater the exposure to social media marketing (longer experience, more time spent per week) the higher the perceived advantages on various areas.

Facebook (92%), Twitter (80%), LinkedIn (70%), blogging (55%), and YouTube (56%) have been the top five platforms used by marketers in 2013, confirming the ranking measured in the past year. The relative preferences remain substantially consistent across categories measuring both the experience and time spent in social media marketing. However, Pinterest, Google+, Instagram, and YouTube get more preferences from committed users with respect to less involved ones, whereas B2B and business-to-consumer (B2C) companies are more focused in LinkedIn/blogging and Facebook, respectively.

Expectations for social media usage in the near future have been always in the direction of a further increase, and the percentage of marketers reporting amplified benefits with respect to the previous year has increased for all activities for which social media are deemed to be helpful. In the 2013 report, in particular, marketers say they plan to increase their use of YouTube, Facebook, blogs, LinkedIn, and Twitter, in that order.

However, the 2011 data indicated that clear directions and strategies for social media to be integrated in the overall company objectives were still not entirely set (only 25% of respondents reported about a clearly defined strategy), "gut feel" remaining one of the main "tool" to determine the approach to use instead of more formal techniques.

This lack of governance and policies was reflected in the absence, in most cases (78%), of specific budget entries for social media and the deficiency of both dedicated social media marketing managers (only 6% respondents having appointed one) and consensus about how to manage social media, whether by one individual or a team. Funds spent in social media was less than 10% of the overall marketing budget for more than 75% of interviewed companies, with a slightly greater inclination to invest time rather than money (50% allocating less than 10% of total marketing time resources, 4% dedicating more than 50% time). These facts can be explained by the relative novelty of social media platforms, which may still be holding back their adoption, and the rather low level of investment required for basic social media activity—such as blog posting, tweeting, and others—when compared with almost every other kind of marketing activity.

The 2011 findings indicated that use of commercial social services was still in its infancy in the B2B arena at that time, with 62% of respondents declaring not to have used any advertising or commercial services from social networks at all to the interview date. The key reasons cited for this—"other priorities," "don't understand," "not good value,"—seemed to indicate that the social networks themselves had not been good enough at developing and selling their commercial propositions.

These facts are confirmed by a more recent report by Nielsen revealing that most of the surveyed advertisers and agencies had been using social media for less than 3 years. Indeed, 20% of respondents declared they had only started in the last year, and for 70% of them the share with respect to overall online advertising was less than 10%.

As social media continues to become more sophisticated and increases its user bases and engagement, the opportunities for advertising are believed to expand significantly. The Nielsen report indicates that the majority (64%) of surveyed advertisers expected to increase their budget for social media advertising in 2013, many of them at the expenses of other channels, even if those increases would have been modest (between 1% and 10%).

Social media are just a part of tactics that also include online display, online video, and mobile, as well as offline means like print and TV. The primary purpose for advertising is branding-related, such as raising awareness and influencing brand options. Free LS (e.g. P2P) offers a very helpful tool towards these goals.

Despite difficulties in demonstrating its effectiveness, social media marketing is more and more perceived as a way reach out potential stakeholders on the Internet, in particular the younger generation.

References

[1] Efthymiopoulou M., Efthymiopoulos N., Christakidis A., Athanasopoulos N., Denazis S., Koufopavlou O. 'Scalable playback rate control in P2P live streaming systems'. *Peer-to-peer Networking and Applications*. 2016; 9(6):1162–1176.

[2] Efthymiopoulou M., Efthymiopoulos N., Christakidis A., Denazis S., Koufopavlou O. 'Scalable control of bandwidth resource in P2P live streaming'. *Proceedings of the 22nd Mediterranean Conference of Control and Automation*; Palermo, Italy, June, 2014, pp. 792–797.

[3] Efthymiopoulou M., Efthymiopoulos N. 'Enabling QoS in peer to peer live streaming through simultaneous control of bandwidth and playback rate'. *Proceedings of the Fourth International Conference on Communications, Computation, Networks and Technologies*; Barcelona, Spain, November, 2015.

[4] Efthymiopoulos N., Christakidis A., Denazis S., Koufopavlou O. 'Liquidstream – network dependent dynamic P2P live streaming'. *Peer-to-Peer Networking and Applications*. 2011;4(1):50–62.

[5] Christakidis A., Efthymiopoulos N., Fiedler J., *et al*. 'VITAL++, a new communication paradigm: embedding P2P technology in next generation networks'. *IEEE Communications Magazine*. 2011;49(1):84–91.

[6] Deltouzos K., Gkortsilas I., Efthymiopoulos N., Denazis S. 'Liquidstream II – scalable P2P overlay optimization with adaptive minimal server assistance for stable and efficient video on demand'. *Peer-to-Peer Networking and Applications*. 2015;8(2):260–275.

[7] Deltouzos K., Gkortsilas I., Efthymiopoulos N., Denazis S. 'Liquidstream II – scalable P2P overlay optimization with adaptive minimal server assistance for stable and efficient video on demand'. *Proceedings of the 32nd Computer Communications Workshops (INFOCOM WKSHPS) IEEE Conference*; Turin, Italy, April, 2013, pp. 57–58.

[8] Efthymiopoulos N., Christakidis A., Efthymiopoulou M., Corazza L., Denazis S., Koufopavlou O. 'Congestion control for P2P live streaming'. *International Journal of Peer to Peer Networks (IJP2P)*. 2015;6(2):1–21.

[9] Slotine J.J.E., Li W. *Applied Nonlinear Control.* Prentice Hall, Englewood Cliffs, NJ; 1991.

Chapter 11

Hybrid resource sharing for QoS preservation in virtual wireless networks

Dimitrios N. Skoutas, Nikolaos Nomikos, Demosthenes Vouyioukas, Charalabos Skianis and Angelos Antonopoulos

An enabling technology for the fifth generation (5G) of wireless communications is the virtualisation of network resources. 5G networks are expected to consist of dense deployments and efficient resource allocation is a top priority [1]. At present, there is a growing number of research projects, investigating network virtualisation at different levels. The virtualisation of core network functionalities is the topic of the T-NOVA project [2] where a virtualised cloud infrastructure is responsible to provide them, thus achieving elasticity and flexibility in network deployment. Moreover, various tools such as OpenFlow [3] have been developed allowing the dynamic and scalable adaptation of the core network regarding the routing of data. Also, open-source solutions such as OpenStack have been adopted for the management of public and private clouds as well as the well-timed creation and effective maintenance of virtual machines [4].

11.1 Wireless network virtualisation

At another level, wireless network virtualisation (WNV) proposes the abstraction of physical resources, including the infrastructure and the spectrum owned by one or more physical network operators (PNOs) in order to improve resource utilisation. Slicing spectral resources results in flexibility in service provisioning to the end-users and in network deployment [5]. In WNV, one or more PNOs allow the leasing of their physical resources to multiple virtual network operators (VNOs) based on various resource sharing schemes, ranging from fixed to complete sharing (CS). Several works [6] investigate the relationships between the PNO, VNOs and service providers (SPs) and describe their roles within the WNV context. The METIS project [7] examines different scenarios of spectrum virtualisation involving exclusive or shared access to spectral resources. Furthermore, project iJoin [8] proposes the concept of radio access network-as-a-service (RANaaS), where RAN functionalities are virtualised through an open cloud infrastructure for both the access and backhaul parts of the network.

11.1.1 Benefits of wireless network virtualisation

The benefits of WNV derive from the efficient resource usage. First, capital expenditure and *operational expenditure* can be reduced through the sharing of infrastructures of the PNO by multiple VNOs [6,9]. In this way, the PNO harvests revenues through leasing, whereas VNOs are not required to invest on acquiring and installing network infrastructure. Moreover, spectrum efficiency can be achieved, as long as dynamic algorithms for spectrum sharing are developed. To this end, combining spectrum pools from one or multiple wireless technologies and providing tailor-made algorithms to specific use cases can lead to significant capacity gains as proposed in [10].

Such flexibility enables the formation of novel business models between the PNOs and the VNOs and allows new players to enter the market providing their services to the end-users. In [11], spectrum and network management trends are discussed paving the way for the consideration of such aspects in the forthcoming WNV setting. It must be noted that in order to achieve spectrum virtualisation, accurate estimation of the available resources is required. In [12], it is proposed to employ users and user-deployed devices such as femtocells to identify vacant spectrum. In this way, wireless network operators (called PNOs in the context of this book chapter) can provide incentives to end-users, thus turning them into wireless prosumers targeting an overall network optimisation. Similarly, in many studies, cognitive radio is shown as an efficient way to maximise spectrum utilisation [13–15]. In addition, power reduction and energy efficiency can be achieved when the spectrum is allocated in an optimal way to provide the requested services by the end-users.

11.1.2 WNV in the future networking environment

As 5G networks are envisioned to support the connectivity of billions of nodes and the provision of services with diversified requirements [16–18], virtual service providers (VSPs) will play a major role in achieving these targets. VSPs are defined as the entities that directly lease resources from the PNOs specialising in providing specific services to the end-users. So, in this context, the role of VNOs is identical to the VSPs with the added characteristic of attracting end-users through novel 5G services such as high-definition video, machine-to-machine (M2M) [19], e-health, vehicle-to-vehicle, smart energy grids and others. This approach can significantly reduce the complexity of the WNV as the VSP has complete knowledge of the required capacity for the specific service and, thus, can perform direct demands to the PNO for dynamic resource allocation or release. As a result, optimal spectrum and backhaul capacity usage can be achieved, while the required quality of service (QoS) is preserved. In this chapter, we adopt the case where multiple VSPs request resources from the PNO in order to provide distinctive services.

Although VSPs are directly related to specific services, the heterogeneity of 5G services demands usage of infrastructure and spectrum for varying time periods. Thus, each VSP should be able to dynamically acquire the necessary resources and towards this end, hybrid algorithms can facilitate the coordination among the VSPs and the PNO. Typical resource-sharing schemes, such as fixed and CS, are not as efficient as a dynamic virtualised networking environment requires, and thus, we present a hybrid resource sharing (HRS) scheme. HRS enables the dynamic sharing of resources among the VSPs, taking into account the requirements of each service in terms of QoS and quality of experience (QoE). In this way, the total incoming traffic, consisting of the different services can be efficiently handled and blocking is avoided. The efficiency of HRS is shown through performance evaluation and comparisons with fixed and complete resource-sharing schemes in terms of blocking probability under different traffic-load conditions and resource-sharing factors. Furthermore, looking at the big picture, we should also emphasise the need for a dynamic mechanism able to coordinate the wireless coverage of multiple VSPs leasing resources at different PNOs that are deployed within the heterogeneous networking (HetNet) environment of the same geographic area. In this chapter, we give a high-level description of such a mechanism that is based on the prosuming concept and is able to enhance the overall HetNet coordination.

The structure of this chapter is as follows. Section 11.2 presents a high-level network planning and optimisation framework, able to improve the coordination of multiple VSPs/PNO sets, towards the efficient wireless coverage of a wide geographical area. Subsequently, in Section 11.3, we focus on a single PNO, and on the basis of specific business models, we discuss on the PNO–VSPs interaction as well as on the interaction among VSPs that lease resources of the same PNO. Section 11.4 includes a detailed description of two mainstream resource-sharing schemes, while Section 11.5 presents the HRS enabling the controlled sharing of resources. Next, Section 11.6 provides the performance evaluation and the comparisons with other schemes. Finally, open issues in the area of WNV are discussed in Section 11.7, while Section 11.8 concludes this chapter.

11.2 Wide area coordination of multiple PNOs/VSPs

The benefits that WNV can offer in terms of efficient management of wireless resources, as noted previously, are obvious. However, in order these benefits to be achievable in real networks, one must also take into account the adverse effect of an uncoordinated coverage of large geographical areas by multiple VSP's able to operate not only in-band but also out-of-band relatively to the PNO's initially intended operation.

Thus, a spectrum monitoring and control system, able to provide the means for the effective coordination among different VSPs and PNOs as they operate within the dynamic HetNet environment of the future, is required. The employment of wireless prosumer's devices as mobile multi-band spectrum sensors should be a key feature of such a system, if we aim for a cost-effective deployment of a dense monitoring system.

11.2.1 Ubiquitous spectrum monitoring based on wireless prosuming

According to the wireless prosumer concept, the majority of user-deployed network devices that consume resources of the wireless network can be utilised, at the same time, to perform spectrum measurements which subsequently could be gathered and fed to a central spectrum controller. Beside the users smartphones, one could also consider other user-deployed devices such as femtocells, Wi-Fi access points and relays, which could also be employed in order to gather the required spectrum data.

Thus, by utilising multiple radio access technologies as spectrum sensors, the monitoring of the occupancy and state of multiple spectrum bands will be made possible in an economically feasible manner. On the other hand, we also have to consider that the end-users, who are asked to offer their resources (e.g. central processing unit, network access and battery power), should be at the same time motivated to adopt the presuming behaviour. Towards this goal, the network operators could define specific incentives for the prosumers, such as better network access, lower billing rates and others.

11.2.2 Forming overall network planning policies

Figure 11.1 shows a high-level framework that allows the monitoring of various spectrum bands, enables cognitive real-time corrective/optimisation actions and provides a solid ground for the formation of long-term network planning policies. Specifically, as discussed previously, the end-user prosuming devices (PDs) will act as spectrum sensors which periodically take measurements and send reports to the local "Spectrum Data Gathering" (SDG) Gateway.

- *PDs* are organised in clusters which should contain the proper density and type of PDs that will allow the acquirement of dense and spectrally accurate data for a wide range of frequencies. Regarding the frequency of measurements, it depends on the variability of the networking environment, and therefore, the time intervals between subsequent measurements may vary from a few minutes to as long as 1 h. Optionally, when considered as appropriate, selective radiation metres with narrow-band capabilities, able of high precision measurements could also assist the measurement process. Such devices will be deployed by the PNOs and they can be either placed at specific locations or they can be mobile (e.g. mounted on public transportation vehicles).
- *The SDG gateway* is responsible for gathering the spectrum reports while it also performs an initial processing of data. Then, the useful information is forwarded to the Virtual Network Planning and Optimisation (VNPO) module. PNOs and secondarily the VSPs, may also have direct access to the spectrum data provided by each SDG in order to enhance the efficiency of their own

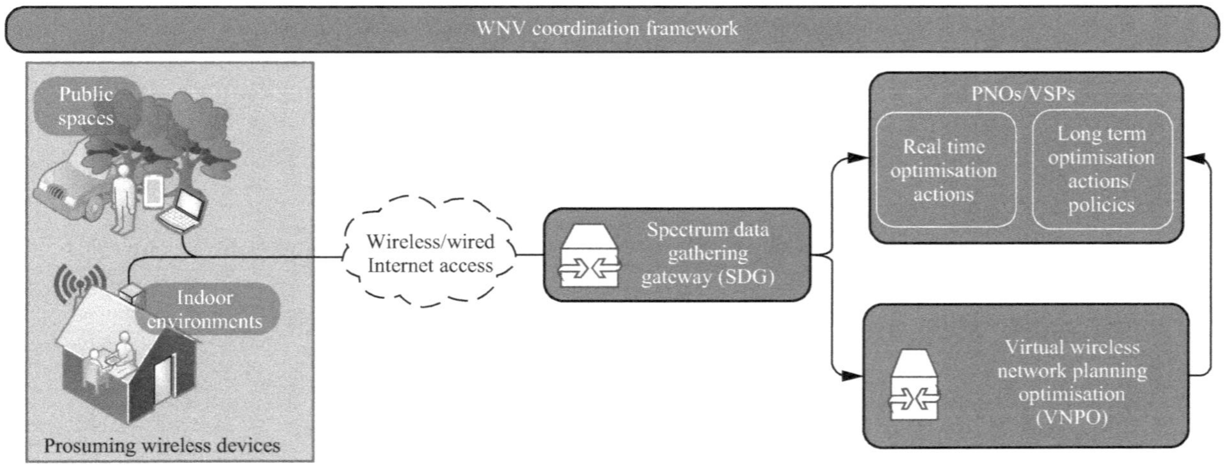

Figure 11.1 High-level network planning and optimisation framework

self-optimising capabilities. Therefore, it is evident, that SDGs should be placed close to each monitoring cluster in order to enable the timely collection of spectral data.

- *At VNPO*, through the use of simulation and specifically developed indoor/outdoor propagation models, possible deployment issues as well as inefficient spectrum allocations are identified. Considering various optimisation factors, VNPO will suggest optimisation actions to both PNOs and VSPs regarding the overall setting of the heterogeneous wireless coverage. These suggested that actions will be assessed and classified in order to gradually form a long-term resource allocation policy.

Provided that the proposed coordination framework is realised (Figure 11.1), each PNO will be aware of the impact of his actions (e.g. leasing his resources to a particular VSP) to the overall heterogeneous wireless coverage as well as to the utilisation of his own resources.

11.3 Emerging business models for sharing the resources of a PNO

In this section, various business models are presented considering the sharing schemes between the PNO and the VSPs of the WNV. These business models assume that end-users are motivated to act as prosumers providing feedback on the spectrum state. In this way, the necessary information to trigger resource sharing among VSPs is acquired. The topic of business models is also discussed in a number of research projects, such as [7,20,21] where the roles of the owner of the physical resources and of other entities such VNOs and VSPs are described.

11.3.1 The role of PNOs and VSPs

In order to give the details for the different business models, a clear description of the roles of PNOs and VSPs has to be provided. Starting with the PNOs, it is considered that they are the owners of the network infrastructure including the backhaul link to the core network. However, different cases of resource ownership exist where an entity owns the network infrastructure and a different one, the spectral resources [6]. It must be emphasised that the interests of the PNOs lie in the optimal utilisation of their resources and towards this end, they can follow various approaches for revenue maximisation. For example, in the network sharing paradigm [9], PNOs can share their infrastructure with other PNOs through different schemes, such as leasing for a specific time period or dynamic sharing. Moreover, PNOs can share their spectrum in order to improve its utilisation. However, the relationship between PNOs is beyond the scope of this chapter, and we focus on the synergy between PNOs and VSPs. In this setting, the PNOs aim to attract VSPs by providing advanced network infrastructure and sufficient spectral resources that can be exploited for service provisioning through various business schemes.

From their point of view, the VSPs are interested in acquiring infrastructure and, most commonly, spectrum from the PNOs aiming to provide services to the end-users with the optimal amount of resources, while avoiding excessive charging from the PNOs. We can also consider the case where a VSP may request to utilise PNO's infrastructure but needs to operate at a spectrum band, that is (close but) different to the one that PNO normally operates. In this case, PNO, before deciding to lease his resources to the specific VSP, needs to have an accurate knowledge of the total spectrum utilisation. This can only be achieved through the realisation of the monitoring framework of Figure 11.1.

Moreover, different categories of VSPs appear, and one may observe that a VSP can offer a set of services such as video streaming, voice and M2M. In this case, the VSP is responsible to demand resources from the PNO and then, allocate them efficiently to different service classes. A different category consists of VSPs specialising in a specific service and so, one or more VSPs can be responsible for the provision of each service class performing optimal utilisation of the PNOs' resources. It is obvious that a VSP performs a trade-off between

1. Demanding resources to satisfy services with different QoS levels, thus incurring complexity but might result in attractive service packages for the end-users.
2. Specialising in a specific service where in this case, it is easier to acquire the optimal amount of resources since the QoS levels of only one service has to be maintained.

11.3.2 Interaction between PNO and VSPs

Here, business models between a PNO and various VSPs are discussed. The basic model consists of a PNO who provides exclusive access to a partition of the resources to the VSPs based on service level agreements (SLAs). In this case, the VSPs are responsible to provide their services by exploiting only the dedicated chunk of the PNO's infrastructure resources. It must be noted that although this case incurs the least complexity, it has many drawbacks in the sense of resource utilisation. For example, a VSP that does not attract many end-users can be given to resources that would otherwise be used by other VSPs that are starving.

A more complex model involves the simultaneous access to all the portion of the PNO's resources from the VSPs. In this way, enhanced-resource utilisation can be achieved when the VSPs' demands are timely given to the PNO. The main disadvantage of this case is that resource-demanding services might experience significantly increased blocking probability. Also, a PNO might apply different pricing on the resources that are allocated to VSPs, for example extra charging for VSPs that use lower chunks of resources in order to motivate a higher level of resource occupancy.

To avoid such shortcomings, novel business models should address the concerns of the PNOs regarding resource utilisation, while maintaining low blocking probability for the services of the VSPs, at the requested QoS from the end-users. So, hybrid models should be developed combining the exclusive access to resources with dynamically allocated partitions according to the needs of each VSP. More specifically, in hybrid-sharing schemes, one part of the VSP's resources is provided by the PNO through SLAs that grant exclusive access to a partition of the resources, and within this chunk, common resource pools can be formed according to *sharing factors* that are chosen by the VSPs.

11.3.3 Interaction between VSPs

According to the SLAs between the PNOs and the VSPs, interesting interplays between VSPs arise. When exclusive licences are granted to each VSP, there is the possibility that a VSP which experiences resource underutilisation, might be interested in leasing a partition of its resources to another VSP which is in shortage. So, SLAs between VSPs can be formed, thus leading to greater revenues for the VSPs and improved resource utilisation.

Enabling the concurrent acquisition of the PNO's resources will allow VSPs to form coalitions and merge their resources that are available for service provisioning. In this way, VSPs which specialise in different services can form diverse service sets that are attractive to end-users. In addition, for such cases, where only shared resources are available, two or more VSPs could sign SLAs that result in prioritising a VSP to access the common resource pool and achieve the desired QoS level. Moreover, for hybrid models where resource partitions that are exclusively allocated coexist with a common resource pool, VSPs can select different strategies which rely on the *sharing factor*. For example, a VSP might be interested in acquiring a larger chunk of exclusive resources that can be merged or leased in agreement with other VSPs. On the contrary, a VSP might choose to hold smaller portions of dedicated resources and offer a bigger portion to the common pool (CP) where other VSPs have access, in order to avoid excessive charging from the PNO.

11.4 PNO's main resource sharing approaches

As mentioned previously, a VSP does not always require to utilise the same spectrum as PNO. Out-of-band transmissions in respect to the PNO's spectrum band are possible as long as it is permitted by the capabilities of the PNO's infrastructure and provided that an overall HetNet coordination mechanism is employed. Therefore, in order to take into account the general case, in the following we will not refer to the sharing of spectrum but instead we will refer equivalently to the sharing of the available backhaul capacity, up to the extent that the backhaul capacity does not exceeds the maximum wireless capacity. Two of the main approaches to share the PNO's resources [7,10] are CS and fixed sharing (FS). In the following sections, the details of these schemes are given.

11.4.1 Complete sharing

As shown in Figure 11.2, the incoming service requests (SRs) are identified and forwarded by the PNO to the appropriate VSP. Subsequently, the service admission

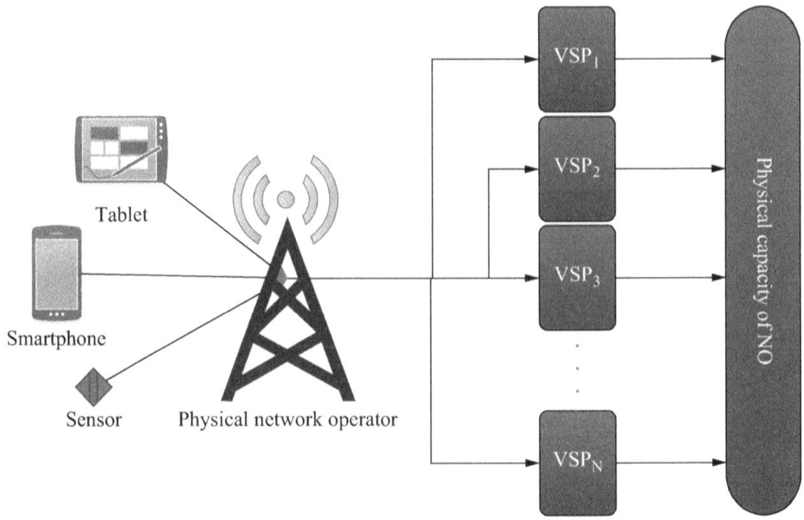

Figure 11.2 The complete sharing scheme where the PNO's capacity is equally accessible to all VSPs

control (SAC) mechanism of the respective VSP decides whether the SR will be accepted or rejected based on the available capacity.

According to the CS sharing mechanism, the VSPs can utilise all the available physical capacity of the PNO in a CP manner. This approach has the advantage of maximising the total throughput and the utilisation of the physical capacity. However, CS cannot provide definite SLAs between the PNO and the VSPs. As long as a VSP has increased incoming traffic load, it can occupy a large fraction of the total capacity, leaving the rest of the VSPs to starve.

Furthermore, following a CS of the available capacity, it is expected to benefit services that require low data rates such as voice over IP (VoIP), contrary to high data rate services such as video conferencing (VC). In other words, CS cannot provide QoS differentiation, and for the case where the VSPs are directly related to specific services, this characteristic can result in significant degradation of the network's performance and affect the required SLAs.

11.4.2 Fixed sharing

One possible solution in order to provide the required SLAs to the various VSPs is to employ the FS approach. In Figure 11.3, each VSP can utilise only a specific fraction (partition) of the total capacity ensuring that a misbehaving VSP cannot lead the other VSPs to starvation.

Therefore, the main advantage of FS is that it has low complexity and can provide definite SLAs to the VSPs, as well as QoS differentiation if a VSP specialises in a specific service. Nevertheless, FS has significant drawbacks, such as offering reduced utilisation of the available capacity and reduced total throughput.

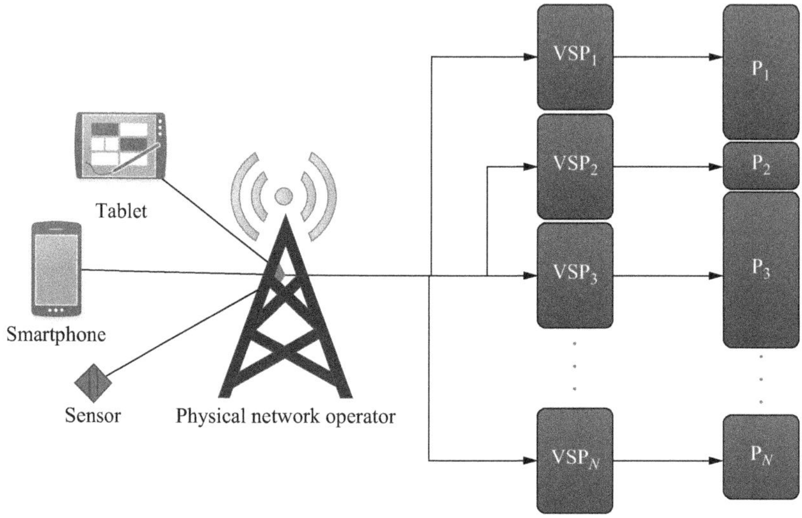

Figure 11.3 The fixed sharing scheme where each VSP is allocated a predetermined capacity partition

Furthermore, FS is not capable of confronting any variations of the composition of the incoming traffic load. This is due to the fact that, if there is available capacity at one of the VSPs, then according to the FS sharing discipline, this capacity cannot be utilised by any of other VSPs, thus leading to the aforementioned weaknesses of FS.

11.5 Hybrid-controlled sharing of resources

Aiming to combine the merits of FS and CS, and at the same time to address their weaknesses, the concept of hybrid-controlled physical-resource sharing is introduced.

11.5.1 The formation of physical capacity partitions

Initially, based on the SLA agreements between the VSPs and the PNO, a number of physical capacity partitions are formed. Each partition P_j consists of two areas: the shared area (SA_j) and the protected area (PA_i). An example of the formation of the partitions is shown in Figure 11.4.

Based on this formation, the concept of native and non-native service calls is defined as follows:

- A service call admitted by VSP_j is considered as native for the respective partition P_j.
- A service call admitted by VSP_j is considered as non-native for the other partitions.

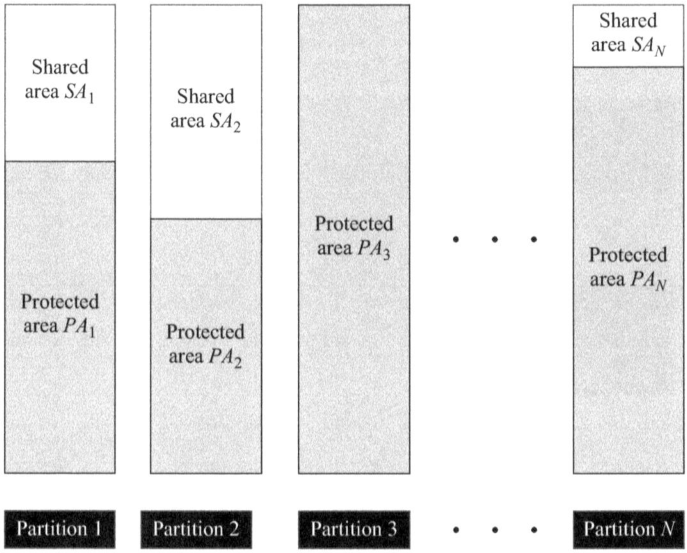

Figure 11.4 An example of the formation of the resource partitions

In greater detail, the SA of all the partitions can be utilised by native and non-native calls. In other words, a call that is native to a partition can utilise both the *PA* and *SA* areas, whereas a non-native call can utilise only the *SA* area. However, the admission of service calls at the *SA* of partitions where they are not native should be performed in a controlled manner to avoid being flooded with non-native calls. Denoting by CP_j the capacity of the jth partition, then the *PA* area can be defined as follows:

$$PA_j = \left(1 - r_j\right) \cdot CP_j, \quad r_j \in [0, 1] \tag{11.1}$$

Where r_i is the sharing factor of the specific partition. Based on Equation (11.1), the SA can be expressed as follows:

$$SA_j = CP_j - PA_j = r_j \cdot CP_j \tag{11.2}$$

Consequently, the SA_j area increases with the increase of the sharing factor r_j and can be equal to the whole partition when r_j becomes equal to 1. Then, the partition acts as a CP of resources and is available to all the VSPs. On the contrary, when r_j decreases, the SA decreases as well, and reduces to zero when r_j reaches zero. In this case, the partition is fully protected and does not accept non-native calls.

11.5.2 Service admission control and capacity allocation

Through the utilisation of the sharing factor r_j, the behaviour of hybrid sharing can be fine-tuned between the two extremes of the CS ($r = 1$) and FS ($r = 0$) of the physical capacity. From the point of view of the VSP, the available capacity that can be utilised is a combination of its own fixed private partition PA, as well as a

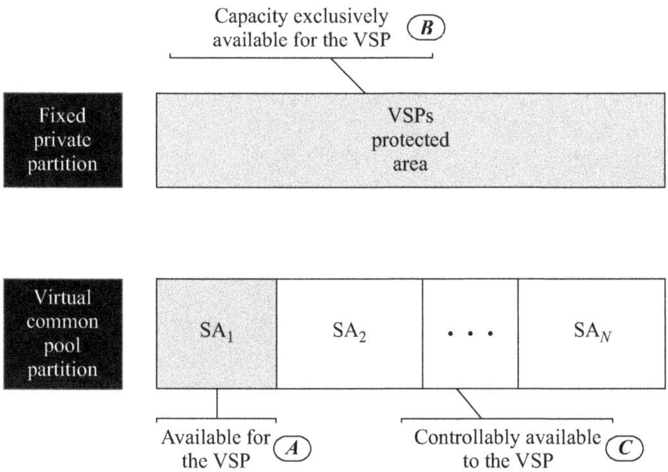

Figure 11.5 The VSP's view of resources

virtual CP partition which is derived by aggregating the SAs of all the physical partitions as shown in Figure 11.5.

Each VSP has its own SAC policy which could be based on multiple criteria (QoE classes, QoS differentiation, user profiling etc.). On the arrival of an SR, the admission control module of the VSP analyses the relative context and identifies its service class. On the basis of the available context information and the admission policy of the VSP, an SR may be temporarily postponed from being served, thus having a 'Policy-Based Rejection'. If this is not the case, the VSP initially admits the SR, as long as there is available capacity, either at the private or at the CP partition and forwards this decision to the PNO's capacity allocation mechanism (CAM) in order to have the final acceptance confirmation.

Upon the acceptance of an SR by the VSP, the underlying CAM, illustrated in Figure 11.6, successively checks if the required capacity can be allocated to

1. The respective SA of the CP partition (area A at Figure 11.5).
2. The respective PA of the private partition area (area B at Figure 11.5).
3. The rest of the CP area (area C in Figure 11.5).

If there are/are not available resources at any of these areas, the call is accepted/rejected and the VSP is notified accordingly.

By following this order of operation, CAM which belongs to the PNO, guarantees that each VSP will offer capacity from its own SA partition to other VSPs only if it is undoubtedly underutilised. Thus, CAM ensures the SLA between the VSP and the PNO, while at the same time, it allows the SA to be utilised as a common resource pool provided that the QoS offered to the respective VSP is not downgraded. Finally, although the VSP has a view of the available capacity, the admission decision of an SR is confirmed by the PNO in order to avoid

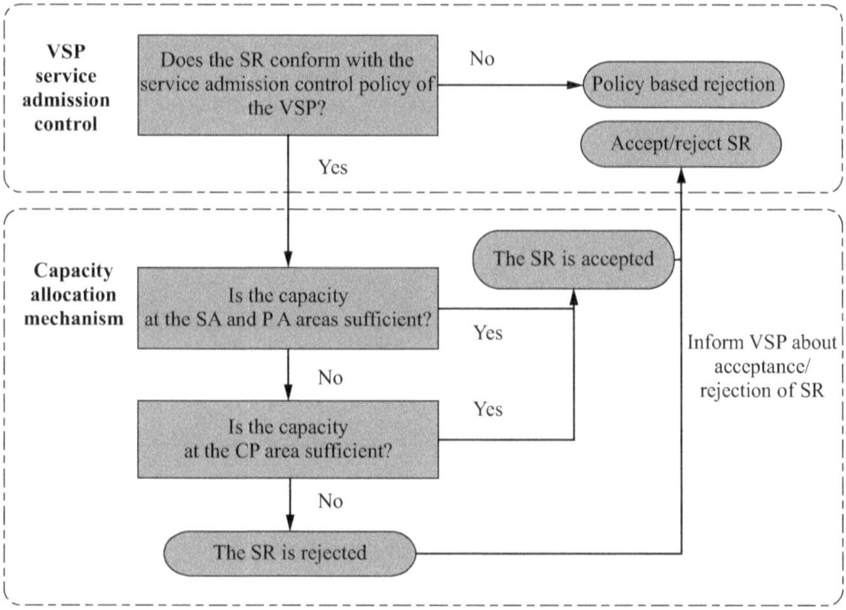

Figure 11.6 Service admission control and capacity allocation

synchronisation issues (i.e. two or more VSPs admitting simultaneously SRs for the same shared resources).

11.6 Performance evaluation

The evaluation of HRS is performed through event-driven simulation using a custom-made C++ simulator. SRs follow a Poisson arrival process and their duration is exponentially distributed. All the parameters of the proposed service admission and capacity allocation framework, as they are analysed at Section 11.5 are defined in the simulations. Consequently, according to the studied scenario, two or more VSPs can be defined whereas the backhaul capacity can be partitioned according to considered SLAs and the composition of the offered traffic load, while a sharing factor r can also be specified for each VSP.

In addition, HRS is compared to the other two typical approaches for the sharing of the available PNO capacity, which are the CS and the FS schemes. As described previously, according to the CS sharing scheme, the VSPs can utilise all the available physical capacity of the PNO as a common resource pool. On the contrary, in the FS scheme, each VSP is allocated a fixed fraction of the PNO's capacity, aiming at a specific QoS for a given traffic load distribution. Following the business models described at Section 11.3, we consider VSPs that are specialised and serve only specific service classes whereas the PNO uses a global sharing

Table 11.1 Services' description

Service/QoS characteristic	VoIP	MTC	VC	BE-IA
Average data rate (kbps)	24	256	256	120
Average duration (s)	900	60	1,200	1,200
Target Block. Prob. (%)	1	0.1	1	BE

Table 11.2 Traffic load distribution per service

Name	VoIP	MTC	VC	BE-IA
Scenario A (%)	50	10	10	30
Scenario B (%)	70	10	0	20

factor common for all VSPs. These assumptions lead to less complicated simulation scenarios that facilitate the evaluation of the features of the proposed resource sharing framework.

Table 11.1 contains the characteristics of the services that are considered in the performance evaluation, namely VoIP, machine-type communications (MTC) for alarm video surveillance, VC and best-effort internet access (BE-IA). Therefore, each admitted service call by a VSP is allocated a predefined data rate that is assumed to correspond to a particular level of QoE. In the following, two scenarios are examined and the respective distributions of the incoming traffic load among the aforementioned services are depicted in Table 11.2.

11.6.1 Scenario A – providing different service level agreements

In this section, a scenario is examined where specific and diverse SLAs are required. Thus, CS is compared to HRS with low sharing factor $r = 0.1$ in a simulation scenario where four VSPs share the capacity of one PNO and each one of them is specialised in a specific service class. Thus, VSP_1 serves only low data rate real-time services (i.e. VoIP) with a low blocking probability of 1%, VSP_2 serves MTC alarm services which require an extremely low blocking probability of 0.1%, VSP_3 serves high data rate real-time services (i.e. VC calls), also with a low blocking probability of 1%, while VSP_4 serves best-effort traffic with no specific QoS requirements. The traffic distribution of the incoming traffic load between the four VSPs is shown in Table 11.2.

Figure 11.7 depicts the blocking probability that is achieved when a VSP uses CS or HRS with a low sharing factor $r = 0.1$, for increasing traffic-load conditions. Both schemes achieve the goal of 1% blocking probability. However, one may observe that, as the traffic load increases, HRS provides lower blocking probability. Therefore, VoIP calls are adequately served under both sharing schemes, but in the case of CS, this is mainly due to the ability of low data rate VoIP calls to utilise efficiently commonly shared resources and not due to a tunable feature of CS.

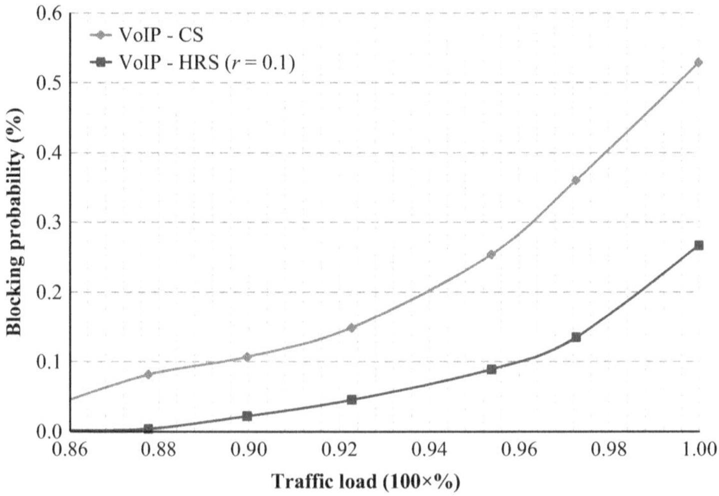

Figure 11.7 The blocking probability comparison of the proposed hybrid and the complete sharing scheme for increasing traffic load for the VoIP service

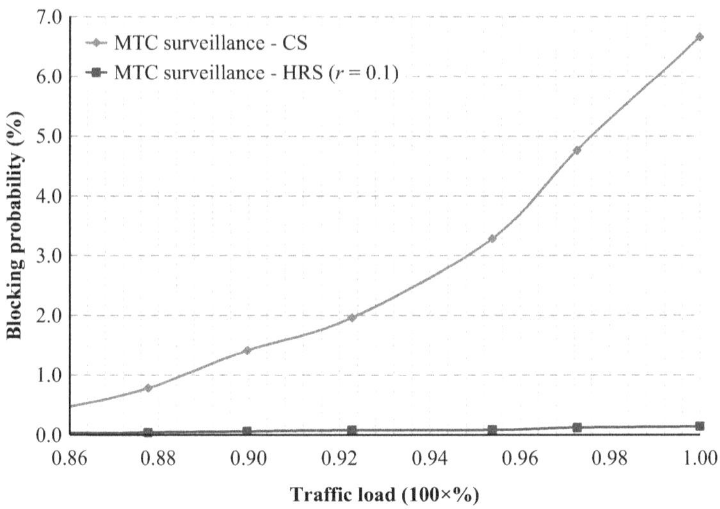

Figure 11.8 The blocking probability comparison of the proposed and the complete sharing scheme for increasing traffic load for the MTC surveillance service

This becomes apparent in Figure 11.8 which shows the MTC surveillance results for the CS and the proposed scheme with $r = 0.1$, for increasing traffic load. In this comparison, it is observed that the hybrid scheme has a clear advantage and offers a blocking probability below 1% for the whole traffic-load range.

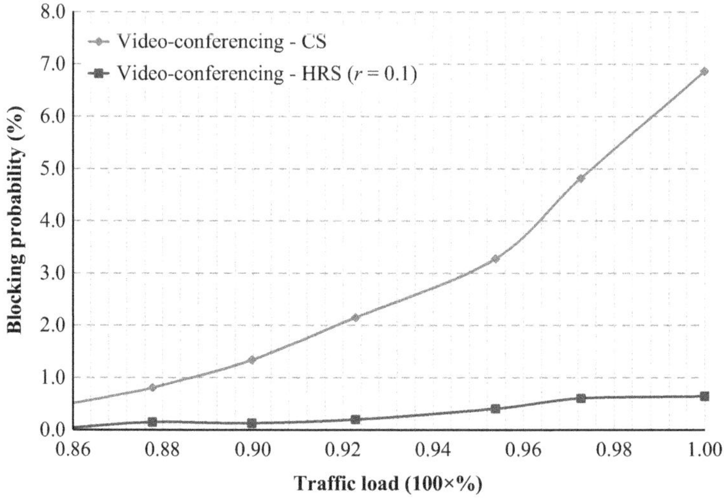

Figure 11.9 The blocking probability comparison of the proposed and the complete sharing scheme for increasing traffic load for the video conferencing service

On the contrary, when the CS scheme is adopted by the VSP of the MTC service, the blocking probability rises nearly up to 7% and the required SLA (QoS) level cannot be achieved. Although usually, the MTC alarm surveillance services have short duration, they demand relatively high data rate, and as a result, they have great difficulties when capacity has to be acquired from a CP. On the other hand, the hybrid scheme offers to the VSP of the MTC services a highly protected (i.e. low sharing factor of 0.1) capacity partition which ensures the required SLA.

The same conclusion is reach in the third comparison which relates to the blocking probability of the VC service for CS and HRS with $r = 0.1$ for different values of traffic. These results are included in Figure 11.9. The improved performance of HRS is obvious in this figure, as the blocking probability is maintained below 1%, thus providing significant benefits for the VSP of the VC service.

Furthermore, HRS treats the BE VSP as such and consequently allocates only the capacity that is left after ensuring the SLAs of the other VSPs. On the contrary, CS offers to the BE VSP relatively low blockages at the expense of the provided SLAs to VSPs 2 and 3. The blocking probability results for the BE-IA service are illustrated in Figure 11.10 for increasing traffic load.

Concluding the first part of the comparisons, it is derived that SLAs between the VSPs and the PNO cannot be ensured when a CS scheme is adopted by the PNO to share its physical resources. The provisioning of resources through a partitioning scheme seems to be the only way to offer guaranteed SLAs.

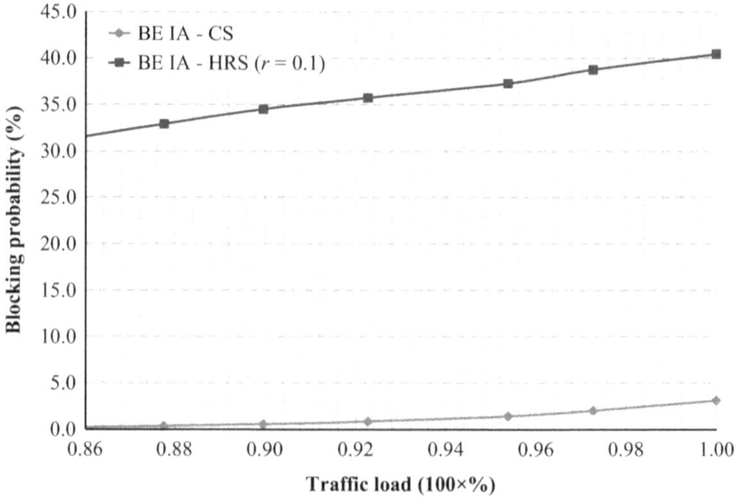

Figure 11.10 *The blocking probability comparison of the proposed and the complete sharing scheme for increasing traffic load for the best-effort Internet access service*

11.6.2 Scenario B – flexible vs inflexible partitioning

However, an FS scheme is inflexible and that can also be the cause of the SLA reduction when short-term variations of the traffic-load distribution occur. In this scenario, it is assumed that the VSPs negotiate with the PNO for specific SLAs based on the same traffic-load distribution as in the previous scenario. Nevertheless, it is considered that the actual traffic-load distribution suddenly changes from (50, 10, 10, 30) to (70, 10, 0, 20) as shown in Table 11.2.

Figure 11.11 investigates the performance improvement offered by the hybrid scheme with $r = 0.3$, compared to the FS scheme when VoIP service is provided. It is easy to observe that FS cannot adapt to the traffic load's variations and the VSP might experience difficulties in achieving the desired QoS level. On the other hand, HRS gives the opportunity to the VSP_1's calls to be served through the shared capacity of VSP_3 which for this short period has not any ongoing traffic flows. As a result, HRS constitutes a better choice as blocking remains well below 1% in this case.

11.6.3 Varying value of the sharing factor (r)

Following the traffic-load distribution of the first scenario (see Table 11.2), the performance of the hybrid scheme is evaluated when different values of the sharing factor are adopted, as shown in Figure 11.12. It must be emphasised that by adjusting r, the performance of the hybrid scheme can be fine-tuned between an FS-like behaviour ($r = 0.1$) to a CS-like behaviour ($r = 0.7$).

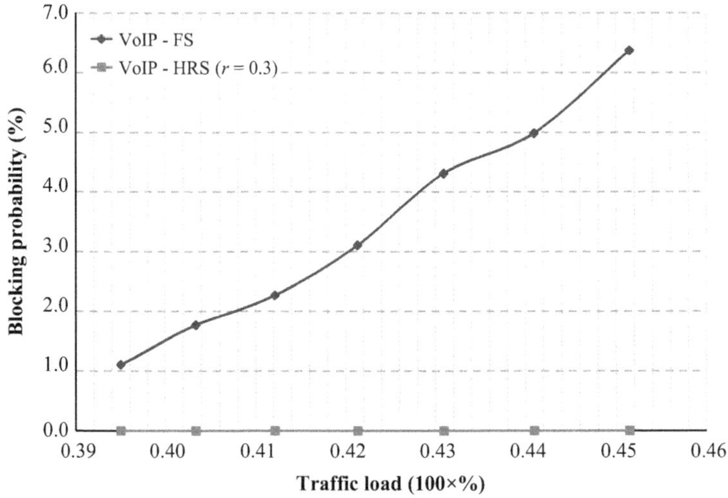

Figure 11.11 The blocking probability comparison of the proposed and the fixed scheme for increasing traffic load for the VoIP service

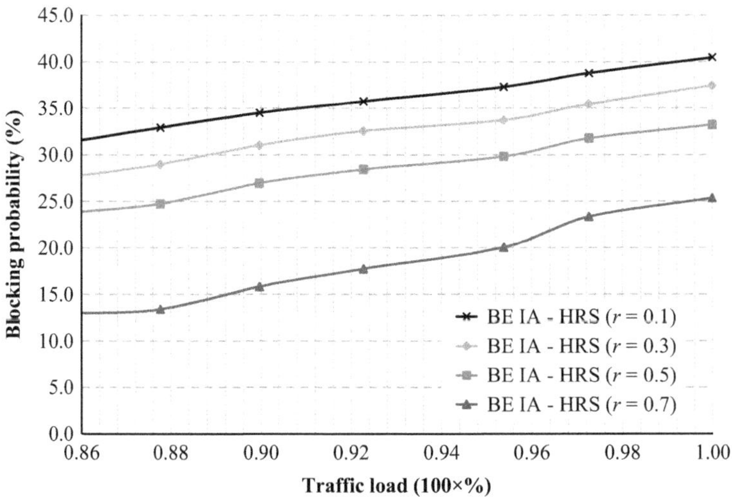

Figure 11.12 The blocking probability for increasing traffic load for the best-effort Internet access service for different sharing factors

As discussed in Sections 11.3 and 11.4, the use of a CS scheme can be useful in cases where all VSPs have similar requirements or when no specific SLAs are needed. In such a context, CS is able to provide higher throughput and capacity utilisation than partitioning sharing schemes. Accordingly in this case, the increase of r, gradually provides a more elastic resource sharing between VSPs and reduces the blockages of BE service calls of VSP_4.

11.7 Open issues

In this chapter, various resource sharing mechanisms and aspects of WNV are discussed. Moreover, wireless prosuming has been envisioned at the heart of WNV due to the dense and distributed spectrum state reports that can be provided by the end-users. However, as the research on WNV is at an early stage, various open issues arise. In greater detail, the following topics should be addressed in order to consolidate WNV as an enabling technology for 5G communications:

- Interference is a serious concern as intra- and inter-cell interference arise due to the frequent spectrum reuse among cell of the same and different tiers, respectively [22–24]. Moreover, user-deployed base stations such as femtocells have led to uncontrolled network deployment from the PNOs' perspective. Thus, efficient interference coordination algorithms should be deployed (e.g. as optimisation functions in the VNPO module of Figure 11.1) in order to enable the use of heterogeneous spectrum bands.
- Also, in 5G communications, the *Device-to-Device* paradigm is considered a technology that can provide significant off-loading from the PNO's network. Through D2D, devices communicate directly and avoid the usage of the base stations' resources. However, in many cases, D2D can be facilitated through signalling from the overlay infrastructure-based network. To enable efficient D2D communication, a VSP can coordinate devices that desire to communicate directly since it has the available data on the utilisation of the virtualised resources.
- The integration of cloud computing elements in enabling WNV is considered an additional challenge. Since the timely and accurate abstraction of resources requires powerful computing capabilities, the PNOs might experience difficulties in satisfying this constraint. Thus, cloud-based approaches could relax this requirement and provide scalable solutions through the formation of local cloudlets as proposed in [12] that will be provided by cloud SPs. These cloudlets will store the information on the state of wireless resources and will perform the allocation of the virtualised resource to VSPs.
- Another concern lies in the level of security that has to be assured in the forthcoming wireless networks [25] and more specifically, in WNV settings. In greater detail, the virtualisation of physical resources should be performed by trusted entities of the network in order to ensure a robust operation. Also, since cloud-based and user-centric approaches, such as prosuming can be integrated into WNV, malicious behaviours could threaten the validity of the information on the availability of resources. To this end, authentication mechanisms and cryptographic algorithms should be applied to ensure data integrity, whereas physical-layer security algorithms can be adopted to ensure secure delivery of spectrum estimation data [26–28].
- Finally, the formation of novel business models must be investigated in order to promote the added value that WNV offers to various stakeholders. In greater detail, WNV's business models should allow new players to enter the market

such as VSPs in the form of SMEs that are specialised in novel services, cloud SPs that bring advanced cloud computing elements with high processing power and users that adopt the role of prosumers of wireless resources. In addition, concepts such as spectrum marketplaces could be examined in cases where PNOs desire to exchange wireless resources and VSPs bid for the acquisition of the available resources.

11.8 Conclusions

This chapter presented the topic of prosumer-based WNV for optimised resource utilisation in 5G networks. The concept of wireless prosuming was presented and a framework where prosumers aid in the acquisition of spectrum state information was given in detail. Then, the role of the PNO and the VSPs within the WNV framework were investigated and various business models focusing on the interplay between PNOs and VSPS, as well as among VSPs were discussed. After, the complete and FS schemes of the PNO's spectral resources by the VSPs were discussed, and an HRS scheme improving upon such approaches was described. The efficiency of HRS was shown through performance evaluation for the blocking probability for different services and traffic load conditions. Results indicate that a controlled sharing between the VSPs coupled with a shared chunk of resources that constitutes a common resource pool can provide elasticity in service delivery. In this way, blockages are avoided and the QoS is improved compared to the fixed and CS schemes. Finally, various open issues were listed that require solutions in order to achieve robust and efficient prosumer-based WNV in the 5G era.

References

[1] N. Bhushan, J. Li, D. Malladi, *et al.*, "Network densification: the dominant theme for wireless evolution into 5G," *IEEE Commun. Mag.*, vol. 52, no. 2, pp. 82–89, 2014.

[2] G. Xilouris, E. Trouva, F. Lobillo, "T-NOVA: a marketplace for virtualized network functions," *2014 European Conference on Networks and Communications (EuCNC)*; Bologna, IEEE, 2014, pp. 1–5.

[3] F. Hu, Q. Hao, K. Bao, "A survey on software-defined network and OpenFlow: from concept to implementation," *IEEE Commun. Surv. Tutorials*, vol. 16, no. 4, pp. 2181–2206, Fourth quarter 2014.

[4] A. Beloglazov, R. Buyya, "OpenStack Neat: a framework for dynamic and energy-efficient consolidation of virtual machines in OpenStack clouds," *Concurr. Comput. Pract. Exp.*, vol. 27, no. 5, pp. 1310–1333, Apr. 2015.

[5] R. Kokku, R. Mahindra, H. Zhang, S. Rangarajan, "NVS: a substrate for virtualizing wireless resources in cellular networks," *IEEE/ACM Trans. Netw.*, vol. 20, no. 5, pp. 1333–1346, 2012.

[6] C. Liang, F. R. Yu, "Wireless network virtualisation: a survey, some research issues and challenges," *IEEE Commun. Surv. Tutorials*, vol. 17, no. 1, pp. 358–380, 2014.

[7] A. Osseiran, F. Boccardi, V. Braun, "Scenarios for 5G mobile and wireless communications: the vision of the METIS project," *IEEE Commun. Mag.*, vol. 52, no. 5, pp. 26–35, May 2014.
[8] P. Rost, C. J. Bernardos, A. De Domenico, "Cloud technologies for flexible 5G radio access networks," *IEEE Commun. Mag.*, vol. 52, no. 5, pp. 68–76, May 2014.
[9] X. Costa-Perez, J. Swetina, T. Guo, R. Mahindra, S. Rangarajan, "Radio access network virtualisation for future mobile carrier networks," *IEEE Commun. Mag.*, vol. 51, no. 7, pp. 27–35, 2013.
[10] A. Apostolidis, L. Campoy, K. Chatzikokolakis, *et al.*, Mobile and wireless communications Enablers for the Twenty-twenty Information Society (METIS), "Intermediate description of the spectrum needs and usage principles," METIS Del. ICT-317669-METIS/D5.1 v.1.0, Aug. 2013.
[11] F. Beltran, J. Gutierrez, J. Melus, "Technology and market conditions toward a new competitive landscape in the wireless access market," *IEEE Commun. Mag.*, vol. 48, no. 6, pp. 46–52, 2010.
[12] N. Nomikos, P. Makris, D. N. Skoutas, D. Vouyioukas, C. Skianis, "Enabling wireless prosuming in 5G networks," *Int. Conf. Telecommun. Multimedia (TEMU)*, Heraklion, pp. 190–195, 2014. doi: 10.1109/TEMU.2014.6917759.
[13] L. Musavian, S. Assa, S. Lambotharan, "Effective capacity for interference and delay constrained cognitive radio relay channels," *IEEE Trans. Wireless Commun.*, vol. 9, no. 5, pp. 1698–1707, 2010.
[14] S. Pandit, G. Singh, "Throughput maximisation with reduced data loss rate in cognitive radio network," *Springer Telecom. Systems*, DOI: 10.1007/s11235-013-9858-z, Aug. 2013.
[15] A. Bourdena, P. Makris, D. N. Skoutas, *et al.*, "Joint radio resource management in cognitive networks", in *Evolution of Cognitive Networks and Self-Adaptive Communication Systems*, vol. 38, no. 3, IGI Global, 1AD, pp. 50–80, 2013. doi:10.4018/978-1-4666-4189-1.ch003.
[16] R. W. Heath, R. Heath, A. Lozano, T. L. Marzetta, P. Popovski, "Five disruptive technology directions for 5G," *IEEE Commun. Mag.*, vol. 52, no. 2, pp. 74–80, 2014.
[17] P. Demestichas, A. Georgakopoulos, D. Karvounas, *et al.*, "5G on the horizon: key challenges for the radio-access network," *IEEE Veh. Tech. Mag.*, vol. 8, no. 3, pp. 47–53, 2013.
[18] D. Niyato, P. Wang, "Cooperative transmission for meter data collection in smart grid," *IEEE Commun. Mag.*, vol. 50, no. 4, pp. 90–97, 2012.
[19] K. Zheng, F. Hu, W. Wang, W. Xiang, M. Dohler, "Radio resource allocation in LTE-advanced cellular networks with M2M communications," *IEEE Commun. Mag.*, vol. 50, no. 7, pp. 184–192, 2012.
[20] N. Feamster, L. Gao, J. Rexford, "How to lease the internet in your spare time," *ACM SIGCOMM Comp. Commun. Rev. (CCR)*, vol. 37, no. 1, pp. 61–64, 2007.
[21] L. M. Correia, H. Abramowicz, M. Johnsson, K. Wünstel, eds. Architecture and design for the future internet: 4WARD project. Springer Science & Business Media, Berlin, 2011.

[22] P. Makris, N. Nomikos, D. N. Skoutas, *et al.*, "A context-aware framework for the efficient integration of femtocells in IP and cellular infrastructures," *EURASIP J. Wireless. Commun. Netw.*, vol. 2013, no. 1, p. 1–15, March 2013.

[23] N. Nomikos, T. Charalambous, I. Krikidis, D. Vouyioukas, M. Johansson, "Hybrid cooperation through full-duplex opportunistic relaying and max-link relay selection with transmit power adaptation," *IEEE Int. Conf. Comm. (ICC)*, Sydney, NSW, pp. 5706–5711, June 2014. doi: 10.1109/ICC.2014.6884231.

[24] N. Nomikos, P. Makris, D. N. Skoutas, D. Vouyioukas, C. Skianis, "A cooperation framework for LTE femtocells' efficient integration in cellular infrastructures based on femto relay concept," *Proceeding of the 17th IEEE Computer Aided Modeling and Design of Communication Links and Networks (CAMAD)*; Barcelona, Spain, Sep 2012, pp. 318–322.

[25] A. Nieto, J. Lopez, "Analysis and taxonomy of security/QoS tradeoff solutions for the future internet," *Secur. Commun. Networks*, vol. 7, no. 12, pp. 2778–2803, Dec. 2014.

[26] I. Krikidis, J. S. Thompson, S. McLaughlin, "Relay selection for secure cooperative networks with jamming," *IEEE Trans. Wireless Commun.*, vol. 8, no. 10, pp. 5003–5011, 2009.

[27] R. Bassily, S. Ulukus, "Deaf cooperation and relay selection strategies for secure communication in multiple relay networks," *IEEE Trans. Signal Process.*, vol. 61, no. 6, pp. 1544–1554, 2013.

[28] N. Nomikos, A. Nieto, P. Makris, *et al.*, "Relay selection secure 5G green communications," *Telecommun. Syst.*, vol. 59, no. 1, pp. 169–187, May 2015.

Chapter 12

Energy efficiency gains through opportunistic cooperative schemes in cognitive radio networks

Abdelaali Chaoub[1], Ali Kamouch[2] and Zouhair Guennoun[2]

12.1 Contribution of the chapter

It is widely known that each decade experiences the emergence of a new generation of mobile services. From the birth of the first generation during the 1980s to the in-progress fourth generation, it is commonly believed that 5G will be soon a reality.

Several players in telecommunication industry have not yet obtained a consensus on what could be the roadmap of the new generation technologies. Yet, certainly, it is the generation of higher data rates and selfish adaptation to surrounding environment, as well as an infinite number of ubiquitously connected devices with energy savings. Cooperation and unlicensed access to spectrum are among the potential candidates to be considered in the context of next generation networks.

This chapter investigates how cognitive radio (CR) and cooperation would be an ideal combination, as cooperative systems can efficiently exploit the connectivity offered by spectrum cognition to ensure better spectrum usage. Approaches based on collaboration between cognitive and noncognitive nodes are actively pursued today to support the explosive growth of bandwidth-consuming services and drive the development of license-exempt applications. However, basic cooperative protocols often make use of relay resources in a deterministic fashion regardless of network randomness. As such, opportunistic cooperation through opportunistically scheduling the relay transmission is vital for reaping the performance benefit of cooperative communication. Little research efforts have developed practical proposals to demonstrate the system capabilities of hybrid architectures where the decision of cooperation is made depending on whether cooperation is beneficial or not. In this chapter, we introduce a novel cooperation selection scheme for dynamic spectrum access networks. When the direct link

[1]Communications Systems Department, National Institute of Posts and Telecommunications (INPT), Rabat, Morocco
[2]Laboratory of Electronic and Communication, Mohammadia School of Engineers, Mohammed V-Agdal University, Rabat, Morocco

between the source and the destination is reliable enough to ensure safe delivery of the licensed content, the direct-transmission (DT) is more suited. However, cooperation takes place when the direct link does not support the data rate of the primary communication; some intermediate nodes called secondary relays are engaged to relay the signal of the source in a many-to-one fashion while simultaneously renting special spectrum rights. Thus, the sender actively picks up the most appropriate strategy contingent on the instantaneous network observations. Afterward, we developed a theoretical model with the aim of finding the optimal power allocation needed by the opportunistic decode and forward (ODF) mechanism to attain a transmission rate target, which converges into a minimization problem of a given utility function, namely the overall power budget devoted to both primary and secondary networks. Numerical simulations are performed in terms of the total and combined power constraint on both the source and the relay to prove that by applying a proper relaying policy the proposed scheme can be more energy efficient than the conventional cooperation.

The remainder of this chapter is structured as follows: Section 12.2 describes the background and the motivations of this research. Further, a concise overview of cooperative diversity in CR networks is presented. Later, several contributions made on this research area are consolidated and summarized. Section 12.3 details the system model and analytically assesses the performance of the proposed opportunistic cooperative framework in terms of the average required total power, as power allocation is a major hindrance to the implementation of cooperative diversity. The key ingredient of this theoretical part is to minimize the average power required by any of the two policies, either direct or relay-aided, to achieve a target transmission rate from the primary link perspective. Throughout Section 12.4, a basic transmission scenario in CR-based cooperative networks is considered, and results of extensive Monte Carlo simulations are plotted to prove the effectiveness of the proposed system. Throughout this section, interesting observations are made along with some insightful comments to underline the benefits of such dynamic strategies compared to the conventional deterministic protocols, as cooperation is adopted only from time to time when it is advantageous. Then, the conclusion follows in Section 12.5.

12.2 Cognitive radio and cooperation: preliminaries

Last years are witnessing a huge demand for high-quality mobile services over wireless networks. The promise of high-bandwidth and low-latency wireless applications with the ability to deliver rich contents whenever and wherever the consumer wants has placed unprecedented stress on the limited spectrum and the available frequencies are becoming increasingly exhausted.

Several measurement campaigns for spectrum occupancy statistics conducted by the Federal Communications Commission as well as its counterparts around the world at many locations and during many time intervals have underlined the inefficient use of frequencies due to the intermittent connectivity of licensed users, referred to as primary users (PUs) [1]. On the contrary, the exclusive and static spectrum governance left over very little bands for fixed frequency assignments, in addition to the

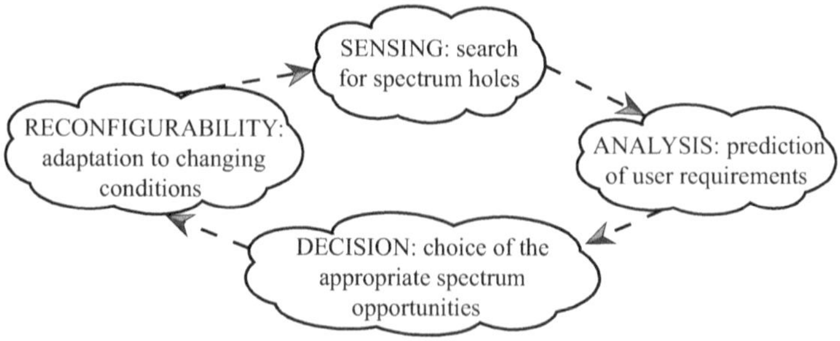

Figure 12.1 Cognitive radio cycle

mandatory auction process which is slow and expensive (billions of dollars). It is clearly seen that there will be no more frequency ranges available in the near future, whereas the already allocated spectrum is not optimally and efficiently used. These two contradictory findings regarding radio frequency behavior are among the most pressing challenges for telecommunication industry to face the prospects of a spectrum crunch.

As the radio spectrum is the lifeblood of the telecommunication market, research community has intensively worked on ways to provide new frequency allocation policies. The debate centers mainly focuses on how to allow new entrants to the wireless market, whilst radio frequencies are limited and naturally finite. As a result, the cognition in spectrum sharing [2] has arisen as a novel concept to change the rules for spectrum bands (Figure 12.1). The central idea is considering opening up the spectral resource to a new type of unlicensed devices, otherwise known as secondary users (SUs), so that licensed and unlicensed users can coexist with each other in a noninterfering mode [3].

Nowadays, CR is arousing immense interest among academics and industry. On 21 November 2013, the European Telecommunications Standards Institute (ETSI) Future Mobile Summit [4] has cited a number of key recommendations and concluding messages to forecast what the beyond-4G is envisioned to be. The unlicensed access to spectrum was one of the raised points. The concept of CR is among the promising technologies subject of research and investigations in the context of next generation networks [5], this technology allows the optimization of spectrum resources utilization and thus the spectral efficiency is maximized which leads to higher amount of bits transmitted per second and increased throughput.

Last years have also witnessed a recent strong interest in CR technical standardization [6] to bridge the gap between theory and practical achievements in this research area. For instance, numerous attempts are made to establish universal standards for CR, namely IEEE 802.22, IEEE 802.11af, and IEEE 802.19.1. IEEE 802.22 aims at deploying wireless regional area network infrastructures in rural and remote areas, and it is able to attain an aggregate data rate of up to 23 Mbps. It is the most mature CR standard at this time as it has started the process of standardization since October 2004. The IEEE 802.11af task group is the IEEE standard working on the issue of providing WiFi services over TV bands to achieve higher throughput data rates and faster connections, this task group was formed during

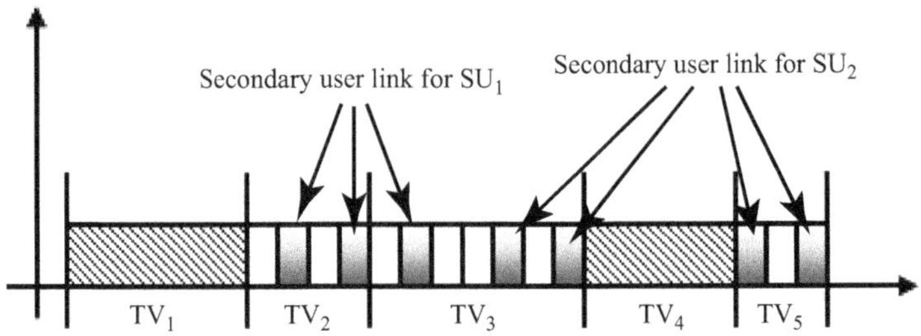

Figure 12.2 Coexistence between several SUs

December 2009. While using a certain band of spectrum, the secondary system must not only avoid disturbing the neighboring licensed network but also obtaining awareness about other coexisting secondary systems. Therefore, standards like IEEE 802.19.1 have been proposed to address the coexistence issue between several secondary technologies operating over TV bands (Figure 12.2), this project was approved on December 2009. Further, the ECMA-392 standard tackles the medium access control (MAC) and the physical (PHY) parts to enable the operation of portable devices over TV white spaces, it was published by the end of 2009. Moreover, IEEE has recently (September 2011) launched the task group IEEE 802.15.4m (802.15.4 over Television White Spaces (TVWS)) with the goal of enabling Wireless Personal Area Network (WPAN) infrastructures over TV bands. All the aforementioned standards are operating over TV white spaces (54–60 MHz, TV channel 2; 76–88 MHz, TV channels 5 and 6; 174–216 MHz, TV channels 7–13; 470–608 MHz, TV channels 14–36; and 614–698 MHz, TV channels 38–51). The basic idea behind is that the transition of TV channels from analog to digital has freed up a considerable amount of unused portions of spectrum. It is commonly known that TV bands have good propagation conditions, excellent building penetration, and high spectrum efficiency.

12.2.1 Interaction between primary and secondary users

Cognitive users need accurate and up-to-microsecond information about the radio frequency (RF) environment in vicinity to guarantee a peaceful coexistence between primary and nearby secondary devices. Spectrum rooms should be actively identified and permanently monitored. Strategies based on spectrum sensing have been suggested as innovative solutions to explore conventional and unconventional degrees of freedom (DoF) to locate the abundant spectrum gaps with high-detection probability [7]. Legacy sensing algorithms supervise the spectrum through three conventional dimensions: frequency, time, and space. However, other degrees of freedom such as the used code and the angle of arrival can be examined. Another alternative mechanism to identify the vacant gaps is the use of a geographic coordination through a central database [8], which is either a substitute or a complementary solution to selfish spectrum sensing. Despite the increasing maturity of these two techniques, the strong heterogeneity and the nonlinearity of wireless environments dilute the RF signals and raise some technical challenges such as the hidden node problem (Figure 12.3), which makes sensing approach prone to

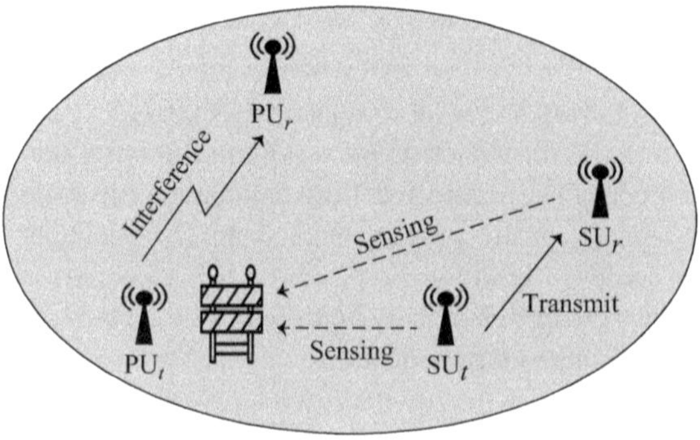

Figure 12.3 An example of hidden node situations

detection errors. Besides, the massive number of devices interconnected through wireless networks, their ever-changing characteristics is a limiting factor in the database solution. For future research, this is a challenging direction that opened up new horizons for this research area [9].

Depending on the degree of knowledge of the primary signal at the secondary network, spectrum gaps can be shared according to three levels of cognition, namely interweave, underlay, and overlay [10].

The interweave-based coexistence approach has been proposed with the objective of enabling devices to occupy the spectrum rooms that have been left vacant by noncognitive users. The surrounding environment should be observed to be able to predict the state of each portion of the frequency spectrum, portions of spectrum considered as being underutilized may be accessed by SUs as long as the primary activity remains idle. In order to facilitate the coexistence of both primary and secondary traffics within the same network in an opportunistic transmission mode, spectrum opportunities should be actively identified and monitored. Cognitive users may conduct sensing operations permanently and reliably, and different dimensions need to be explored to find the abundant spectrum gaps. Legacy sensing algorithms monitor and supervise the spectrum through three conventional dimensions: frequency, time, and space. However, other degrees of freedom such as the used code and the angle of arrival may be examined. Another alternative mechanism to identify the vacant gaps is the use of a geographic coordination through a central database, which is either a substitute or a complementary solution to selfish spectrum sensing.

In underlay systems, simultaneous cognitive and noncognitive transmissions are allowed as long as the interference level at the PU side remains confined within the interference limit imposed to cognitive devices. In recent literature, many advanced signal processing techniques have proven to be very efficient for interference avoidance, among which we find the beamforming and the spread spectrum techniques to be excellent. Beamforming consists of exploiting the superposition concept of waves to guide the signal toward a specific receiver using multiple antennas. More importantly, in spectrum-sharing contexts, constructive or destructive interference is provoked at the intended cognitive receiver to lessen the interference caused to noncognitive users while focusing the signal energy toward the desired direction. Using the spread spectrum technique, the secondary signal is multiplied by a spreading code to obtain a weaker signal with wider band. The resulting spread signal causes lower interference level to noncognitive users. The original secondary signal is recovered at the receiver side by simply multiplying the input signal with the same spreading code. The spread spectrum technique is also useful for alleviating the interference caused by the primary signals to the secondary ones. Another common solution could be limiting the power of the secondary signal to keep the interference level at the primary side bounded, albeit restricting the secondary transmissions to short range communications.

Using the overlay approach, cognitive users are able to track the PU messages or codebooks to be able to either null out the interference caused by the primary signal at the secondary receiving end or to strengthen the primary signal through serving as a relay for the licensed traffic. Unlike the underlay scheme, there is no interference temperature constraint enforced on the secondary signal power. To coexist with the licensed network without any interference, SUs seek the best compromise between the interference induced and the improvement brought to the primary signal to achieve a stagnant signal-to-noise ratio (SNR).

Advanced levels of cognition can be obtained using a combination of the above-cited frameworks [11]. Later, we will show that increased efficiency may also be achieved by occasionally and not permanently exploiting the available side information.

12.2.2 Overview of spatial diversity in cognitive radio networks

CR networks need to be engineered to meet tight constraints in terms of energy, error resilience, and interference avoidance. To this end, the integration of the spatial dimension has been proposed as a means to meet these requirements by creating transmit and/or receive diversity. Recurrent techniques for achieving this spatial diversity include, typically, MIMO (multiple-input–multiple-output) configurations and relaying. The extra diversity offered by the spatial DoF can be exploited to substantially boost the rate and speed of communication systems without the need to purchase extra bandwidth.

To date, MIMO technology is gaining further momentum worldwide [12]. This technique endows wireless devices with the capability of simultaneous transmission and/or reception of multiple independent streams using antenna diversity in a noninterfering mode via interference concealment techniques. However, still a number of challenges arise including the need for miniaturization to stack many antennas in a small volume and how to increase the number of collocated antennas without creating significant inter-antenna interference.

Nowadays, relaying concept has ignited an intensive research as an innovative technique to achieve reliable connectivity and guarantee favorable conditions for quality of service (QoS)-sensitive secondary communications [13]. The concept of relay channel characterizes a class of three-device communication channels and was examined the first time by van der Meulen [14,15] and later by Cover and

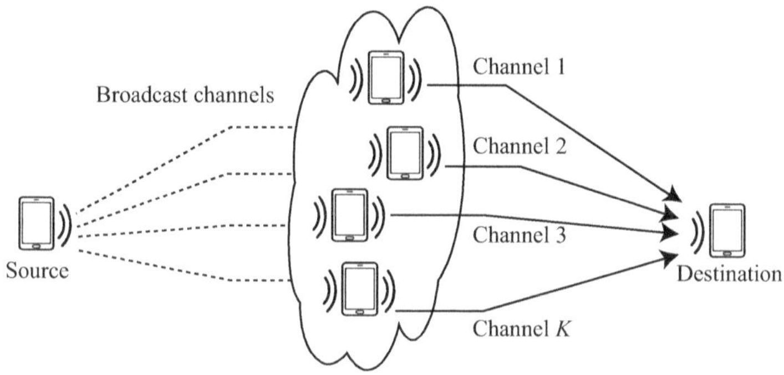

Figure 12.4 Dual-hop multirelay CR network

Gamal [16]. In particular, the source and destination nodes may not always be able to proceed with a DT especially in harsh environments. Thus, the link between source and destination can be bridged by a third party, called relay, with rich spectrum opportunities when there is no available spectrum gaps to establish the direct source–destination link (Figure 12.4). During the past few years, because of their ability to deliver streams over longer distances with extended network range and faster speeds, the applications of relaying have become quite popular, such as Wireless Sensor Networks [17], Vehicular Ad Hoc Networks [18], and Wireless Mesh Networks [19]. Given the above promising potentials, cooperative relay-based schemes still need to overcome some major challenges [20] related to security vulnerabilities, resource control and management, synchronization protocols, implementation issues, delay overheads, and the number of nodes available and ready for cooperation.

12.2.3 To cooperate or not? That is the question!

Most of the research on cooperation protocols is based on the approach of 'always cooperate when possible'. That said, each transmitting source recruits a number of relays and solicit their help in stream delivery regardless of the network state. One defect of such schemes is that the recruited nodes will always cooperate even when it is not necessarily gainful for the entire communication. It can be also noticed that a plenty of works that tackled the principle of cooperation via relays have focused on the goal of maximizing spectral efficiency, minimizing bit error rate (BER), or maximizing transmission rate. Nevertheless, energy-efficient cooperative designs have not been paid sufficient attention. Further, the QoS of relay-aided networks hinges upon the varying network state and need to be subsequently adapted to. These findings together revealed a number of shortcomings related to fixed cooperative scenarios and emanate the need for truly flexible and opportunistic cooperative strategies [21].

It becomes increasingly imperative to develop intelligent systems, wherein the source–destination pair cooperates with the candidate relays only when it is in its self-interest based upon local decisions according to optimum energy allocation.

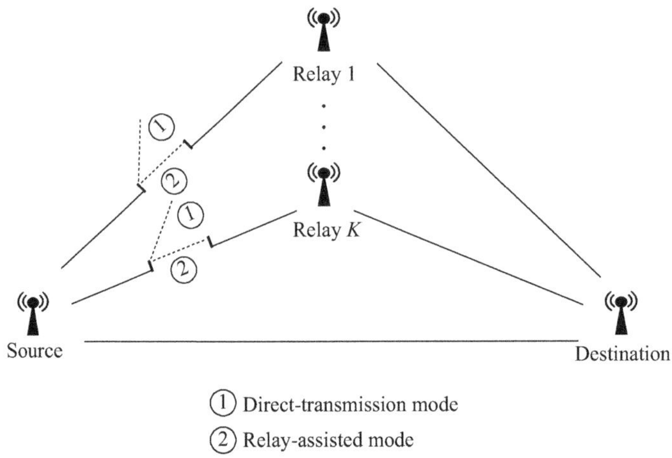

Figure 12.5 Hybrid scheme combining direct-transmission and relay-assisted modes

The sender chooses to make use of a two-hop high data rate path to the destination via a relay when the source–destination channel is not advantaged, whereas it transmits data straight to the intended receiver in case of better conditions on the direct channel (Figure 12.5); bearing in mind that channel conditions are closely related to energy consumption. Such dynamic strategies enjoy high-error resilience against channel fluctuations in RF environments and offer better resource exploitation. More importantly, such schemes 'only cooperate when beneficial' achieve low-power consumption which prolong the battery life of wireless devices. In fact, the tremendous proliferation of smartphones accompanied by the booming growth of energy-consuming applications urged the industry and academia around the world to stimulate research on innovative means for longer power autonomy [22].

12.2.4 A literature survey on opportunistic cooperation protocols

In recent studies, the non-fixed type cooperative schemes have raised interest among researchers.

Li *et al.* [23] have used some opportunistic cooperative protocols for distributed sensing activities to attain significant improvements in terms of detection probability and energy consumption. Unlike earlier schemes, the proposed solution has the property that dedicated fusion infrastructures are no longer needed as spectrum decisions are basically dependent on local observations.

Gunduz and Erkip have designed an ODF scheme subject to delay and power constraints and analyzed its outage behavior in [24] and its delay-limited capacity in [25]. The relay's intervention is governed by a local decision regarding the mode of transmission, that is, direct or relay-assisted (RA), providing the least total power. Depending on the amount of channel knowledge available at the transmitter side, the authors have investigated the ODF model under a complete [25] and a partial [24] channel state information (CSI).

The same authors had consolidated the above findings in [26] and proved that with partial CSI at the transmitter and perfect CSI at the receiver a non-zero delay-limited capacity can be achieved regardless of the average power constraint. The authors also showed the merits of the ODF strategy in terms of both delay-limited capacity and maximum outage probability compared to the classical multihop (MH) protocol.

In [27], El-Bakoury *et al.* derived the closed form expressions for the average power needed for both MH and ODF strategies to achieve a target rate through a one-bit feedback on the relay–destination channel status and a maximum power constraint on each participating node. The authors argued that a limited information about the status of the relay–destination channel suffices for the source node at the cost of a negligible degradation in terms of the minimum required average power.

A different approach has been adopted in [28], wherein the appropriate transmission mode is chosen according to the presence of an outage event. A BER analysis has been conducted for the proposed scheme and for the non-cooperation and the coded cooperation cases as well, whereby insightful conclusions are reported regarding the superiority of the newly designed scheme over the conventional strategies.

Urgaonkar and Neely [29] explore the cognitive femtocell concept and investigates the improvement provided by the opportunistic cooperation to such contexts. More precisely, cognitive devices may enhance the noncognitive transmission and negotiate to rent special spectrum rights instead of vying for dynamic access. These interactions have been modeled using a constrained Markov Decision Problem to converge to an optimal solution.

In [30], a communication system is presented where a threshold-based criterion is used to switch between the cooperative and the noncooperative modes under the assumption of partial channel information. The authors demonstrated that the adequate choice of the thresholds permits to compensate for the effects of imperfect CSI knowledge.

Urgaonkar and Neely have investigated the concept of dynamic cooperation in mobile ad hoc network under tight delay constraints and random packet arrivals [31]. The user-level outage behavior is studied according to optimal power allocation goals.

A more generic opportunistic solution has been proposed in [32] where the transmission rate is maximized by using a hybrid scheme that switches dynamically among three transmission modes, which are decode-and-forward, compress-and-forward, and direct strategy.

As shown in the previous works on the topic, opportunistic cooperative strategies and their potential to enhance the design of emerging CR networks have not been fully investigated. In addition, most previous works have omitted the multiple relay scenario for complexity avoidance.

In the following, we address the question of whether cooperation would always produce higher levels of reliability than the direct mode wherein no assistance from any intermediate node is provided. To study this question, we present our arguments in favor of a new type of cooperative relaying schemes based on a certain energy selection criteria that use the available CSI collected over multiple links. We will introduce an opportunistic cooperation model where the source–destination pair acquires the available relay resources only for irregular intervals depending on channel impairments.

12.3 Proposed work

12.3.1 General analysis

In this section, we examine a CR network collocated with a primary link. Time is divided into frames of fixed duration T. Consider the practical scenario in which a common noncognitive source PT_x is sending a single stream to a common noncognitive destination PR_x during one time frame. Simultaneously, K secondary transmitter-receiver pairs $ST_x^i - SR_x^i$ $(i \in \{1, \ldots, K\})$ operate on the same wireless channel, that is, in the range of the ongoing primary communication. All the links between different nodes are supposed to be Gaussian and each link suffer from a Rayleigh fading process. Denote the channel gain from the primary transmitter to the primary receiver as h_{PP}. $h_{TT,i}$ is the channel gain between the primary transmitter and the ith secondary transmitter and $h_{SP,i}$ is the channel gain between the ith secondary transmitter and the primary receiver. The channel gain of the secondary link $ST_x^i - SR_x^i$ is denoted as $h_{SS,i}$ and the channel gain between the secondary transmitter j and the secondary receiver i is denoted as $h_{SS,ij}$ with $i \neq j$ (Figure 12.6). We assume a frequency-flat block Rayleigh fading which means that the channel gain is invariant during each frame but changes from one frame to the other. We assume also that noise processes over various links are zero mean and have identical variance N_0.

In what follows, we suppose that the fading conditions are slowly varying so that the receiver can have an accurate estimate of the state of its associated physical resources and may disseminate these measurements to the transmitter using a dedicated feedback channel. Hence, CSI is perfectly known to the transmitter. Another alternative is periodically sending some pilot signals and collecting feedback. It is interesting to note that with only a partial information of the channel at the transmitter side, good results can be obtained [24]. Besides, we assume a half-duplex constraint and we suppose that each antenna can ensure the transmission as well as the reception of radio signals (i.e. transceiver).

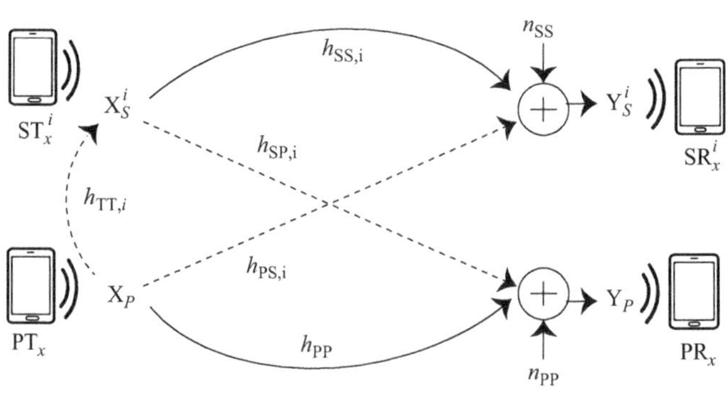

Figure 12.6 Concurrent cognitive and noncognitive transmissions

We use the notation $|.|$ for the absolute value in the case of a scalar and the Frobenius norm in the case of a vector.

12.3.2 Opportunistic cooperative schemes

Due to channel irregularities in radio environments, the primary link cannot be always connected in the common channels. The primary source can use some intermediate secondary nodes that experience relatively good channels with the source and the destination. In return, the chosen relays can overlay their secondary signals with that of the PUs. However, the involvement of a third party can also be viewed as another concern. The cognitive and noncognitive users need to balance the desire to cooperate with the need for maintaining optimal energy levels.

Hybrid schemes, using a combination of point-to-point and cooperative transmissions, have a great potential to ensure optimal service to the final users. The primary pair cooperates with some chosen SUs only when the direct channel is not advantaged. Concretely, at each network state s, the proposed scheme picks the transmission mode that requires the least total power for the same preassigned rate, denoted R. In the following, we assume that the source broadcasts its decision, whether to use the DT or the RA mode, through a control channel to the destination and the relays. We assume also that the messaging volume resulting from the transmission of the source decision causes no significant overhead.

To provide a performance analysis of the described scheme, we characterize the end-to-end instantaneous capacity at the receiving end for both primary and secondary networks. The instantaneous capacities of the two modes DT and RA are denoted, respectively, as $\eta_{\mathrm{DT}}^{\mathrm{p}}$ and $\eta_{\mathrm{RA}}^{\mathrm{p}}$. Define $P_{\mathrm{p}}(s)$ and $P_{\mathrm{s}}^{i}(s)$ as the transmit power of the primary transmitter and the ith secondary transmitter, respectively, for a given realization of channel gains s. The power allocation between the primary and secondary networks is subject to a long-term maximum allowable power $P_{\max}$. Let C be the set of channel realizations where the DT mode is preferred and $\overline{C}$ is the set of network states where relay assistance is sought. The optimal transmission policy is the one with the lowest energy consumption while guaranteeing a certain rate R at any channel state.

12.3.2.1 Direct transmission mode: $s \in C$

The instantaneous capacity of the direct mode is defined as the throughput of a straight communication between the primary transmitter and the primary receiver during the duration of one-time frame

$$\eta_{\mathrm{DT}}^{p} = \log_2\left(1 + \frac{|h_{\mathrm{PP}}|^2 P_p(s)}{N_0}\right). \tag{12.1}$$

The minimum required total power of the direct mode can be written as follows:

$$P_{\min}^{\mathrm{DT}} = \min\{P_p(s), \eta_{\mathrm{DT}}^{p} \geq R\}. \tag{12.2}$$

The inequality above means that the transmission rate cannot fall below the threshold value R because this will occur an outage event.

Substituting (12.21) with (12.2), we get

$$P_{\min}^{\mathrm{DT}} = N_0 \frac{\left(2^R - 1\right)}{\left|h_{\mathrm{PP}}\right|^2}. \tag{12.3}$$

12.3.2.2 Relay assisted mode: $s \in \overline{C}$

Popular overlay schemes proposed in research literature allow multiple coexisting cognitive and noncognitive nodes to share the same infrastructure through the use of time splitting. SUs are allowed to rent some portions of frequency bands for a negotiated power cost and following a predetermined time schedule subject to an explicit agreement with the license holders. Concretely, the time frame structure is decomposed into three blocks according to two parameters α and β $(0 < \alpha, \beta < 1)$ as illustrated in Figure 12.7. During the first phase, the noncognitive source keeps broadcasting the initial stream to both the relays and the destination. These decode-and-forward relays will collaborate in the second-phase transmission to help accumulating enough data at the noncognitive destination to recover the original message with the help of the signal received from the source. The third phase is granted to the cognitive devices for secondary content delivery as a reward for their cooperation and assistance (Figure 12.7). A robust approach to select the appropriate set of relays is to sort the paths connecting the primary source and the available secondary transmitters according to the channel gain and select those with the K highest values (Figure 12.8).

Based on the model described above, the primary source PT_x transmits continuously a given stream X_P until the K relaying nodes $\left(\mathrm{ST}_x^i\right)_{i\in\{1,\ldots,K\}}$ are able to successfully decode the transmitted message. Then, the decoding relays perform

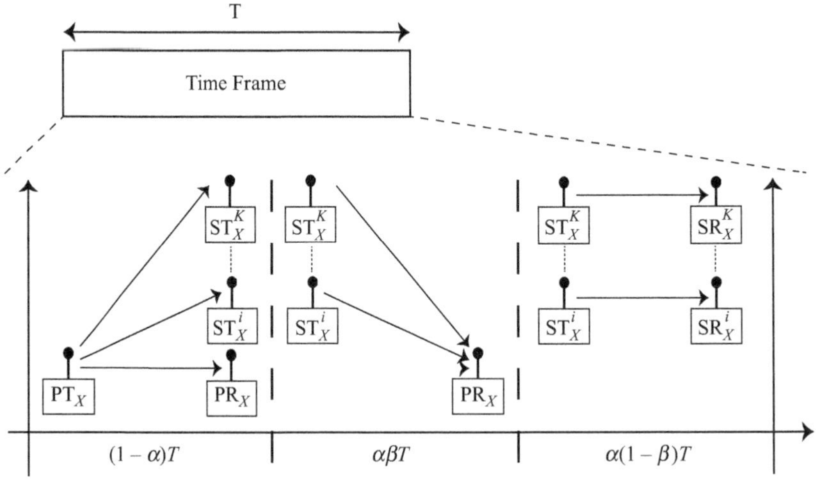

Figure 12.7 Three-phase transmission scheme

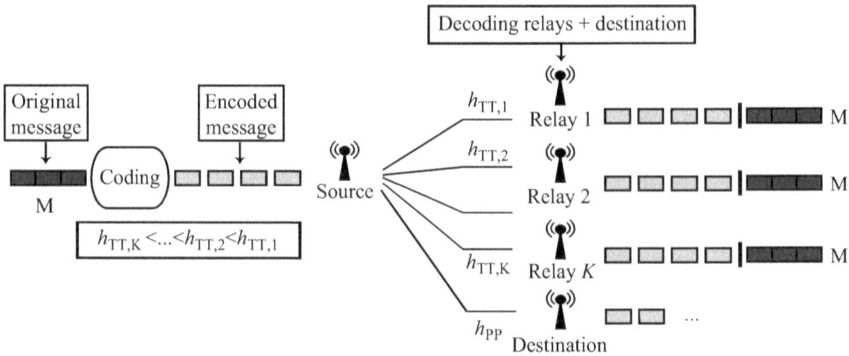

Figure 12.8 Decode-and-forward relays sorted in decreasing order of the channel gain

parallel transmissions of the same stream over the second hop to the primary destination. Due to the broadcast nature of the wireless transmission, a copy of the source signal transmitted during phase 1 is also overheard by the destination and combined with relays signals to increase diversity. Obviously, the final signal Y_P is dominated by the worst hop among the first two stages. During the third phase, each cognitive user ST_x^i conveys its private stream X_S^i to the destination SR_x^i. This signal encounters a Rayleigh channel fading $h_{\mathrm{SS},i}$ and an additive white Gaussian noise n_{SS} with zero mean and power N_0, along the opportunistic path from the source to the destination. Besides, the secondary pairs sharing the same primary link can harmfully interfere with each other. The output signal Y_S^i subject to fading, noise, and mutual interference can be expressed as follows:

$$Y_{\mathrm{S}}^i = h_{\mathrm{SS},i}X_{\mathrm{S}}^i + \sum_{j=1,j\neq i}^{K} h_{\mathrm{SS},ij}X_{\mathrm{S}}^j + n_{\mathrm{SS}}. \tag{12.4}$$

For a simple derivation of the closed-form expression of the throughput, we assume a constant transmit rate which is equal to the ergodic capacity. The set of relaying nodes may be tackled in a holistic manner as all the members behave as a single destination dominated by the worst member, from the source node viewpoint. The same set of relays in addition to the source act as an unified structure with accumulated energy at the destination node. Accordingly, the throughput of the primary communication is the minimum of the throughput in the two first phases and can be derived as follows:

$$\eta_{\mathrm{RA}}^p = \min\{C_{\mathrm{RA}}^{\mathrm{TT}}, C_{\mathrm{RA}}^{\mathrm{SP}}\}. \tag{12.5}$$

with capacities $C_{\mathrm{RA}}^{\mathrm{TT}}$ and $C_{\mathrm{RA}}^{\mathrm{SP}}$ formulated as follows:

$$C_{\mathrm{RA}}^{\mathrm{TT}} = (1-\alpha)\log_2\left(1 + \frac{\min_{i\in\{1,\ldots,K\}}\left|h_{\mathrm{TT},i}\right|^2 P_p(s)}{N_0}\right). \tag{12.6}$$

and

$$C_{\mathrm{RA}}^{\mathrm{SP}} = (1-\alpha)\log_2\left(1+\frac{|h_{\mathrm{PP}}|^2 P_p(s)}{N_0}\right) + \alpha\beta\log_2\left(1+\frac{\sum_{i=1}^{K}|h_{\mathrm{SP},i}|^2 P_s^i(s)}{N_0}\right). \quad (12.7)$$

In this case, the minimum required total power of the relay-aided transmission is defined as follows:

$$P_{\min}^{\mathrm{RA}} = \min\left\{(1-\alpha)P_p(s)+\alpha\left(\sum_{i=1}^{K}P_s^i(s)\right), \eta_{\mathrm{RA}}^p \geq R\right\}. \quad (12.8)$$

We need to determine the lowest cost in terms of energy for the primary link in accordance with the throughput of the first hop defined in (12.6). The throughput from source to relays is a decreasing function of α and the relays-to-destination throughput is monotonically increasing as the value of α increases, thereby the optimal $P_p(s)$ (P_p^*) is obtained when both uplink and downlink achieve a same throughput value.

$$C_{\mathrm{RA}}^{\mathrm{TT}} = C_{\mathrm{RA}}^{\mathrm{SP}}. \quad (12.9)$$

Under optimal power allocation, which achieves the lower bound of the transmission rate, we have

$$C_{\mathrm{RA}}^{\mathrm{TT}} = R. \quad (12.10)$$

based on which the minimum required power of the primary link can be calculated by

$$P_p^* = N_0\frac{\left(2^{R/(1-\alpha)}-1\right)}{\min_{i\in\{1,\ldots,K\}}|h_{\mathrm{TT},i}|^2}. \quad (12.11)$$

On the other side, the vector of optimal powers $P_s^* = \left(P_{s,i}^*\right)_{1\leq i\leq K}$ picked up by relays can be set while keeping in mind that each SU should strive to maximize its achievable rate toward its intended receiver during the fraction of time $\alpha(1-\beta)$. More importantly, the PU has full rights on its spectrum, and it is responsible for deciding the appropriate parameters in view of time splitting (α,β) and the number of cooperating nodes K that maximize its utility in terms of the achieved throughput. Secondary links observing the decision made by the primary link act subsequently and choose the optimal power levels P_s^* in return for the leased fraction of time β, while seeking to maximize their individual utilities too that is, throughput. In either case, different players attempt to find a good revenue/payment trade-off. Both the selection of the appropriate strategy and the optimization of the expected utility issues become more tractable by transforming the multidimensional problem into a Stackelberg game and a Nash equilibrium with the primary link as a leader and the followers are the secondary links [33]. This approach is widely adopted in literature. However, a major handicap of such frameworks is the fact that the factor β is imposed by the primary network, which has substantial impacts on the secondary network reliability especially in the case of bandwidth-consuming applications. The primary link has tendency to choose a large β which reduces the time fraction dedicated to secondary communications and as a result secondary devices may become reluctant to cooperate, whereas cognitive users are incapable of providing high-transmit powers to gain access to a larger portion of time (smaller β values) because they have tight power constraints.

In the current chapter, to circumvent this bottleneck we proceed differently by assuming a negotiated spectrum sharing approach in which licensed users conclude explicit arrangements with interested parties to grant them spectrum access rights in exchange for monetary and nonmonetary compensations. This option allows the unlicensed users to trade their power requirements and their utility target for better spectrum sharing. The licensees may be very interested in allowing high-transmit power levels at the unlicensed network side, but only with strong technical assurances that cognitive devices guarantee an interference-free environment and as long as a part of the secondary power will be used to assist the primary traffic. Moreover, the additional monetary revenues will encourage the licensed operators to satisfy and deal with the unlicensed traffic demands; in such case, these monetary incomes can be invested in strengthening the resilience of the licensed infrastructure against potential interference incidents.

Motivated by above discussions, the following formulation is proposed. Each SU targets at optimizing an utility function that represents its transmission rate and is equal to the achievable throughput minus the energy cost during the time duration of the allocated subslot. This utility is given by

$$U_s^i = \alpha\left(\eta_{\mathrm{RA}}^s - cP_s^i(s)\right). \tag{12.12}$$

with c is the cost per unit transmission energy and η_{RA}^s is the normalized throughput at the secondary receiver, obtained from (12.4).

$$\eta_{\mathrm{RA}}^s = \alpha(1-\beta)\log_2\left(1 + \frac{|h_{\mathrm{SS},i}|^2 P_s^i(s)}{N_0 + \sum\limits_{j=1,j\neq i}^{K} |h_{\mathrm{SS},ij}|^2 P_s^j(s)}\right). \tag{12.13}$$

According to Lee *et al.* [22], this utility function is concave and thus the maximum achieved utility is attained at the point where the first derivative is null. By computing the first-order derivative of U_s^i with respect to $P_s^i(s)$, the above utility function reaches its maximum at the following stable point:

$$P_{s,i}^* = \begin{cases} P_{\max}, & \text{if } P_{s,i}^* \geq P_{\max} \\ \dfrac{1-\beta}{c} - \dfrac{N_0}{|h_{\mathrm{SS},i}|^2} - \sum\limits_{j=1,j\neq i}^{K} \dfrac{|h_{\mathrm{SS},ij}|^2}{|h_{\mathrm{SS},i}|^2} P_{s,j}^*, & \text{if } 0 < P_{s,i}^* < P_{\max} \\ 0, & \text{if } P_{s,i}^* \leq 0 \end{cases} \tag{12.14}$$

This system of K linear equations has a unique solution: $P_s^*(\beta)$.

Subsequently, we conclude from results in (12.11) and (12.14) that the minimum required power $P^{\mathrm{RA}}_{\min}$ (12.8) depends closely and only on the values of α and β: $P^{\mathrm{RA}}_{\min} = P^{\mathrm{RA}}_{\min}(\alpha, \beta)$.

12.3.2.3 Power minimization framework for opportunistic cooperation

The overall average power required by the system to achieve a certain rate R can be expressed as follows:

$$E[P_T] = E[P_T/s \in C] \times P(s \in C) + E\left[P_T/s \in \overline{C}\right] \times P\left(s \in \overline{C}\right). \quad (12.15)$$

$E[P_T]$ is calculated according to two decision regions, the first one is defined as the region where the DT mode is more energy efficient than the RA mode and the opposite happens in the second region.

Algorithm 12.1 Calculate the optimal average power numerically for a given rate R

Generate N network states $S = \left(h_{\mathrm{PP}}, (h_{\mathrm{TT},i})_{1\le i\le K}, (h_{\mathrm{SP},i})_{1\le i\le K}\right)$
$p_{\mathrm{b}} = 0$
$av_{\mathrm{DT}} = 0$
$av_{\mathrm{RA}} = 0$
for all $s \in S$ **do**

$$P^{\mathrm{DT}}_{\min} = \min_{\log_2\left(1+\frac{|h_{\mathrm{PP}}|^2 P_p}{N_0}\right)\ge R} P_p$$

$$P^{\mathrm{DT}}_{\min} = N_0 \frac{\left(2^R - 1\right)}{|h_{\mathrm{PP}}|^2}$$

if $h_{\mathrm{PP}} \ge \max_{1\le i\le K} h_{\mathrm{TT},i}$ **or** $h_{\mathrm{PP}} \ge \max_{1\le i\le K} h_{\mathrm{SP},i}$ **then**

$p_b \leftarrow p_b + 1$
$av_{\mathrm{DT}} \leftarrow av_{\mathrm{DT}} + P^{\mathrm{DT}}_{\min}$

else

$$P^{\mathrm{RA}}_{\min} = \min_{0<\alpha,\beta<1} (1-\alpha)P^*_p(\alpha, R) + \alpha\left(\sum_{i=1}^{K} P^*_{s,i}(\alpha,\beta)\right)$$ {P^*_p and $P^*_{\mathrm{s},i}$ are given in (12.11) and (12.14), respectively.}

if $P^{\mathrm{RA}}_{\min} > P^{\mathrm{DT}}_{\min}$ **then**

$p_b \leftarrow p_b + 1$
$av_{\mathrm{DT}} \leftarrow av_{\mathrm{DT}} + P^{\mathrm{DT}}_{\min}$

else
$av_{\mathrm{RA}} \leftarrow av_{\mathrm{RA}} + P_{\min}^{\mathrm{RA}}$
end if
end if
end for
$P(s \in C) = \frac{p_b}{N}$
$P(s \in \overline{C}) = 1 - P(s \in C)$
$P_{\mathrm{av}}^{\mathrm{DT}} = \frac{av_{\mathrm{DT}}}{p_b}$
$P_{\mathrm{av}}^{\mathrm{RA}} = \frac{av_{\mathrm{RA}}}{N - p_b}$
$E[P_T] = P_{\mathrm{av}}^{\mathrm{DT}} \times P(s \in C) + P_{\mathrm{av}}^{\mathrm{RA}} \times P(s \in \overline{C})$

We can write the first term in (12.15) as

$$E[P_T/s \in C] = E\left[P_p^*/s \in C\right] \tag{12.16}$$

while the second term is given by

$$E\left[P_T/s \in \overline{C}\right] = E\left[(1-\alpha)P_p^* + \alpha \sum_{i=1}^{K} P_{s,i}^*/s \in C\right] \tag{12.17}$$

whereas the weight of each region can be inferred using the below useful observation

$$s \in C \Leftrightarrow P_{\min}^{\mathrm{DT}} \leq P_{\min}^{\mathrm{RA}}. \tag{12.18}$$

and

$$P(s \in \overline{C}) = 1 - P(s \in C). \tag{12.19}$$

Hence, the theoretic formulation of the power minimization problem (12.20) is completely characterized and can be easily numerically solved (see Algorithm 12.1).

$$\min_{0<\alpha,\beta<1} E[P_T]. \quad \text{for all network states } s \tag{12.20}$$

To avoid needless calculations, an intuitive case is when the primary link is better than all the available source-to-relay or relay-to-destination links in terms of the channel gain, in which case the source sends its data directly to the destination without any external assistance.

$$\left\{h_{\mathrm{PP}} \geq \max_{1 \leq i \leq K} h_{\mathrm{TT},i} \text{ or } h_{\mathrm{PP}} \geq \max_{1 \leq i \leq K} h_{\mathrm{SP},i}\right\} \subseteq C. \tag{12.21}$$

12.4 Numerical results and discussions

To corroborate the above theoretical design, some experimental simulations have been undertaken below to investigate the throughput benefits of the examined schemes. These simulations have been performed using a large number of random fading samples generated through Monte Carlo experiments.

For these experiments and unless otherwise stated, a primary transmitter–receiver pair selects a set of $K \in \{3, 4, 5\}$ competing secondary transmitter–receiver pairs whose transmitter is willing to cooperate with the primary network. In order to achieve optimal cooperation, we assume that all the chosen relays are located at the same position on the straight line between the primary transmitter and the primary receiver, at a normalized distance $d = 0.2$ from the primary transmitter. For the relaying operation, we assume a large-scale path loss model with a coefficient $\zeta = 2$, as a result the average power gains on the source-to-relay and the relay-to-destination channels are given as: $E\left[|h_{\mathrm{TT},i}|^2\right] = 1/d^{\zeta}$ and $E\left[|h_{\mathrm{SP},i}|^2\right] = 1/\left(1-d\right)^{\zeta}$ $(\forall\, i \neq j \in \{1,\ldots,K\})$, respectively. For the secondary competing links, the small scale Rayleigh fading gives rise to the following average channel gains: $E\left[|h_{\mathrm{SS},i}|^2\right] = \mathrm{E}\left[|h_{\mathrm{SS},ij}|^2\right] = 10$ dB $(\forall\, i \neq j \in \{1,\ldots,K\})$. Likewise, the average channel gain on the primary link is $E\left[|h_{\mathrm{PP}}|^2\right] = 1$. The maximum tolerated power for the overall system is fixed to $P_{p,\mathrm{dB}} = P_{\max,\mathrm{dB}} = 40$. The variance of noise is $N_0 = 1$ and the cost per unit energy is $c = 0.2$.

Figure 12.9 illustrates the primary link achieved rate versus the average power of the overall system. We surprisingly observe that the capacity of the primary network is strongly insensitive to the size of the secondary network. This is due to the fact that only the best cognitive nodes; the one, with the highest channel gain from the primary source, intervene in the delivery of the licensed stream. Moreover, system resources in terms of time and energy are dynamically allocated in such way to attain a desired rate regardless of the number of cooperating nodes.

The probability of DT mode under different values of the average total power is plotted in Figure 12.10 while varying the number of available relays. In low power regime, data packets are more likely to be promptly accumulated at the receiver side using the extra diversity offered by the spatial dimension than the straight communication and thus the cooperative mode becomes more gainful. Conversely, for higher transmit powers, the direct link may support the data rate of the primary communication without any external assistance. Hence, the opportunistic system will favor the DT over cooperation to avoid spending the power budget in simultaneous transmitting and relaying. We also notice that the direct mode occurs most of time with probability greater than 0.5, which implies that, unlike the conventional systems, the fixed cooperation strategy does not necessarily works in CR contexts.

At low and high total average power, relays are listening most of the time as it can be seen in Figure 12.11. In this case, the cooperation is not beneficial and most of packets are transmitted through the direct link. Yet at moderate power levels,

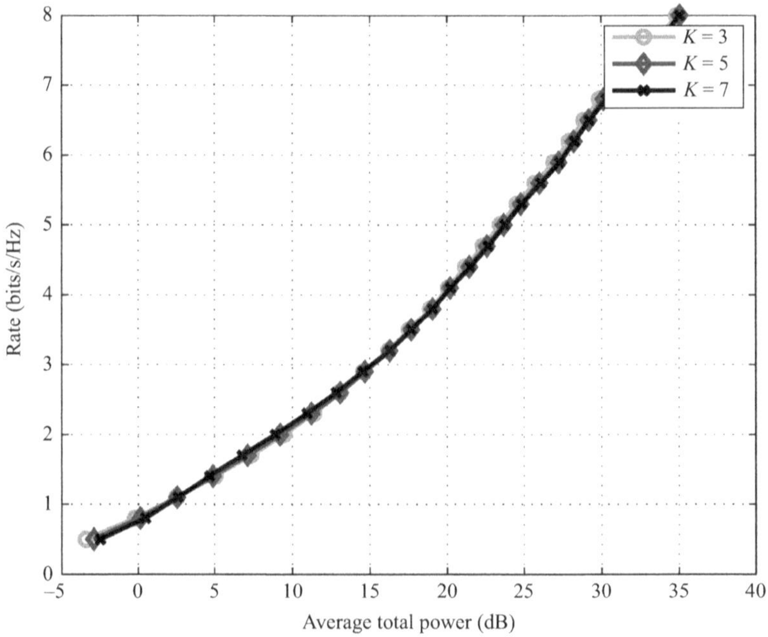

Figure 12.9 Transmission rate versus average total power for different K values

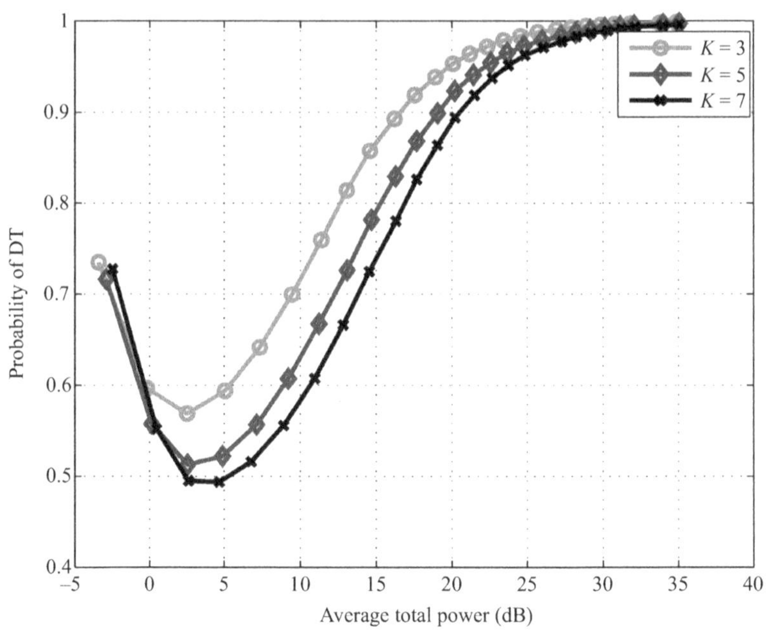

Figure 12.10 Probability of direct transmission versus average total power for different K values

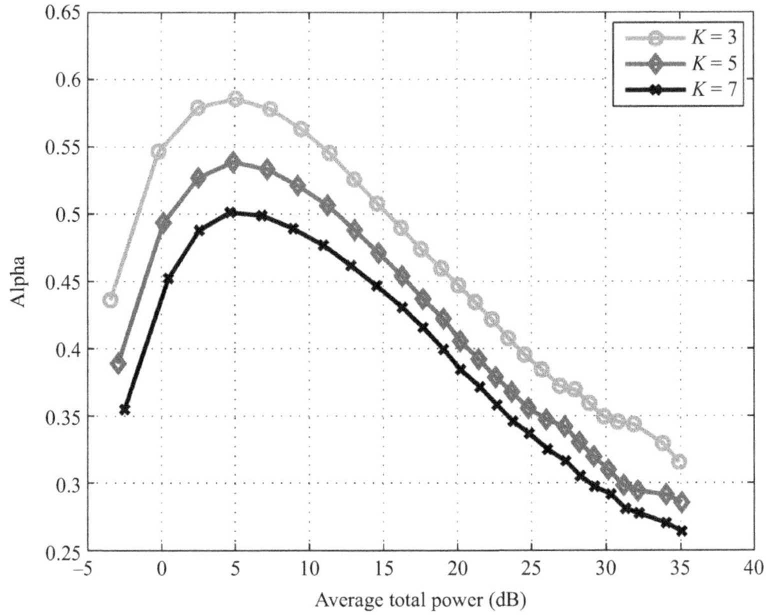

Figure 12.11 Fraction of time α versus average total power for different K values

relays start to be increasingly involved in the transmission process. Intuitively, by increasing the secondary transmit power a larger time need to be leased to the cognitive transmission and thus the duration of the relaying phase becomes shorter. However, higher secondary transmit powers cut down the time needed by the destination to gather necessary packets because the energy is summed over all the intermediate nodes, and as a direct consequence more time may be granted to all relays to accomplish receiving enough data packets from the primary source. Extending the duration of the third phase constitutes an incentive for secondary devices to cooperate, but shortens the duration of the primary communication. This observation translates into a trade-off between the throughput gained by SUs and the overhead added to the licensed communication.

Figure 12.12 shows that most of the cooperation time is used to maximize the utility for SUs. This situation is still beneficial for the primary network as the portion of time slot reserved for cooperation is very small. As the total power is increasing, the relaying nodes show an increased tendency toward cooperation as the increase in the source power increases the chances of successful decode-and-forward operations at the relays side. Furthermore, the portion of the time where the band is leased for secondary usage gets decreased. This is essential for the PU to remain interested by cooperation while ensuring maximum utility for SUs. At high-power regime, the source power is higher enough to strengthen the direct link communication. Under these conditions, the probability of DT is increasing, whereas cooperation time fraction gets decreased and it is dedicated essentially to secondary utilization.

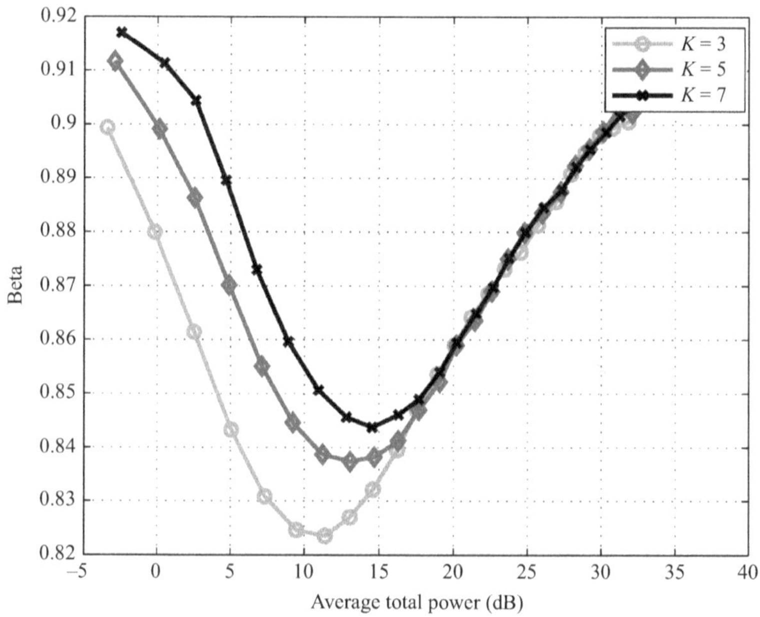

Figure 12.12 Fraction of time β versus average total power for different K values

It is worth noting that other metrics like capacity and outage probability are also recurrent performance indicators in communication systems. Concretely, for some applications where CSI is not available, there is no minimum power that guarantees reliable transmission for all possible channel states and thus the preassigned transmission rate cannot be achieved. In this case, the outage probability is non-zero and an outage minimization problem can be easily formulated similarly to (12.20). Additional metrics such as delay, throughput, and jitter can be readily computed from the outage probability estimation.

12.5 Conclusions

Opportunistic cooperation is emerging as a promising solution overcoming the environmental uncertainties characterizing wireless CR networks and ensuring a peaceful spectrum sharing between incumbent and rental users as well. The system under consideration is designed to achieve the maximum possible rate for the PU, whereas ensuring an optimal utility for the SUs in power-constrained environments. The decision about the need and the amount of cooperation (how and when) is taken by the PU while considering the impact it will have on the SU behavior. For further insight into the proposed framework performance, deep numerical simulations have been conducted. It has been shown that the decision to choose between DT or cooperation implicitly includes an optimal and adaptive power allocation mechanism. More importantly, under limited power levels, the system will adopt a

DT policy and surrounding relays will not be involved. This prevents unnecessary energy consumption. For moderate power values, the relay can successfully decode the sent packet and cooperate with the source thus forming a virtual antenna array toward destination. At high SNR regime, the straight link is enough to accommodate the primary traffic volume and the relays help becomes useless and can be discarded. The suggested cooperation scheme allows all the stakeholders to benefit from a fear spectrum allocation and an effective power management.

Lastly, we firmly believe that the fifth generation represents a big opportunity for a real and tangible commitment from various telecom actors from industry and academia to boost and leverage the widespread deployment of CR-based infrastructures, especially with the recent successful achievements in terms of hardware and pilot trials around the world to achieve universal service provision [34]. Before being an opportunity in the context of 5G networks, the concept of CR brings a paradigm shift as a key solution for various problems encountered by operators such as band saturation, weak deployments in rural zones, and high quality of service demands.

References

[1] FCC Spectrum Policy Task Force, Report of the spectrum efficiency working group, *FCC, Tech. Rep.*, November 2002.

[2] Mitola III, J. and Maguire, G. Q., Cognitive radio: Making software radios more personal, *IEEE Personal Communications*, 6(4), 13–18, 1999.

[3] Weiss, T. and Jondral, F., Spectrum pooling: An innovative strategy for the enhancement of spectrum efficiency, *IEEE Communications Magazine*, 42, 8–14, 2004.

[4] Summit Presentations. Available at: <http://www.etsi.org/news-events/events/682-2013-etsi-future-mobile-summit> [Accessed 18 November 2016].

[5] Hong, X., Wang, J., Wang, C. X., and Shi, J., Cognitive radio in 5G: A perspective on energy-spectral efficiency trade-off, *IEEE Communications Magazine*, 52(7), 46–53, 2014.

[6] Sherman, M., Mody, A. N., Martinez, R., Rodriguez, C., and Reddy, R., IEEE standards supporting cognitive radio and networks, dynamic spectrum access, and coexistence, *IEEE Communications Magazine*, 46(7), 72–79, 2008.

[7] Sun, H., Nallanathan, A., Wang, C. X., and Chen, Y., Wideband spectrum sensing for cognitive radio networks: A survey, *IEEE Wireless Communications*, 20(2), 74–81, 2013.

[8] Database administrators. Available at: <https://www.fcc.gov/general/white-space-database-administration> [Accessed 18 November 2016].

[9] Kamil, A. S. and Khider, I., Open research issues in cognitive radio, in *Proceedings of the 16th Telecommunications forum TELFOR*, Belgrade, Serbia, November 2008, pp. 250–253.

[10] Goldsmith, A., Jafar, S. A., Maric, I., and Srinivasa, S., Breaking spectrum gridlock with cognitive radios: An information theoretic perspective, *Proceedings of the IEEE*, 97(5), 894–914, 2009.

[11] Hesammohseni, S. A., Moshksar, K., and Khandani, A. K., A combined underlay and interweave strategy for cognitive radios, in *Proceedings of the IEEE International Symposium on Information Theory (ISIT)*, Honolulu, HI, USA, June 2014, pp. 1396–1400.
[12] Biglieri, E., Calderbank, R., Constantinides, A., Goldsmith, A., Paulraj, A., and Poor, H. V., *MIMO Wireless Communications*, Cambridge University Press, Cambridge, 2007.
[13] Sendonaris, A., Erkip, E., and Aazhang, B., User cooperation diversity. Part I. System description, *IEEE Transactions on Communications*, 51(11), 1927–1938, 2003.
[14] Van der Meulen, E. C., Transmission of information in a T-terminal discrete memoryless channel, *Ph.D. dissertation*, Department of Statistics, University of California, Berkeley, CA, June 1968.
[15] Van Der Meulen, E. C., Three-terminal communication channels, *Advances in Applied Probability*, 3, 120–154, 1971.
[16] Cover, T. M. and Gamal, A. E., Capacity theorems for the relay channel, *IEEE Transactions on Information Theory*, 25(5), 572–584, 1979.
[17] Akyildiz, I. F., Su, W., Sankarasubramaniam, Y., and Cayirci, E., A survey on sensor networks, *IEEE Communications Magazine*, 40(8), 102–114, 2002.
[18] Hartenstein, H. and Laberteaux, K. P., A tutorial survey on vehicular ad hoc networks, *IEEE Communications Magazine*, 46(6), 164–171, 2008.
[19] Akyildiz, I. F. and Wang, X., A survey on wireless mesh networks, *IEEE Communications Magazine*, 43(9), S23–S30, 2005.
[20] Ahmed, M. H. and Ikki, S. S., To cooperate or not to cooperate? That is the question!, in *Cooperative Networking*, eds. M. S. Obaidat and S. Misra, John Wiley & Sons, Ltd, Chichester, 2011, pp. 21–33.
[21] Gong, X., Chandrashekhar, T. P., Zhang, J., and Poor, H. V., Opportunistic cooperative networking: To relay or not to relay? *IEEE Journal on Selected Areas in Communications*, 30(2), 307–314, 2012.
[22] Lee, W., Wang, Y., Shin, D., Chang, N., and Pedram, M., Optimizing the power delivery network in a smartphone platform, *IEEE Transactions on Computer-Aided Design of Integrated Circuits and Systems*, 33(1), 36–49, 2014.
[23] Li, Y., Nosratinia, A., and Zhang, W., Opportunistic cooperation for distributed spectrum sensing in cognitive radio, in *Proceedings of IEEE International Conference on Communications (ICC)*, Kyoto, Japan, June 2011, pp. 1–5.
[24] Gunduz, D. and Erkip, E., Outage minimization by opportunistic cooperation, in *Proceedings of International Conference on Wireless Networks, Communications and Mobile Computing*, Maui, Hawaii, June 2005, vol. 2, pp. 1436–1442.
[25] Gunduz, D. and Erkip, E., Opportunistic cooperation and power control strategies for delay-limited capacity, in *Proceedings of 39th Annual Conference on Information Sciences and Systems (CISS)*, Baltimore, MD, USA, March 2005.

[26] Gunduz, D. and Erkip, E., Opportunistic cooperation by dynamic resource allocation, *IEEE Transactions on Wireless Communications*, 6(4), 1446–1454, 2007.
[27] El-Bakoury, I., Seddik, K. G., and Elezabi, A., Opportunistic relaying with partial CSI and dynamic resource allocation, in *Proceedings of Global Conference on Signal and Information Processing (GlobalSIP)*, Austin, TX, USA, December 2013, pp. 899–902.
[28] Zou, Y., Zheng, B., and Zhu, W. P., An opportunistic cooperation scheme and its BER analysis, *IEEE Transactions on Wireless Communications*, 8(9), 4492–4497, 2009.
[29] Urgaonkar, R. and Neely, M. J., Opportunistic cooperation in cognitive femtocell networks, *IEEE Journal on Selected Areas in Communications*, 30(3), 607–616, 2012.
[30] Jeong, D. K. and Kim, D., Outage-optimal threshold-based opportunistic cooperation in AF relaying systems with staying link information, in *Proceedings of IEEE Asia Pacific Conference on Wireless and Mobile*, Bali, Indonesia, August 2014, pp. 264–268.
[31] Urgaonkar, R. and Neely, M. J., Delay-limited cooperative communication with reliability constraints in wireless networks, *IEEE Transactions on Information Theory*, 60(3), 1869–1882, 2014.
[32] Shaqfeh, M., Zafar, A., Alnuweiri, H., and Alouini, M. S., Hybrid DF-CF-DT for buffer-aided relaying, in *Proceedings of International Zurich Seminar on Communications*, Zurich, Switzerland, March 2016, pp. 84–88.
[33] Wang, B., Wu, Y., and Liu, K. J., Game theory for cognitive radio networks: An overview, *Computer Networks*, 54(14), 2537–2561, 2010.
[34] Harada, H., White space communication systems: An overview of regulation, standardization and trial, *IEICE Transactions on Communications*, 97(2), 261–274, 2014.

Chapter 13

The role of edge computing in future 5G mobile networks: concept and challenges

Pouria Sayyad Khodashenas[1], Cristina Ruiz[1], Muhammad Shuaib Siddiqui[1], August Betzler[1] and Jordi Ferrer Riera[1]

Abstract

Future 5G technologies are expected to overcome the challenges of next generation networks aiming to tackle the novel and manifold business requirements associated to different vertical sectors. Extraordinarily high speed and capacity, multi-tenancy, heterogeneous technologies convergence, on-demand service-oriented resource allocation or even coordinated, automated management of resources are only few examples of the complex demands 5G aims to undertake. The shift from centralized cloud computing-based services towards data processing at the edge is becoming one of the fundamental components envisaged to enable those future 5G technologies. Edge computing is focused on pushing processing to the network edge where all the actual interactions in the access networks takes place and the critical low-latency processing occurs. Combination of network functions virtualization (NFV) and edge-computing technologies and mechanisms provides a wide range of novel opportunities for value-added service provisioning covering different features required in future access networks, such as Quality of Service (QoS), security, multi-tenancy, and low latency. This chapter provides an overview of edge-computing technologies, from supporting heterogeneous infrastructure up to service provisioning methodologies related to the application-specific requirements. It describes the role of edge computing and NFV in future 5G mobile networks. It also provides an insight into how edge computing can potentially facilitate and expedite provisioning of security in 5G networks. The manuscript analyses the role of the networking resources in edge-computing-based provisioning, where the demands of 5G mobile networks are to be met with wireless-networking technologies, which in essence are different to wired technologies present in core data centers. Initial results obtained from the evaluations of wireless fog networking

[1]Fundació i2CAT, Internet i Innovació Digital a Catalunya, C/Gran Capità 2-4, 08034, Barcelona, Catalunya, Spain

backhauls are presented, and the challenges ahead of the actual implementation of those technologies are also analyzed in the chapter.

13.1 Introduction

5G technologies are envisaged as a new networking paradigm in order to overcome the challenges and requirements associated with the future communication systems. In fact, 5G systems aim to support extraordinarily high speed and capacity, multi-tenancy, fixed and wireless network convergence, self-X, unconventional resource virtualization, on-demand service-oriented resource allocation and automated management and orchestration [1]. As a consequence, those 5G upcoming technologies will facilitate both the development and materialization of novel vertical sectors, such as Industry 4.0, Smart Grids, Smart Cities and e-Health. 5G deployment is envisaged therefore to accelerate industrial environments, facilitating them to enter the value chain and increase revenue generation. It is expected that, besides improving the citizen's digital experience, such a market revolution contributes billions to the EU economy each year and creates hundreds of thousands of new jobs, mostly for the small and medium-sized businesses [2].

The landscape of future communication is reshaped and redefined significantly by the ongoing digitalization trends [3]. One of the main pillars of such revolution is the way that new network functions are introduced to the value chain. Traditionally, such a process demands deployment of specialized devices with "hard-wired" functionalities. It implies that any adaptation to the ever increasing and heterogeneous market requirements demands a huge investment to change/deploy hardware. Thanks to the advent of cloud computing, software-defined networking (SDN) [4] and network function virtualization (NFV) [5], the idea of having general-purpose computing and storage assets at networks has been realized along with the virtualization of networks and network functions, which enables the automation of network service (NS) provisioning and management [6].

This approach paves the way for various benefits both on network performance and network control and management aspects, including: (i) efficient management of hardware resources, (ii) rapid introduction of new network functions and services to the market, (iii) ease of upgrades and maintenance, (iv) exploitation of existing virtualization and cloud management technologies for virtual network function (VNF) deployments, (v) reduction in capital expenditure (CAPEX) and operating expenses (OPEX) expenditures, (vi) enabling a more diverse ecosystem, and (vii) encouraging openness within the ecosystem, as defined by the European Telecommunication [7].

This chapter provides an overall and schematic review of the concept of virtualization of networks and network functions, and then reviews their role to realize 5G challenging features, for example, multi-tenancy and end-to-end security over Evolved Universal Terrestrial Radio Access [8], that is, radio access network (RAN) and backhauling network. In essence, in terms of multi-tenancy, the challenge remains on how the isolation among different tenants, utilizing virtual networks on top of, possibly, the same physical infrastructure, can be ensured at all the levels. For ensuring end-to-end security in 5G networks, a holistic approach is required, which takes into account not only the physical but also the virtual resources of the network. These challenges require a substantial change on the network devices, from being only network equipment to cloud-enabled devices enhanced with, for example, novel processor architectures.

13.2 Multi-tenancy over the cloud-RAN

Traditionally, to provide coverage in one point of presence (PoP), actual installation of physical infrastructure, for example, small cell (SC), is required. Despite the fact that mounting equipment in one place may not be possible (e.g. dense areas), such an ownership increases operators' CAPEX and significantly hampers business agility, particularly when considering the high degree of cell densification needed to deal with the 5G requirements. Moreover, the static nature of physical ownership makes it difficult (impossible in some cases) to handle scenarios with dynamic capacity requirements. For example, a flash crowd event at a venue (e.g. stadium, urban area, etc.) cannot be well served without overprovisioning of the underlying physical infrastructure. It can be easily translated to more operators' expenses (CAPEX and OPEX), which, in turn, increases the service cost for the end users. To address this issue, the idea of multi-tenancy has been initiated in 3GPP [9], and it is expected to play a vital role in 5G networks [10]. In a multi-tenant scenario, a third party owns the underlying infrastructure and provides access to the actors of the telecom scene like network operators, service providers (SPs), over-the-top players, and others. Such a sharing increases service dynamicity and reduces the overall cost and energy consumption.

Furthermore, thanks to the advent of cloud computing, SDN and NFV, stakeholders can enter the network value chain without deploying specialized "hard-wired" devices. It relaxes the ever-increasing cost of adaptation to heterogeneous market needs. This new concept calls for a substantial change on the architecture of current RAN nodes, from being only a wireless head to a cloud-enabled device equipped with, for example, novel processor architectures, graphics processing units, digital signal processors, and/or field-programmable gate arrays. In this line, new industry initiatives have already introduced the concept of Mobile-Edge Computing (MEC) [11] and the related key market drivers [12]. The resulting solution will allow several operators/SPs to engage in new sharing models of both access capacity and edge-computing capabilities, that is, the logical partitioning of the localized network infrastructure in one or more PoPs.

In this section, we review the implementation of cloud-enabled small cells (CESCs) as an example of future cloud-enabled 5G RAN node, able to support edge cloud computing in a multi-tenant, multi-service ecosystem. Then, some of its potential benefits and challenges are presented.

13.2.1 Enabling technologies

In this subsection, we introduce the basic concept of CESC and Light Data Centre (Light DC). The design originates from the Small cEllS coordination for Multi-tenancy and Edge services (SESAME) project [13].

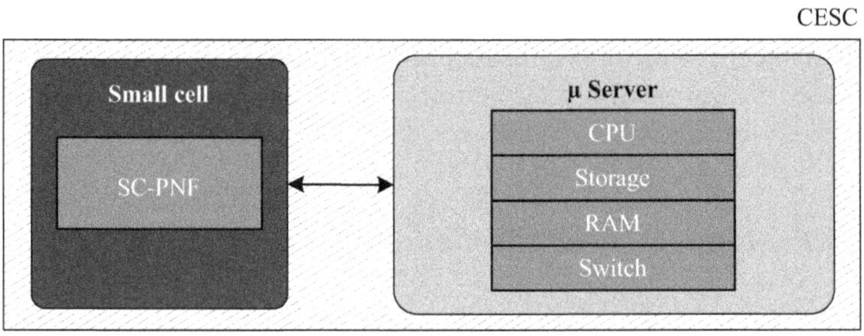

Figure 13.1 Cloud-enabled small cell (CESC) consisting of the small cell radio head (SC–PNF) and the micro-server

13.2.1.1 Cloud-enabled small cell (CESC)

The concept of CESC, shown in Figure 13.1, is the key enabler in order to form a cloud-enabled network that supports both radio access and edge-computational services at one PoP. It consists of the union of a SC, although it can be extended and applied to any other access network system, and a micro-server in one single device. SC functionalities are foreseen to be split into physical network functions (PNFs) and SC VNFs [14]. SC VNFs may represent different layers of the Evolved Universal Terrestrial Radio Access (E-UTRA) protocol stack, while the rest of protocol entities will remain as PNF (e.g. below Packet Data Convergence Protocol (PDCP) at data plane protocol stack). The use of SC VNFs provides the required support for splitting the SC resources in different virtual slices. Therefore, CESC is able support multi-tenancy in a cloud-RAN environment by instantiating a dedicated SC VNF per tenant, also called virtual network operator (VNO) from now, while keeping PNFs common for all.

SC VNFs, as well as service VNFs, are going to be hosted by the micro-server, whose architecture and characteristics are optimized for the MEC environment. Instantiation of service VNFs (e.g. virtual firewall and virtual caching) over the CESC micro-server aims to satisfy the identified requirements of 5G (e.g. low latency). CESC becomes the main serving node, it allows overall edge service provision avoiding data transfer to the core network.

13.2.1.2 Light Data Centre (Light DC)

As it may be inferred, resources on a single micro-server (i.e. RAM, CPU, storage, HWA) might not be enough to support the MEC services of all tenants. CESC clustering enables the creation of a micro-scale virtualized execution infrastructure in the form of a distributed DC, denominated Light DC (Figure 13.2), enhancing the virtualization capabilities and processing power at the edge. The hardware architecture of the Light DC envisages that each micro-server in a CESC will be able to communicate with all others via a dedicated network, guaranteeing the latency and bandwidth requirements needed for sharing resources. The resulting solution will allow VNOs/tenants not only to support connectivity but also to provide added value mobile edge services (e.g. caching, video transcoding, etc.) in a PoP.

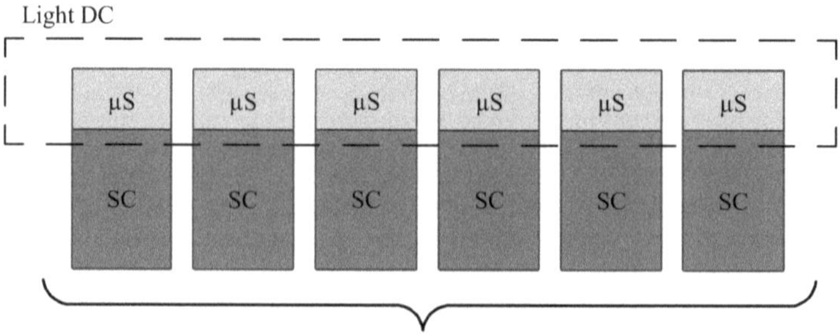

Figure 13.2 Light DC overview formed by all the micro-server in a CESC cluster

Like any DC, Light DC is a pool of resources (computational, storage, network) interconnected using a communication network. Such a clustering can be achieved with different network topologies (e.g. star, tree, mesh, etc.) and technologies (e.g. Ethernet over copper or fiber, wireless radio, etc.). The network topology plays a pivotal role in the architecture, since it determines the scalability, robustness, performance (e.g. throughput and latency), efficiency and in simple words viability of a solution. Light DC might adopt and adapt any of well-known DC topologies such as three-tier, DCell, or even BCube [15]. However, considering metrics like performance, complexity and the number of network elements (i.e. network interface card) required per micro-server, the three-tier topology stands out as the most appropriate topology for the Light DC [16]. In small areas with no geographical barriers, such as a building or a hospital, the topology might be simplified to a flat tree or even a star topology, as shown in Figure 13.3. There are many (physical transport layer) technological options to form the selected network topology, such as Ethernet over copper or fiber, or even wireless radio. Section 13.5 will review these alternatives focusing on describing a wireless backhauling solution with more details.

13.2.2 Multi-tenant multi-service management and orchestration

The Light DC concept offers a virtualization platform to meet 5G requirements, however, management and orchestration of this uniform virtualized environment, able to support both radio connectivity and edge services, is a challenging task itself. The most clearly highlighted specific challenges that a cloud-RAN environment entails are the dynamic composition of the Light DC resources based on the current status of CESC cluster(s), the coordination of specific type of resources (radio-related resources, service-related HW accelerators, etc.) and the isolation of dedicated network slices to each tenant.

Considering the aforementioned challenges, the SESAME project [17] proposes a solution to consolidate multi-tenancy and orchestration in SC cloud-enabled mobile communication infrastructure. The consolidation is built over the virtualization and NFV pillars. Figure 13.4 better illustrates that consolidation as

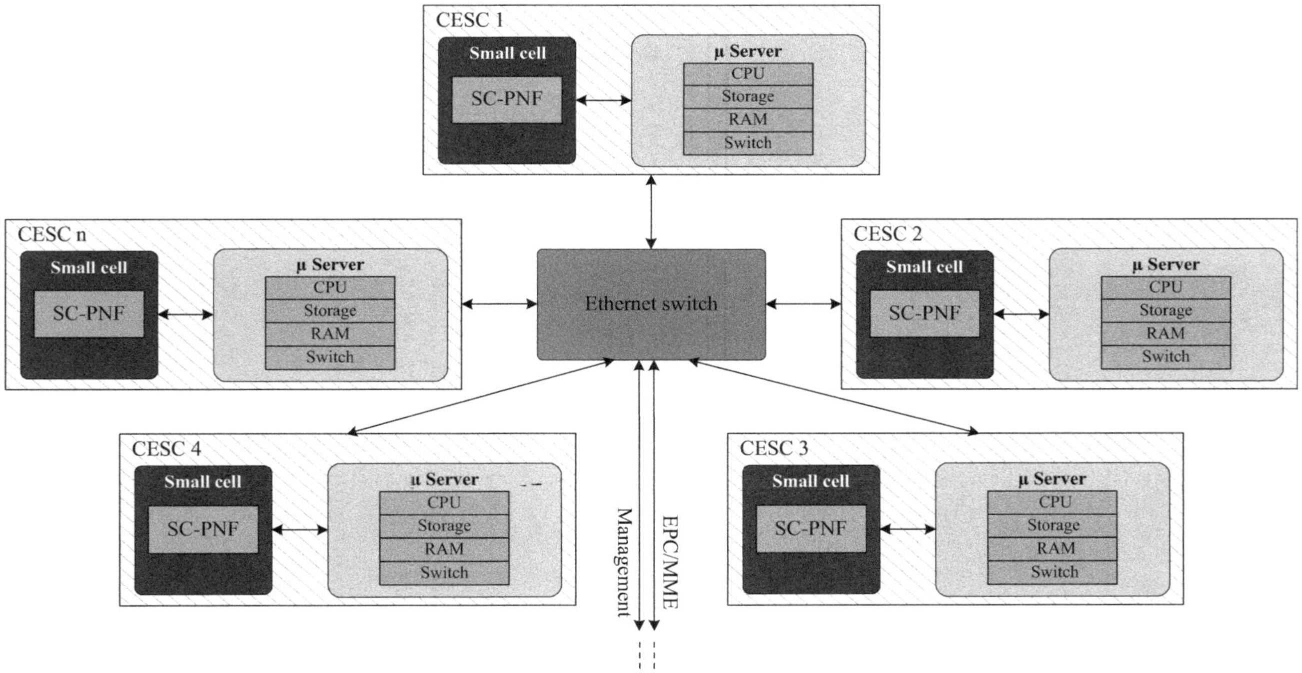

Figure 13.3 CESC cluster physical architecture

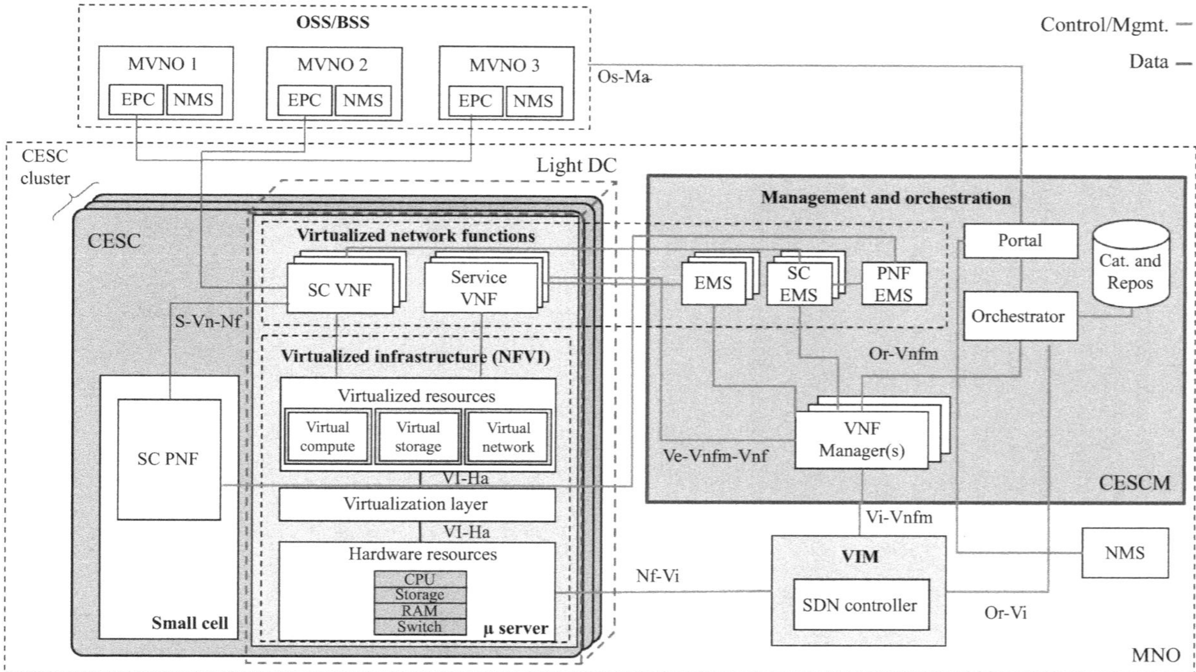

CESC cloud-enabled small cell | **CESCM** CESC manager | **EMS** element management system
MNO mobile network operator | **SC VNF** SC virtual network function | **NMS** network management system
MVNO mobile virtual network operator | **SC PNF** SC physical network function | **VIM** virtualized infrastructure manager

Figure 13.4 Possible cloud-RAN multitenant architecture based on ETSI MANO framework

envisaged by SESAME through the high-level architecture. The diagram depicts the most relevant building blocks of the proposed system, as well as both their internal and external inter-connections.

On the right side of the figure, the CESC Manager (CESCM) is the central service management and orchestration component in the architecture. This component, in essence, is responsible for the integration of all the traditional 3GPP network management elements; at the same time, it adds the novel recommended functional blocks in order to materialize the NFV paradigm [18]. With respect to multiplicities during deployment and operation time, the system has been designed to allow a single instance of a CESCM to control, monitor, and operate several CESC clusters, whereby each cluster constitutes a Light DC, as described above. The clusters are independently controlled by a dedicated Virtual Infrastructure Manager (VIM), whose responsibilities are bound within a single cluster. The CESCM holds a holistic view of the physical infrastructure, so to optimize its management.

There exists the possibility to chain distinct VNFs over the provided virtualized execution environment, that is, a given Light DC, in order to meet the requirements of a given NS as requested by a tenant. It is worth to mention that, in this context, a NS is understood as a collection of VNFs that jointly supports data transmission between user equipment (UE) and operators' Evolved Packet Core network, with the possibility to involve one or several service VNFs in the data path. Therefore, each NS is deployed as a chain of PNF, SC VNFs, and Service VNFs. Again, due to the distributed nature of the Light DC, the proposed VIM requires data packet extraction (from the traditional 3GPP data path) and a forwarding rule implementation to guarantee possible communication between SC VNFs, and Service VNFs, which may reside in different CESCs. SDN principles are used to provide the system with the required scalability. In this way, the CESCM instructs the embedded SDN controller at VIM with the specific VNF forwarding rules, and the SDN controller in return applies them to support the desired connectivity within the Light DC.

We consider an element management system (EMS) deployed in the CESCM for each instantiated VNF. The EMS is responsible to perform fundamental management actions, such as fault monitoring, configuration, accounting, performance monitoring, and security. Besides, the central management point for the whole network of the mobile operator is the network management system. Thus, the corresponding PNF EMS and SC EMS are respectively the elements accounted for carrying out the management of the physical and virtualized network functions residing at the SC.

The VNF Manager (VNFM) carries out the basic lifecycle management of the deployed VNFs. It is included in the CESCM element by leveraging on the monitoring mechanisms, the CESCM, in conjunction with the VNFM, is able to apply policies for NS-level rescaling and reconfiguration to achieve high resource utilization. It is worth to mention that monitoring mechanisms are dictated by the CESCM service level agreements (SLA) monitoring unit that allows the monitoring of SLAs between different business role players. Two main role players interact in the proposed scenario, the Mobile Network Operator (MNO) is the owner of the radio access (Light DC) and management infrastructure and offers sliced NSs to different Mobile Virtual Network Operators (MVNOs), which act as tenant MNOs.

Another essential component at the heart of CESCM is the NVF Orchestrator (NFVO). Besides management and orchestration of the above-mentioned functionalities, NFVO composes service chains (constituted by two or more VNFs located either in one or several CESCs) and manages the deployment of VNFs over the Light DC. This includes not only the management of a typical NFV Infrastructure (i.e. processing power, storage, and networking), but also assignment of HW accelerators. Besides, to improve the energy efficiency of the proposed solution, NFVO may need to take care of switching on and off resources at CESC level.

The CESCM portal is a control panel web graphical user interface that constitutes the main graphical frontend to access the cloud-RAN management platform. It includes two login procedures, a login for the MVNO to retrieve monitoring information and be able to browse catalogues, request and delete NSs, and another login for the VNO administrator to register extra resources, add new NSs to the catalogues and configure CESCM elements.

13.2.3 Benefits and challenges

As evaluated in [19], it is expected that in the framework of cloud-RAN networks the speed of service delivery significantly increases, since the edge cloud services are executed very near to the end user. It creates an excellent opportunity for stakeholders, for example, operators and cloud based platforms, to serve customers (individuals and businesses) demand for intelligence and complex services in a practical and latency-free manner. Also, multi-tenancy achieved through the virtualization of network resources, allows efficient use of deployed physical infrastructure at one PoP, via the on-demand network topology changes (e.g. add/drop of CESC to a cluster) and elastic per-tenant capacity allocation, aiming to guarantee the quality of experience and reduce the total cost of operation. Having this in mind, compared to the current 4G systems, some of the cloud-RAN merits can be listed as follows [20]:

- Higher wireless area capacity and more diverse service capabilities: by deploying high-density multi-service multi-tenant SC networks higher traffic and capacity per geographical areas are supported.
- Reducing the average service creation time: the flexible design of the CESC platform and the associated management layers promotes a shared virtualized infrastructure, that is, a cloud environment, right at the network's edge which reduces the service deployment time scale.
- Creating a secure, reliable and dependable Internet with a "zero perceived" downtime for services provision: cloud-RAN solutions allow rapid integration of multiple virtual operators sharing the same infrastructure, thus allowing isolated and secure provision of vertical services for massive amount of

connected devices. Also, automated network resource monitoring/optimization allows to provision them where they are needed the most, that is, resource rebalancing/repurposing. It guarantees service reliability and minimizes service downtime.

Despite the potential technical benefits, there are many challenges to address. For example, viability of a solution strongly depends on the service provisioning model on the joint radio and cloud environment. The point is that, even though there is already a good understanding and experience available on the radio access and/or cloud computing service provisioning, there is no clear vision on a joint radio-cloud case which covers both worlds simultaneously. From the business perspective, three major role players are identified [21], as depicted in Figure 13.5: function provider, is the VNF developer which sells/develops VNFs; SP, is the one who composes NS that is, chain of VNFs, PNFs with the available VNFs and offers them to the customer, is the one who purchases NSs. In multi-tenant cloud enabled RAN, there are two main possible ways to form a joint radio-cloud model with the above defined roles, as illustrated in Figure 13.5.

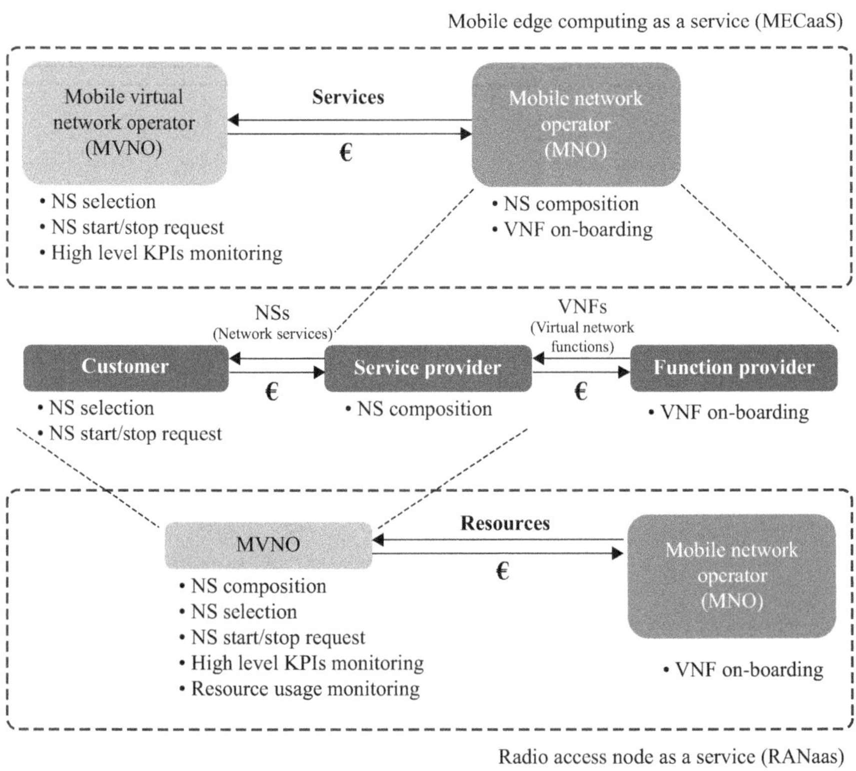

Figure 13.5 Cloud-RAN possible role players and service provisioning schemes

- *Mobile edge computing as a service:* This model (depicted as option 1 in Figure 13.5) has been inspired mainly from the MNO–MVNO business relationship. Briefly, in this model, MVNO relies completely on the infrastructure and other services provided by the MNO. Bearing this in mind, MVNO asks for high level key performance indicator (KPIs) on the SLA, for example, on the radio connectivity at one area with support of a desired downlink/uplink capacity and on the cloud, for example, support for a number of caching hits, transcoding delay less than a threshold, and others. Here, MVNO only has an overall vision of the system and MNO has to provide enough support, that is, both in terms of hardware and number/composition VNF chains (i.e. NS), to meet the agreed KPIs. Performance reports are provided to MVNO on time intervals (even real time). In simple words, with this model, MVNO does not chain VNFs to form a mobile edge service (i.e. VNF1 connected to VNF2 connected to VNF3), and a high level KPI view is enough for it to request a service without going to details.
- *Radio access node as a service:* In this model, shown as option 2 in Figure 13.5, MVNO on SLA asks for connectivity in a certain coverage area with some radio KPIs as above and an aggregated cloud resources on the Light DC, for example, a certain amount of GB of storage, of RAM, and others. This model corresponds with the famous Infrastructure as a Service (IaaS) paradigm, which is one of the three fundamental service models of cloud computing [22]. In an IaaS model, a third-party provider (MVO) hosts hardware (e.g. CESC), software (e.g. Hypervisor, VIM, CESCM, VNFs, etc.), and other required infrastructure components on behalf of its users (MVNO). IaaS providers also host users' applications (i.e. edge/cloud service) and handle tasks including system maintenance, backup, and resiliency planning. With this model in place, MVNO can compose VNF chains on demand, that is, MVNO decides to have VNF1 connected to VNF2 connected to VNF3. As a consequence, any VNF instantiation (depending on the used hardware resources) consumes a portion of available MVNO's aggregated resources. Therefore, the deployment of VNF chains is conditioned to the amount of requested resources by the MVNO. Note that VNF (e.g. vCaching, vTranscoding) hardware resources are fixed and determined by the VNF developer (e.g. 2 GB of storage and 2 GB of RAM). Although, it is possible to have different flavors of one VNF in place, for example, vCaching with extra/less storage capabilities. In this case, MVNO has more choices among one family of VNF.

The discussion above showed only a part of complexity of service provisioning in a joint cloud-radio environment. It is worth to note that, as mentioned above, there are still other open issues to address on a multi-tenant scenario, for example, business related questions such as multi-tenant pricing procedure and technical questions such as VNF placement over the distributed Light DC environment. These call for further discussions and efforts.

13.3 Security in 5G networks

5G enables innovative scenarios and applications making use of ultra-high speed, low-latency telecommunication networks for fixed and mobile users and machine-to-machine communications, as described by the 5G-PPP project CHARISMA (Converged Heterogeneous Advanced 5G Cloud-RAN Architecture for Intelligent and Secure Media Access) [22]. These scenarios together with the introduction of the new paradigm for computing and network infrastructure which decouples the actual functionality from the underlying hardware and middleware functions (Cloud Computing and SDN) further reinforces the need for automated management and control of the telecommunication infrastructure. In particular, since a cloud-based paradigm promotes that infrastructure is highly accessible and shared by multiple users (for instance, VNOs), the concept of a highly secure network gains even more relevance. It is of outmost importance to be able to provide robust, flexible, and proactive mechanisms to detect and prevent security issues and to be able to perform that on real time and in an automated fashion.

13.3.1 Research challenges

Today, we can't foresee the new and ever changing threats that 5G networks will have to protect against, but we do have the basis to create autonomic network management solutions that shall cope with them, being fed with insights from governed real-time analytics systems on the one hand, and actuating on network resources in order to minimize or prevent the effects of the detected threats in real-time on the other hand.

The following research challenges are by no means an exhaustive list of security research challenges facing 5G networks; however, they represent few security aspects that must be considered in 5G networks.

- *NS end-to-end security*
  This refers to all the different mechanisms that can be utilized to ensure confidentiality, integrity, availability, and non-repudiation for a NS. These security mechanisms include authentication, authorization, user privacy/anonymity, anti-jamming, encryption, digital signatures, and others, which can be used, based on priorities and requirements, to mitigate service related vulnerabilities.
- *Tenant isolation*
  Infrastructure sharing by multiple VNOs will require strict isolation at multiple levels in order to ensure absolute security. In particular, different aspects of control-plane, data-plane, and resource isolation must be investigated and guaranteed to ensure zero correlation among different tenant operations. Tenant isolation is ultimately important in order to ensure a reliable and warranted service assurance, together with data and communication integrity and confidentiality.
- *Virtualized security*
  Leveraging on the NFV environment, required network security functions could be deployed, orchestrated, and managed at different locations in the

network as a VNF, also referred as virtual security functions (VSFs). However, the security of VNF in itself as an element, for example, VNF hardening, VNF verification/attestation, VNF code robustness, and others, has to be carefully considered [23].

- *Security management*
  As mentioned earlier, the complexity of security mechanisms grow manifolds in the 5G networks not only due to virtualization of resources but also due to security requirements at different levels or domains such as network slice, NS, and network resource (physical & virtual). Hence, a holistic security management system, guided by a set of defined security policies, is essential to ensure that security mechanisms functions are enforced as planned.
- *Trust management*
  Another vital security challenge is the management of trust among the different modules of the NFV environment to ensure reliable and secure operation of the MANO framework.

13.3.2 A potential approach

As discussed in [24], the already high complexity of securing a network and its services has scaled up another notch with the introduction of SDN and NFV in 5G networks, that is, due to softwarization and virtualization of networks and network functions. In order to build a consistent and robust security ecosystem, network functions, and NSs in SDN/NFV environments require a comprehensive approach to end-to-end security for network resources, both physical and virtual, which ensures automated alignment of security policies to changes in network [25]. The dependence of security functions on monitoring information becomes even more crucial in SDN/NFV-based 5G networks because greater level of security related monitoring is required as compared to the traditional non-SDN/NFV networks. This is mainly due to dynamic and automated provisioning, orchestration and management of networks, network functions, and application services that SDN/NFV-based network deployments allow. In 5G, the difficulty to examine the complete network, both virtual and physical, increases the complexity of security monitoring, in order to ensure consistent automated security monitoring management. Figure 13.6 shows the security management architecture proposed in the 5G CHARISMA project, mapped on an ETSI MANO framework inspired control, management, and orchestration plane for a converged 5G access network.

The proposed security management architecture has two main sub-components, security policy management (SPM) and security & monitoring analytics (SMA). The SPM enables the management of end-to-end security policies at service level. It translates the defined security policy of a service into specific security requirements, for example, a certain VSF, hardening configuration of the VM running a VNF, and others, and initiates their provisioning. The SMA receives monitoring information from all physical and virtual resources. As shown in Figure 13.6, monitoring data is gathered at service level (e.g. VNFs), virtual resource level (e.g. VMs) and physical resource level (e.g. server machine). Being provided with a holistic status of the entire

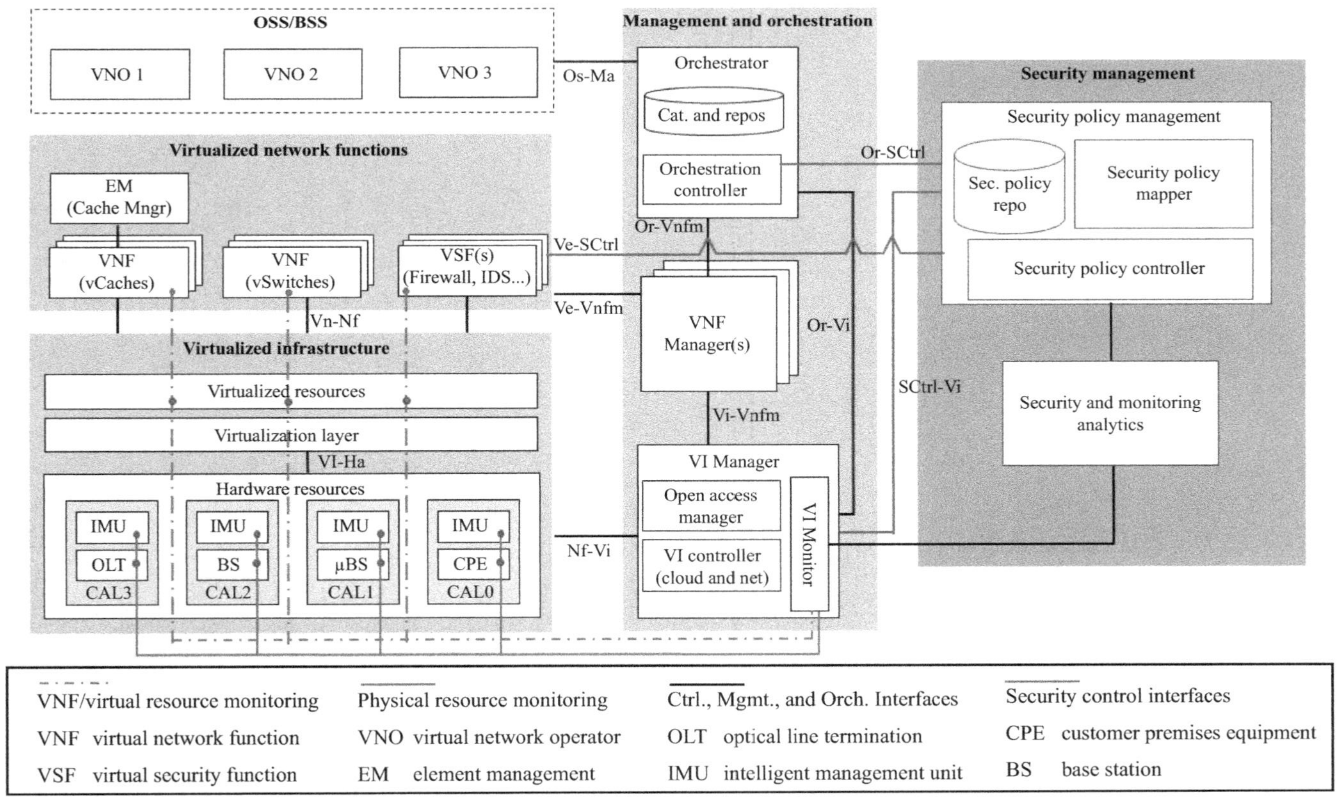

Figure 13.6 Possible security management architecture based on ETSI MANO framework

network, the SMA can extract useable knowledge and recommendations, by running smart analytics/algorithms on the available information, for SPM consumption. Based on the monitoring knowledge/recommendations and defined policy of a service, the SPM can take further actions. It is worth mentioning that the above architecture also considers multi-tenancy; thus, the SPM can also be observed on per tenant basis.

13.4 Wireless backhauling in 5G

One of the main paradigms of 5G networks is to shift tasks originally performed in centralized clouds, such as caching or data processing (video acceleration, encoding, etc.), to the edge. There, a variety of services may run on different physical devices, occasionally requiring data to be moved between devices at the edge in order to provide the required services.

Thus, to deliver a high-performance, the quick-processing and forwarding of access data entering the edge network is required. Further, in cases where service function chaining (SFC) is applied, trespassing several services that may be running on different machines becomes necessary. This is only possible with a flexible and powerful backhauling infrastructure that connects the devices at the edge. A solution that promises to satisfy the requirements of 5G deployments are wireless mesh backhauls [26,27].

In this approach, the access nodes (SCs) are equipped with wireless radio interfaces that allow to establish links to other nodes within transmission range. Figure 13.7 shows an example of topology in a use case where several SCs equipped with additional wireless backhaul interfaces form a mesh network. Each node can act as relay for traffic exchanged between any two points of the network: whenever, access traffic enters the backhaul, it can travel over multiple wireless links, either to reach one of the other nodes for further processing (at the edge), or to reach a gateway node. Gateway nodes dispose the additional network interfaces to establish a connection between the backhaul and the core network.

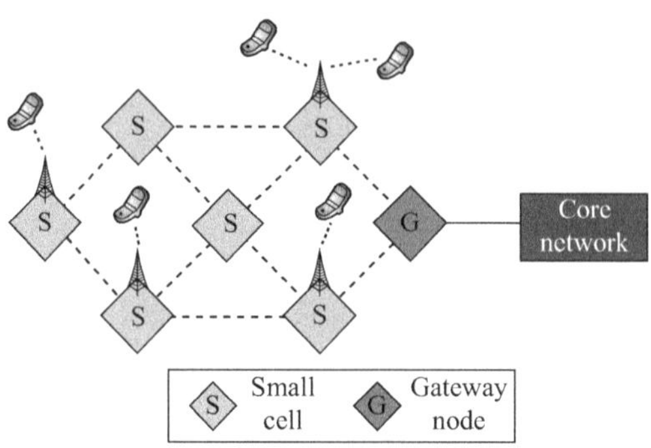

Figure 13.7 A small cell (SC) deployment using wireless interfaces to form a mesh backhaul. The connection to the core network is established via a gateway node (G)

Table 13.1 Typical characteristics of wired technologies used for backhauling in comparison with wireless backhauling technologies

Technology	1000BASE-T	1000BASE-SX	10GBASE-T	WiFi (sub 6 GHz)	WiFi (60 GHz)
Max. data rate	1 Gbit/s	1 Gbit/s	10 Gbit/s	<300 Mbit/s	7 Gbit/s
Range	100 m	200–500 m	55–100 m	100–200 m	50–100 m
Medium	Copper	Fiber	Copper	Radio	Radio

Wireless mesh networks may be composed of a heterogeneous mix of different wireless technologies that differ in terms of the used radio spectrum and communication standards or protocols. Well known standards like IEEE 802.11, operating at the 2.4 and 5 GHz band, can coexist along with other radio technologies, for example, 60 GHz band technologies [28]. Typical characteristics of aforementioned technologies, as well as the characteristics of wired technologies used for wireless backhauls, are shown in Table 13.1.

When comparing key features of wireless backhauls [29] with traditional, wired backhauls, certain advantages can be identified that can have a crucial impact on the performance of future 5G deployments:

- Wireless backhauls profit from a high dynamicity when it comes to the deployment of access nodes in the field. Contrarily to nodes that rely on a wired backhaul connection and require access to a physical infrastructure, nodes equipped with wireless interfaces are not bound to a particular physical location, as long as the connection over radio to at least one of the remaining nodes of the mesh network is guaranteed. Thus, using a wireless backhaul allows for a much more flexible deployment.
- The deployment of wireless backhauls is much less expensive when compared to the deployment of wired backhauls.
- The wireless backhaul architecture allows for a quick integration of new devices into the existing network infrastructure: newly deployed devices can join and leave the existing backhaul infrastructure at any moment. This can be of use whenever the network coverage needs to be increased.
- Since no wired infrastructure is required to interconnect the network nodes with each other and with the core network, the cost of deploying and extending the network is much lower when compared with traditional wired backhaul approaches.

Overall, as strong characteristics of wireless mesh backhauls, the high flexibility and adaptability stand out, which can be useful in a large variety of use cases. The easy deployment and integration of new nodes into the already existing wireless backhaul infrastructure, facilitates the MNO to effectively increase the network coverage and capacity. This is particularly useful whenever during sporadically a large number of UEs needs to be supported in environments without static infrastructure, for example, flash events like music festivals in remote areas or other similarly crowded events. In other scenarios, the easiness of deploying and reshaping the network topology can be crucial. In situations where within a short period of time, the deployed SC infrastructure breaks down, it might be necessary to quickly re-establish network connectivity. A cause for the loss or malfunctioning of network operation can be natural catastrophes that destroy parts of the deployed infrastructure. In such scenarios, a wireless backhaul provides the tools to quickly re-establish basic and emergency communications by reconfiguring the data plane of the backhaul to assure end-to-end user connectivity via alternative paths and by deploying new nodes at critical locations to regain access radio and backhaul connectivity in an area.

In spite of these clear advantages of wireless backhauls, several challenges arise for the design of such wireless architecture. For example, in contrast to wired solutions, noticeable fluctuations in the link qualities between the nodes of the network are possible. This can have a substantial impact on the stability of the backhaul and can also directly affect the achievable throughput.

Yet, it is possible to overcome possible issues since a careful planning of the spectrum to be used by the backhaul can avoid interference issues and in general there is path diversity between the nodes of the network.

Further, for an efficient operation of the backhaul during the network operation, forwarding policies are required to dynamically adapt the routes taken in the data plane of the backhaul. Simple policies can avoid problematic or poor performing links, whereas advanced policies can take into account elaborate interference models, cross-flows performance impacts, and others.

There exist centralized and decentralized approaches on how to determine the data plane routes. Centralized solutions can be SDN based, where a network controller takes routing decisions by monitoring the network and installing forwarding rules in the nodes of the wireless backhaul. In decentralized approaches, routing decisions are taken from the nodes at the edge without a supervisor. In such cases, routing protocols such as OSLR are used.

Features like SFC of VNFs requires an SDN controller, which takes care of installing the forwarding rules in the network, enabling the necessary node-to-node and node-to-core communications.

In a wireless backhaul network, it is possible to extend these SDN controllers by additionally providing the controller with specific information about the status of the wireless backhaul, for example, the availability and the quality of wireless links within the network, as well as the link utilization and ongoing transmissions. This information can be included in the metrics that are used for the calculation of optimal data paths. A wireless backhaul architecture for LTE traffic that implements such a type of logic has been designed and is currently verified [30]. The basics of intelligent, SDN-based wireless backhauls have also been evaluated in the context of fog computing networks, where even a backhaul formed of constrained devices has proven to deliver the required features for wireless backhauling [31]. These preliminary investigations show how a wireless, SDN-based backhaul is successfully deployed in a network of constrained devices. In spite of the

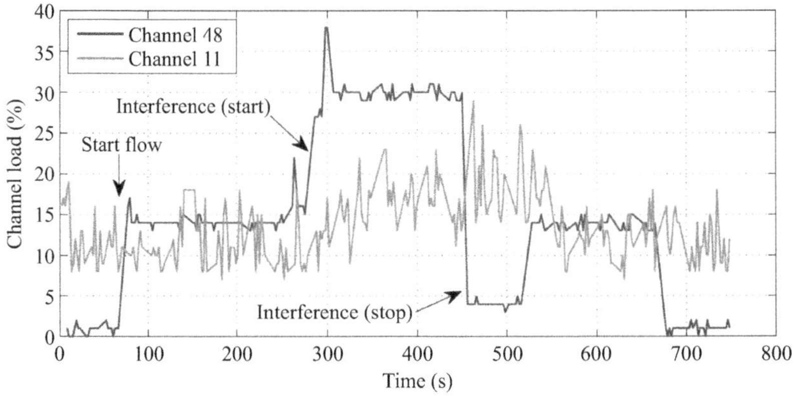

Figure 13.8 Channel load information gathered at a network node, showing how the decisions taken by the SDN-controller affect the cannel load of a wireless backhaul over the course of an experiment in a fog computing use case [31]

constraints of the used hardware, SDN-based wireless backhauling solutions can be applied, which represents a corner stone toward using wireless backhauls in fog networking.

The analyzed use case involves several constrained devices that use wireless transceivers to build a backhaul mesh network that reports channel load statistics to an SDN controller. The controller uses the gathered information to apply load-balancing mechanisms in order to avoid the congestion of wireless links, which can be one of the main causes for performance losses in wireless networks.

Figure 13.8 shows an excerpt from the results, where we can see the channel load statistics gathered by the SDN controller from a network device. The controller observes some background traffic on channel 11 and as soon as a traffic flow is generated by a device from the backhaul ($t \sim 70$ s), the data flow is allocated on the less busy channel 48. The reaction to strong external interference introduced at $t \sim 290$ s on channel 48 then leads the controller to take the decision to rebalance the network load by reassigning the data flow to channel 11. As soon as the external interference stops, the data flow is reallocated to channel 48, again to balance the network load and avoid congestion. The load-balancing paradigm followed by the SDN controller in this example is just one out of many paradigms that can be used to determine how traffic is routed in wireless backhaul, be it for fog computing or other 5G use cases. The investigations performed in this works give a hint of the possibilities and capacities intelligent wireless backhauls might offer to us in future deployments.

Another important aspect of 5G networks is multi-tenancy, which imposes several requirements on the wireless backhaul. While accessing edge services, performing SFC, or accessing the core network, each mobile operator (tenant) may have different types of service subscriptions and SLAs with varying KPIs. Since the wireless backhaul has a direct and crucial impact on some of the KPIs (delay, data rate), it carries a high responsibility of delivering the QoS each tenant expects, making multi-tenancy one of the key design parameters for the wireless backhaul architecture.

One way to handle multi-tenancy in SDN-based solutions is the virtualization of the backhaul network at several layers. At a higher abstraction layer, the backhaul nodes and the wireless links are virtualized and part of the virtualized infrastructure are assigned to each tenant depending on its requirements on network coverage and edge services. On a lower layer of abstraction, the wireless radio resources, that is, the data rate of wireless links, the wireless interfaces, and others, are virtualized and slices of the backhaul are assigned to each tenant [32].

The physical properties and limitations of the radio communications per se are a challenge to overcome. They impose several restrictions when compared to wired solutions. In particular, the limited data rate capacities of wireless links compared to the bandwidths, typical for wired connections and the fact that parallel transmissions on the same frequency may interfere each other. The issue of data rate limitations mainly affects communications in the sub 6 GHz band, for example, the 2.4 and 5 GHz bands used in IEEE 802.11, where data rates of up to 600 Mbit/s are possible. However, the use of the 60 GHz band gives access to data rates of up to 7 Gbit/s, while covering ranges of up to 100 m between two nodes. The recent loosening of the limitations for the use of the 60 GHz spectrum have converted this technology in the most promising and attractive one for future 5G wireless backhaul deployments.

Further, the use of directional antennas in the sub 6 GHz band and a careful planning of the radio spectrum used for communications within the backhaul are two methods to reduce the degree of radio interference and assure a high network performance.

13.5 Conclusion

Needs and requirements for future 5G networks are very diverse and ambitious hence, they pose several important research challenges. In particular, one of the main stated paradigm, is the shift from centralized cloud-computing services toward an edge-computing service provisioning approach. This chapter highlights the most relevant subsequent challenges, including multi-tenancy, network security provisioning, and wireless backhaul implementation. Furthermore, it gathers different ideas and proposals for innovative architectures as potential approaches to address some of the identified 5G requirements.

The presented concepts, along with their corresponding enabling technologies, constitute a promising set of topics under research that are paving the way to successfully accomplish the desired 5G networks requirements in a mobile edge-computing environment.

Acknowledgments

Content presented in this chapter covers work performed in the European H2020 SESAME (no. 671596), H2020 CHARISMA (no. 671704), H2020 5G-XHAUL (no. 671551), and FP7 FLEx (no. 612050) projects.

References

[1] European Commission, "5G Infrastructure Public Private Partnership (PPP): The Next Generation of Communication Networks Will Be Made in EU," *Digital Agenda for Europe*, European commission, 2016 (Retrieved: May 23, 2016).
[2] European Commission, "Digital Agenda for Europe," *Digital Agenda for Europe*, European commission, 2016 (Retrieved: June 10, 2016).
[3] European Commission, "The 5G Infrastructure Public Private Partnership: The Next Generation of Communication Networks and Services," *5G Vision* European commission, 2016 (Retrieved: June 23, 2016).
[4] D. Kreutz, F. M. V. Ramos, P. E. Veríssimo, C. E. Rothenberg, S. Azodolmolky and S. Uhlig, "Software-defined networking: a comprehensive survey," *Proceedings of the IEEE*, vol. 103, no. 1, pp. 14–76, 2015.
[5] European Telecommunications Standards Institute, "Network Functions Virtualisation: An Introduction, Benefits, Enablers, Challenges & Call for Action," *Network Functions Virtualisation—Introductory White Paper,* European Telecommunications Standards Institute (ETSI), 2012.
[6] H. Karl, S. Dräxler, M. Peuster, *et al.*, "DevOps for network function virtualisation: an architectural approach," *Transactions on Emerging Telecommunications Technologies*, DOI: 10.1002/ett.3084, 2016.
[7] European Telecommunications Standards Institute, "Network Functions Virtualisation (NFV): Network Operator Perspectives on Industry Progress," *Network Functions Virtualisation—White Paper,* European Telecommunications Standards Institute (ETSI), 2013.
[8] P. Rost, A. Banchs and I. Berberana, "Mobile network architecture evolution toward 5G," *IEEE Communications Magazine*, vol. 54, no. 5, pp. 84–91, 2016.
[9] 3GPP, "Network sharing: Architecture and functional description," *TS 23.251 v12.0.0, Release 12,* December 2013.
[10] K. Samdanis, X. Costa-Perez and V. Sciancalepore, "From network sharing to multi-tenancy: the 5G network slice broker," *IEEE Communications Magazine*, vol. 54, no. 7, pp. 32–39, 2016.
[11] European Telecommunications Standards Institute (ETSI), "Mobile-Edge Computing: Introductory Technical White Paper," European Telecommunications Standards Institute (ETSI), 2014.
[12] I. Son, D. Lee, J.-N. Lee and Y. B. Chang, "Market perception on cloud computing initiatives in organizations: an extended resource-based view," *Information & Management*, vol. 51, no. 6, pp. 653–669, 2014.

[13] I. Giannoulakis, G. Xylouris, E. Kafetzakis, *et al*., "System architecture and aspects of SESAME: small cells coordination for multi-tenancy and Edge services," *IEEE ICC*, Kuala Lumpur, Malaysia, 2016.
[14] I. Giannoulakis, J. O. Fajardo, J. Garcia, *et al*., "Enabling Technologies and Benefits of Multi-Tenant Multi-Service 5G Small Cells," in *EuCNC*, Athens, Greece, June 2016.
[15] Y. Liu, J. K. Muppala, M. Veeraraghavan, D. Lin, M. Hamdi, "Data Center Networks: Topologies, Architectures and Fault-Tolerance Characteristics," Springer, 2013, ISBN 978-3-319-01949-9.
[16] H2020 SESAME Project, "Deliverable 4.1 'Light DC Architecture Design,'" H2020 SESAME Project, 2016 (Retrieved: March 10, 2016).
[17] H2020 SESAME Project, "Deliverable 2.2 'Overall System Architecture and Interfaces,'" H2020 SESAME Project, 2016 (Retrieved: June 2, 2016).
[18] European Telecommunications Standards Institute (ETSI), "NFV Management and Orchestration—An Overview," *GS NFV-MAN001 V1.1.1*, European Telecommunications Standards Institute (ETSI), 2014.
[19] I. Giannoulakis, E. Kafetzakis, I. Trajkovska, *et al*., "The emergence of operator-neutral small cells as a strong case for cloud computing at the mobile edge," *Transactions on Emerging Telecommunications Technologies*, DOI: 10.1002/ett.3077, 2016.
[20] H2020 SESAME Project, "Deliverable 2.1 'System Use Cases and Requirements,'" H2020 SESAME Project, 2015 (Retrieved: May 10, 2015).
[21] P. S. Khodashenas, C. Ruiz, J. Ferrer Riera, *et al*., "Service Provisioning and Pricing Methods in a Multi-Tenant Cloud Enabled RAN," in *IEEE CSCN*, Berlin, 2016.
[22] S. Sharma, "Evolution of as-a-Service Era in Cloud. A review on as-a-Service Framework (white paper)," *Center for Survey Statistics and Methodology*, *Iowa State University*, Ames, Iowa, USA, 2015.
[23] M. D. Firoozjaei, J. Jeong, H. Ko and H. Kim, "Security challenges with network functions virtualization," *Elsevier Future Generation Computer Systems*, vol 67, pp. 315–324, February 2017, DOI:10.1016/j.future.2016.07.002.
[24] P.S. Khodashenas, J. Aznar, A. Legarrea, *et al*., "5G Network Challenges and Realization Insights," in *ICTON*, Trento (Italy), 2016.
[25] European Telecommunications Standards Institute (ETSI), "Security Management and Monitoring for NFV," *NFV-Sec013, Release* 2, European Telecommunications Standards Institute, 2015.
[26] M. Coldrey, H. Koorapaty, J. Berg, *et al*., "Small-Cell Wireless Backhauling: A Non-Line-of-Sight Approach for Point-to-Point Microwave Links," in *IEEE Vehicular Technology Conference*, Quebec, 2012.
[27] U. Siddique, H. Tabassum, E. Hossain and D. Kim, "Wireless backhauling of 5G small cells: challenges and solution approaches," *IEEE Wireless Communications*, vol. 22, no. 5, pp. 22–31, 2015.
[28] S. Hur, D. Love, J. Krogmeier, T. Thomas and A. Ghosh, "Millimeter wave beamforming for wireless backhaul and access in small cell networks," *IEEE Transactions on Communications*, vol. 61, no. 10, pp. 4391–4403, 2013.

[29] G. Xiaohu, H. Cheng and M. Guizani, "5G wireless backhaul networks: challenges and research advances," *IEEE Network*, vol. 28, no. 6, pp. 6–11, 2014.

[30] FP7 FLEX project, "D5.1 Report on first Open Call activities (SODA-LITE)," FP7 FLEX Project, 2016 (Retrieved: February 12, 2016).

[31] A. Betzler, F. Quer, D. Camps-Mur, I. Demirkol and E. Garcia-Villegas, "On the Benefits of Wireless SDN in Networks of Constrained Edge Devices," in *27th European Conference on Networks and Communications*, Athens, 2016.

[32] H2020 SESAME Project, "Deliverable 3.1 'CESC Prototype Design Specifications and Initial Studies on Self-X and Virtualization Aspects,'" H2020 SESAME Project, 2016 (Retrieved: February 20, 2016).

Chapter 14

A novel marketplace for trading/brokering virtual network functions over cloud infrastructures

George Alexiou[1], Evangelos Pallis[1], Evangelos Markakis[1], Anargyros Sideris[1], Athina Bourdena[1], George Mastorakis[2] and Constandinos X. Mavromoustakis[3]

Abstract

Following up the success story of the OS-Specific App Stores, we present a new business case in network function virtualization (NFV), where function provider (FP) can publish, broke, trade, offer, and advertise their developed functions inside a novel Marketplace for NFV. This novel approach is able to attract new entrants to the networking market, including among other, a Novel Brokerage Platform, allowing Service Providers to transact with the FP. Finally, via the Marketplace, customers can browse and select services and virtual appliances that best match their needs, as well as negotiate Service Level Agreements and be charged under various billing models browse and select the services and virtual appliances that best match their needs.

14.1 Introduction

This chapter elaborates on the research work that was conducted under the framework of T-NOVA ("network functions as-a-service (NFaaS) over Virtualized Infrastructures"), which is a European FP7 Large-scale Integrated Project. The primary aim of this project is the design and implementation of a management/orchestration (MANO) framework for the automated provision, configuration, monitoring, and optimization of NFaaS over virtualized network and IT infrastructures. T-NOVA

[1]Technological Educational Institute of Crete, Heraklion, Crete, Greece
[2]Department of Business Administration, Technological Educational Institute of Crete, Agios Nikolaos, Crete, Greece
[3]Department of Computer Science, University of Nicosia, Nicosia, Cyprus

leverages and enhances cloud management architectures for the elastic provision and (re-) allocation of IT resources hosting network functions (NFs). It also exploits and extends Software Defined Networking platforms for efficient management of the network infrastructure.

The T-NOVA framework allows operators to deploy virtualized NFs, not only for their own needs, but also to offer them to their customers, as value-added services. Virtual network appliances (gateways, proxies, firewalls, transcoders, analyzers, etc.) can be provided on-demand "as-a-Service," eliminating the need to acquire, install, and maintain specialized hardware at customer premises.

Leveraging this NFaaS concept and in order to facilitate the involvement of diverse actors in the network function virtualization (NFV) scene as well as the attraction of new market entrants, T-NOVA introduces a novel concept of "NFV Marketplace," in which network services (NSs) and functions offered by several developers can be published and brokered/traded. The NFV Marketplace enables customers to browse and select services and virtual appliances that best match their needs, as well as negotiate Service Level Agreements (SLAs) and be charged under various billing models. A novel business case for NFV is thus introduced and promoted.

The NF Framework is the conceptual element of the T-NOVA system devoted to the definition of the structure and behavior of the virtual network functions (VNFs). It comprises a NF store, where the VNFs are kept and made available to T-NOVA as building blocks for creating NSs. VNFs and NSs in T-NOVA are described, traded, and offered to the final users by an innovative Marketplace that opens the NFV market to software developers and traditional service providers (SPs) for the benefit of large adoption of NFV solutions.

A VNF is characterized by two attributes: the operational functionalities and the management behavior. The operational part explicitly defines the NFs that are supported, whereas the management part is responsible for the VNF lifecycle. Therefore, a VNF in T-NOVA shall support the application programming interface (APIs) for interacting with the T-NOVA orchestration and the virtualized infrastructure and shall implement the VNF lifecycle described in this report. The VNF metadata is a fundamental part of each VNF. It provides the information for describing how the VNF is composed, which functionalities it provides, and how to manage it. Currently, there are many approaches for implementing this concept that are focused on the technical requirements for making a virtual application running. In T-NOVA, this information is extended with business aspects that allow the registration and trading of a VNF in the marketplace.

The marketplace concept has been introduced by T-NOVA as a novelty in the NFV scheme in order to facilitate the interaction between the different stakeholders that are identified in the NFV business scenarios. On one hand, the VNFs can be implemented by a wide range of developers providing software implementation, and on the other hand, NS providers may want to acquire VNFs to compose NSs to be provided to their own customers [1]. The T-NOVA Marketplace has been designed as a distributed platform placed on top of the overall architecture which, besides of including the users front-end, comprises Operating Support System/Business

Support System (OSS/BSS) components as billing and accounting, and innovative modules as the T-NOVA Brokerage to allow trading functionality.

The virtualization of NFs is being addressed by notable standardization bodies such as European Telecommunication Standard Institute (ETSI) and Internet Engineering Task Force. In particular, ETSI has developed a NFV reference architecture and has provided a common language in this area. Therefore, this report looks at ETSI to build on it. The T-NOVA Marketplace specification relies also on ongoing standardization activities such as business best practices provided by TM Forum [2], as it is for instance the integration of a business service catalog and SLA Management issues in virtualization.

14.1.1 Motivation, objectives, and scope

NFV constitutes a topic of immense interest to the networking community in the research/academic domain but also in industry since it is candidate approach for short-term exploitation. Via the concept of infrastructure "softwarization," NFV has the potential to entirely transform the networking market and open it to new entrants. In this context, T-NOVA introduces a complete open solution for NFV deployment, focusing on the VNF as a service perspective with a strong business orientation.

In order to provide this business orientation to the NFV scheme T-NOVA develops a novel marketplace that will facilitate T-NOVA customers to select virtual appliances by means of a friendly front-end, "plug" them into their existing connectivity services, configure/adapt them according to their needs, and, in the case of NS providers, also allow them to offer NSs composed by several VNFs to their own customers.

The service request will be carried out via a tailored customer front-end/ brokerage platform that is part of the T-NOVA Marketplace. This marketplace will also provide all the T-NOVA stakeholders SLA and billing functionalities. On the other hand, T-NOVA introduces an innovative NF store following the paradigm of already successful OS-specific "App Stores." This NF store contains VNFs by third-party developers, published as independent entities and accompanied with the necessary metadata for both technical and business description of the VNF.

Software developers willing to sell their VNFs through the T-NOVA Marketplace shall extend their implementation of NFs supporting the APIs for interacting with the virtualized infrastructure and the T-NOVA orchestration for the service composition, and the VNF lifecycle.

In this way, thanks to the Marketplace, it is expected that T-NOVA will contribute to expand market opportunities by attracting new entrants to the networking market. This capability will be particularly important for small and medium-sized enterprises and academic institutions that can leverage the T-NOVA architecture by developing innovative cutting-edge NFs as software modules that can be included in the NF store. This will also enable the rapid introduction of VNFs into the market.

14.1.2 T-NOVA Marketplace high-level overview

All features supported by the T-NOVA Marketplace will have to be compliance with the generic T-NOVA business scenario as depicted in Figure 14.1 that reflects

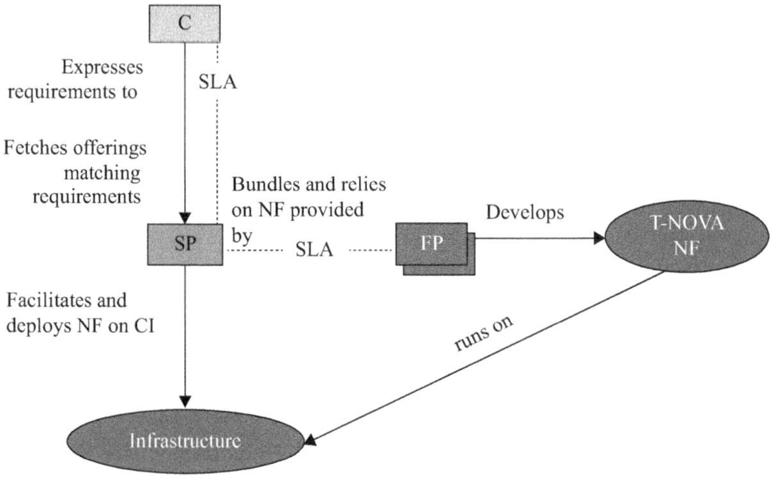

Figure 14.1 Business T-NOVA stakeholders relationships

the two main commercial relationship that are in T-NOVA: one between the SP and Function Providers (FPs) to acquire standalone VNFs to compose a NS and the second one between the SP and the customer (C) who acquire NSs.

The FPs that want to sell their VNFs through T-NOVA Marketplace will enter the system providing their VNFs information: VNF metadata including technical constraints, SLAs, price and others.

The SP that wants to purchase VNFs in order to later sell NSs through T-NOVA enters the system. The SP will be able to compose services acquiring VNFs by means of a brokerage that will facilitate auctioning process among several FPs offering VNFs with similar features in achieve the lowest price offer. Then, the SP will be able to compose NSs bundling VNFs and advertise them creating offerings that will include service description, SLA, and price and will be exposure by means of the T-NOVA Marketplace to the customer. The customer will be able to search for the end-to-end NSs offerings that can be composed by one or several VNFs, and with different SLA level and price in each offering. In the event that there is not any available service offering matching a customer request, a new service composition could to take place triggered by the SP and trading mechanisms [3] will be performed among FPs if several FPs offer similar VNFs dynamically. The customer will be able to select offerings, and the SLA agreement procedure will be initiated: between customer and SP and consequently between SP and FPs; then the service provisioning will start. All the related information SLAs, prices, and others will be stored in the marketplace modules for later billing purposes. Customer, SP, and FP will be able to access their related information by means of the dashboard as it can be the service monitoring information, SLA fulfillment information and billing information.

14.2 Specifications of the T-NOVA Marketplace

The marketplace in the NFV scheme is an innovative concept that T-NOVA introduces with the aim of promoting the VNF service offerings and facilitating the commercial activity and fluent interaction among the different business stakeholders identified in. Besides providing the Graphical User Interface (GUI) for all of the stakeholders, the T-NOVA Marketplace will facilitate all the necessary features related to the market activity, such as trading, SLA negotiations, and billing. The components of the T-NOVA Marketplace high level described in Table 14.1.

14.2.1 State-of-art

In order to design and later implement the marketplace context in T-NOVA, we have first looked at the most relevant ongoing standardization works when applying NSs provision business processes to NFV. In Sep' 14 TM Forum provided some first inputs mapping their standardization document about business process to the ETSI NFV MANO architecture [4]. Analyzing the state-of-art, we have found some solutions from which we can build on in order to develop the T-NOVA Marketplace; however, at this stage it does not exist a proper marketplace to deliver VNF as a Service.

14.2.1.1 European Telecommunication Standard Institute (ETSI) Industry Specification Group (ISG) network function virtualization (NFV)

A network operator-led Industry Specification Group (ISG) was setup in the last quarter of 2012 under the umbrella of ETSI to work through the technical challenges

Table 14.1 Main T-NOVA Marketplace components definitions

Name	Description
SLA management module	The marketplace functional entity that establishes and stores the SLAs among all the involved parties and checking if the SLAs have been fulfilled or not will inform the accounting system for the pertinent billable items (penalties or rewarding)
Accounting module	The marketplace functional entity that stores all the information needed for later billing for each user: usage resources for the different services, SLAs evaluations, etc.
Billing module	The marketplace functional entity that produces the bills based on the information stored in the accounting module
Access control module	The marketplace functional entity that administers security in a multiuser environment, managing and enabling access authorization/control for the different T-NOVA stakeholders considering their roles and permissions
Brokerage module	The marketplace functional entity that enables the interaction among actors for service advertisement, request and brokerage/trading
Dashboard	The marketplace functional entity that provides the user front-end, exposing in a graphical manner all customer-facing services

of NFV. ETSI ISG NFV in its document on global architecture [4] illustrates the high-level NFV framework, where three main working domains can be identified:

- VNF, as the software implementation of a NF that is capable of running over the NFV Infrastructure (NFVI).
- NFVI, which includes the diversity of physical resources and how these can be virtualized. NFVI supports the execution of the VNFs.
- NFV MANO), which covers the orchestration and lifecycle management of physical and/or software resources [5–7] that support the infrastructure virtualization, and the lifecycle management of VNFs. NFV MANO focuses on all virtualization-specific management tasks necessary in the NFV framework.

The T-NOVA Marketplace is designed considering the above solutions, the current NFV ETSI architecture, the ongoing TM Forum best practices for business services delivery and SLA management. T-NOVA introduces the marketplace concept aiming at opening the NFV market to third-party developers for the provision of VNFs, a concept that currently falls outside the technical view of ETSI NFV. On the other hand, ETSI MANO has been deeply explained; ETSI NFV does not provide yet any more insight on the OSS/BSS of the operator besides the definition of an interface. Though OSS/BSS systems are not within the scope of T-NOVA, the proposed marketplace contains partially some OSS/BSS functionalities (i.e., billing, accounting, SLA monitoring, authentication, authorization, and accounting (AAA)), which will be implemented/adapted.

14.2.1.2 TM Forum

The general objective of Tele Management Forum [2], as a global trade association of SPs and suppliers, is the improvement on business agility and the growth of business through knowledge, tools, standard, training, and best practices. The specific TM Forum's Agile Business and IT Program aims at optimizing SPs' operations reducing costs, risks, and time to market by providing a set of integrated offerings that collects the experience and best practices gleaned from the major players within the industry. The TMF's standards are collectively known as Frameworx, which is composed of four underlying components, each aimed at standardizing information models, interfaces, and lexicon:

- Business Process Framework (eTOM, Telecom Operations Map): the industry's common process architecture for both business and functional processes. This framework is meant to aid in the creation of a comprehensive, multi-layered view of all of the business processes necessary for a carrier's operation. It provides both guidelines and process flows and aligns with standards from Information Technology Infrastructure Library and other external bodies.
- Information Framework (SID, Shared Information/Data model): It provides a common reference model for enterprise information that SPs, software providers, and integrators use to describe management information. It is used to develop databases and provide a glossary of terms for business processes. The framework is intended to reduce integration costs and to reduce project

management time and cost by minimizing the number of necessary changes to underlying architecture during the launch of a new product or service offering.
- Application Framework (Telecom Application Map): It provides a common language between SPs and their suppliers to describe systems and their functions, as well as a common way of grouping them. It attempts to group the information and processes defined by the eTOM and the SID into recognizable applications.
- Integration Framework (TM Forum Integration Program): It shows how the business process, information, and application frameworks interact to
  - Create a catalog of business services that define functional and nonfunctional aspects of a service based on service-oriented principles;
  - Develop a platform or domain-based enterprise architecture that provides the business agility required to compete in today's market;
  - Define critical standard interfaces that speed integration.

14.2.1.3 Conclusions

Analyzing the state-of-art, we have found some solutions from which we can build on in order to develop the T-NOVA Marketplace; however, at this stage it does not exist a proper marketplace to deliver VNF as a Service. The T-NOVA Marketplace is designed considering the above solutions, the current NFV ETSI architecture, the ongoing TM Forum best practices for business services delivery and SLA management.

14.2.2 Requirements for T-NOVA Marketplace

The requirements capture process has focused on identifying the desired behavior for the T-NOVA Marketplace and its components, most of which were identified on the basis of the previous requirements analysis performed at T-NOVA system level. The goal of these requirements is to develop an understanding of what the marketplace components need, how they interact between each other and their relationship to the overall T-NOVA architecture.

Requirements were primarily anchored to the existing T-NOVA use cases and the interactions with the whole system both in terms of the actions and requests that would be expected. In addition, the high-level data/information that is required by the marketplace to successfully deploy its functionalities was also identified. Identified requirements were primarily functional as they are related to the behavior that is expected from the marketplace.

Using a systems' engineering approach, the high-level architecture for the marketplace, each component of the overall system was specified in terms of high-level functional blocks. This approach identified the following functional blocks

- Dashboard
- Access control
- Brokerage module
- SLA management module
- Accounting module
- Billing module

Also a Business Service Catalog has been identified to be part of the Marketplace matching TM Forum proposal for business agility.

14.2.3 Specification of the T-NOVA Marketplace architecture

Based on the requirements performed at system level, the requirements gathered for each component in the marketplace including the overall diagram for the T-NOVA Marketplace architecture with both the external and internal interfaces.

Tables 14.2 and 14.3 collect a brief description of the purpose of each external and internal interface depicted in Figure 14.2.

14.2.4 External Interfaces to the T-NOVA Marketplace

The marketplace modules will communicate with other two T-NOVA components: the orchestrator, and the NF store.

14.2.4.1 Orchestrator

The T-NOVA Orchestrator deals with the optimal deployment of NSs instances [8], as requested by the customer or the SP on the marketplace, according to a yet to be designed algorithm, the required SLA, and the current status of the available infrastructure.

Although all NSs instances have been instantiated and are running, it is also the orchestrator's responsibility to follow the available metrics, both from the infrastructure and from the service metrics. In order to meet the agreed SLAs, the orchestrator may scale out or up the supporting infrastructure, communicating such changes to the marketplace, so that a change in accounting is registered and later billed to the customer. Later, if the scaled (out or up) infrastructure [9] is perceived

Table 14.2 Marketplace external interfaces

Marketplace external interface	Description
T-Da-Or	It is used to get monitoring information of the service by the customer and SP
T-Ss-Or	It is used to notify the orchestrator about a new NS instantiation including the service configuration
T-Sl-Or	SLA module requests currently running NS metrics from the monitoring system in the orchestrator
T-Ac-Or	The accounting is notified about any status change of each Network Service (NS) or VNF instances
T-Bsc-Or	The BSC uses this interface to push the Network Service Descriptor (NSD) relevant fields to the orchestrator when a new service offering has been created. Also once the orchestrator validates it, the availability of a service is notified to the BSC to be offered to the customer
T-Br-NFS	The brokerage module will use it to retrieves information about the available VNFs
T-Da-Nfs	It is used to upload VNF and metadata by the FPs to the Function Store

Table 14.3 Marketplace internal interfaces

Marketplace internal interfaces	Description
T-Ac-AA	It is used by the accounting module to access the "user profiles"
T-Ac-Bi	All the information needed for billing is stored in the accounting module
T-Sl-Ac	SLA module is accessed by the accounting module to extract information about SLA violations for penalties to be applied
T-Br-Ss	This interface is used by the Service Selection module to check any potential change in the price as a result of the trading process before creating an entry in the accounting
T-Da-Sl	It is used to introduce SLA templates in the SLA module when a new service of VNF is created and also to provide the dashboard with SLA fulfillment related information
T-Da-AA	It is used to provide and collect all the information necessary to authenticate the T-NOVA users
T-Da-Ss	Once a Customer selects a service in the BSC from the dashboard, it is managed by the service selection module in order to provide the custom service configuration
T-Ss-Ac	It is used to create the entries in the accounting module to track every service or VNF instance created in the orchestrator for billing purposes
T-Da-Bi	The three stakeholders use it to visualize billing information
T-Da-Br	It is used to request VNFs in order to facilitate auctioning among FPs
T-Da-BSC	It is used to publish offerings by the SP, and to browse offerings by the customer

as being more than enough to fulfill the SLA, it can be scaled in or down [10,11]. Through all this process, the orchestrator must provide the marketplace with meaningful metrics showing how NS instances are working.

14.2.4.2 Network function store (NF store)

This T-NOVA component will store the VNFs images and metadata that the marketplace, more concretely the brokerage module, will use to perform trading mechanisms among FPs, to later include those VNFs in the service composition process performed by the orchestrator.

For a VNF to be part of a service composition process, it is necessary that the orchestrator make it available. Whenever a VNF is uploaded, updated or removed from the NF Store, the orchestrator is informed in order to update its internal registers. This process makes the VNF available in the Function Store to be retrieved by the brokerage module.

14.2.5 Marketplace modules specification

14.2.5.1 Dashboard

With the aim of creating a single entry to the T-NOVA system that provides simplicity for the different T-NOVA users or stakeholders, a unified T-NOVA Dashboard will be

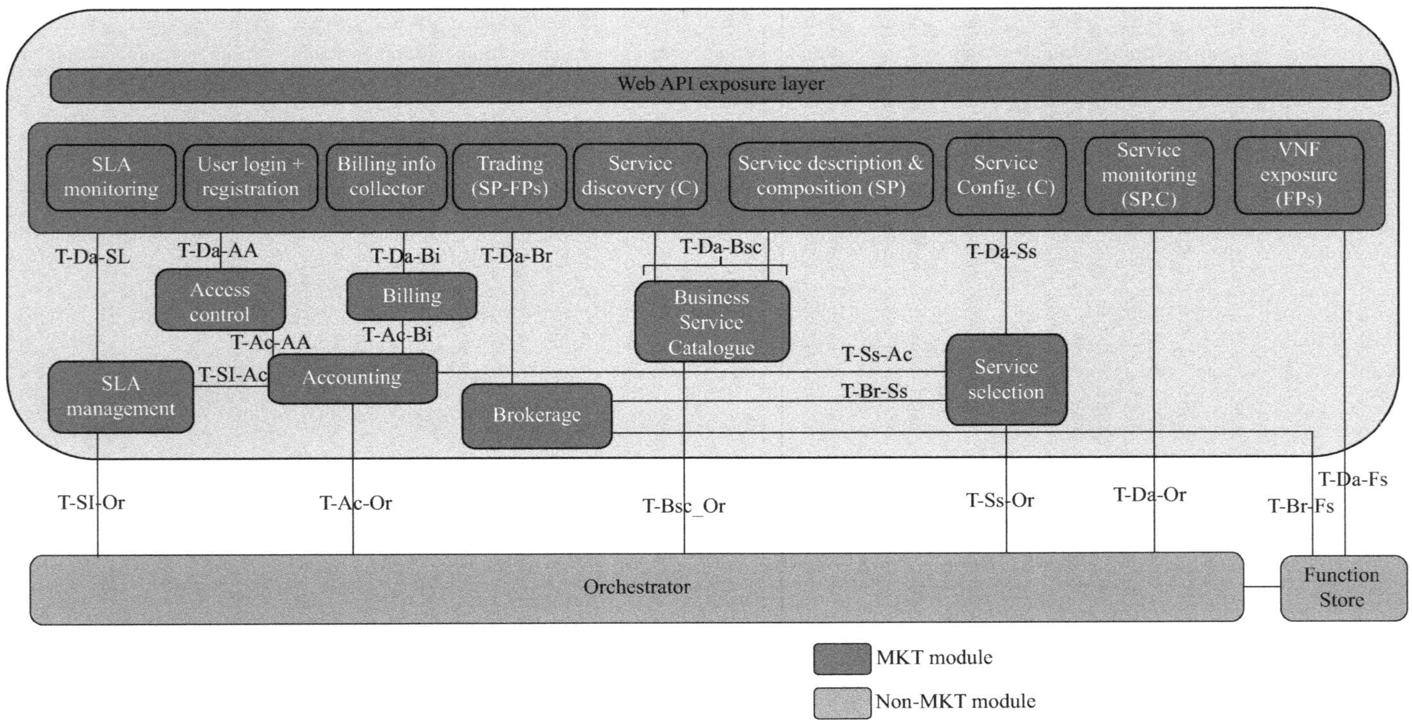

Figure 14.2 Marketplace architecture

designed, taking into account the different roles of the T-NOVA Marketplace. This common dashboard for the whole T-NOVA environment will host three views for the three basic stakeholders that will access the T-NOVA Marketplace: the SP, the FP, and the Customer.

Starting from the dashboard requirements, in this section we include the general description of the information that the dashboard will have to show, first ideas for its design and the general information that will have to be collected by the different APIs of the dashboard coming from the rest of the T-NOVA components.

Functionality

The main features of the dashboard are presented in Figure 14.3. Each view is allocated with specific functionalities stemming from the requirements gathered.

The SP view of the dashboard will allow the SP to perform the functionalities shown in Table 14.4.

The FP view of the dashboard will allow the FPs to perform the functionalities shown in Table 14.5.

The customer view of the dashboard will allow the customer to perform the functionalities shown in Table 14.6.

Design

The dashboard constitutes the T-NOVA system front-end, as offered to the Customer, the SP and the FPs for service consumption, discovery, interaction, publication, and others. In order for the dashboard to be as up-to-date as possible and terminal-agnostic, a web-based design has been selected.

Furthermore, the dashboard shall be able to meet and, if necessary, to adapt to the specific stakeholder's needs/requirements as much as possible providing the best experience to a specific stakeholder. This implies that the implementation shall achieve a flexible service presentation by means of an appropriate choice of technologies and tools. T-NOVA will allow every role to personalize some settings such as interface, appearance, and content according to its profile.

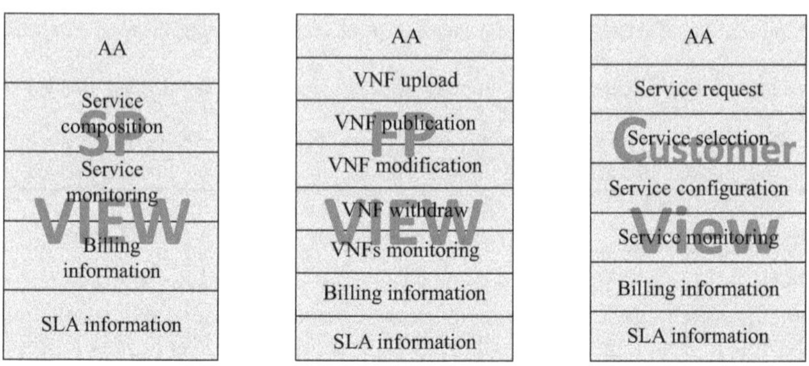

Figure 14.3 Dashboard views

Table 14.4 SP dashboard view

Functionality	Short explanation
AA	Authorization and authentication of the respective role into the T-NOVA dashboard
Service composition	Graphical wizard that will help the SP to compose a new Network Service (NS) starting from the brokerage among the FPs owing the available VNFs
Service monitoring	Graphical representation of all monitoring data for a selected or "consumed" Service
Billing information	Graphical representation of the billing outcomes of selected or "consumed" service. There will be two types of billing information for the SP: • Charges for the SP's customers (BSS functionality) • Invoices on behalf of its own suppliers, the FPs
SLA information	Details of the selected or "consumed" service based on how they respect the agreed SLA. The SP will have accessed to two different kinds of SLA contract and SLA monitoring information: • SLA between SP and its customers (BSS) • SLA agreed with his its suppliers, the FPs

Table 14.5 FP dashboard view

Functionality	Short explanation
AA	Authorization and authentication of the respective role into the T-NOVA dashboard
VNF upload	Graphical wizard that will help the FP to upload his VNF with the necessary parameters
VNF Publication	Graphical representation for the FP to provide the last check in order to publish the uploaded VNF
VNF modification	Small graphical wizard that provides the ability to the FP to modify the uploaded VNF
VNF withdraw	Graphical representation that gives to the FP the ability to remove an already published or uploaded VNF
VNFs monitoring	Graphical representation of all monitoring data for a selected or "consumed" NF
Billing information	Graphical representation of the Billing outcomes for a selected or "consumed" NF
SLA information	Information of the selected or "consumed" NFs based on the agreed SLA and its fulfillment

More specifically, in the authentication stage, all stakeholders share a common layout that displays the generic graphical interface composed by the basic controls that enable stakeholder specific authentication. Once authenticated, every stakeholder will be able to customize the overall experience according to a set of preferences and his profile.

The main design decision has been to have a common dashboard with different customized views based on different roles.

Table 14.6 Customer dashboard view

Functionality	Short explanation
AA	Authorization and authentication of the respective role into the T-NOVA dashboard
Service request	Graphical representation of the services/functions returned by the T-NOVA business service catalog
Service selection	Graphical representation assisted by a check box providing the ability to the customer to select a service for consumption
Service configuration	Small graphical wizard providing to the customer predefined parameters for defining the selected service
Service monitoring	Graphical representation of the data gathered from the monitoring modules
Billing information	Graphical representation of the billing outcomes of selected or "consumed" service

Interfaces

The information that will be collected from the rest of T-NOVA components to be used by dashboard will be provided through the following APIs:

AA: the AA access control system will provide an API to the dashboard to provide and collect all the information necessary to authenticate the T-NOVA users or stakeholders.

SLA management: the goal of this API is to show the users the following information coming from the SLA management module:

- SLA template specification to be filled by the SP and FPs.
- SLA offering to the customer and associated to each service.
- SLA fulfillments by all the stakeholders.

The SLA selection performed by the customer to manage the SLA negotiation process and the SLA contract information will come from the brokerage module that performs the trading. The SLA front-end tool that will be integrated in the dashboard will be also responsible to make the correct request to the SLA API and then gather and show the results.

Brokerage: This interface is exploited for trading issues, among the T-NOVA users (i.e., SP, FP) and the brokerage module. The information that will go through this API will be related to

- VNF request and selection: by means of this API the SP requests and selections will be sent to the brokerage module.
- Advertise VNF: this functionality is exploited for the communication between FP and the brokerage module, as the latter perform the intermediate communication, this is trading.

Orchestrator: The interface between the dashboard and the orchestrator will be used to manage service usage data: through this interface the SP and customer will be able to get the monitoring information of the service.

Billing: The billing API for the dashboard will have to manage the following information between dashboard and billing module:

- Bills charged per user and per service (SP and customer)
- Charges done to SP's customers (BSS functionality to the SP)
- Charges done to FP's customers, which is the SP.

Business Service Catalog: The Business Service Catalog API for the dashboard will be used to

- Publish and on-board service offerings by the SP to T-NOVA
- Browse available service offerings by the customer.

Service Selection: This API will have to manage the information needed to configure each service to be customized for the customer when selecting each service, for instance providing customer's network details for later the orchestrator properly deploy the service. Also used to update on configuration or need to remove a running NS.

NF Store: This interface allows the FPs to publish and manage their VNFs into the NF Store. The publication consists in uploading the VNF image, registering the VNF and its metadata into the function store. The VNFs are versioned allowing the FPs to provide further upgrades. Finally, the FPs can remove their VNFs. In summary, the information managed with this interface is

- VNF image and VNF metadata descriptor
- VNF version
- Upload, upgrade, and delete the VNF package.

14.2.5.2 Access control (AA)

In T-NOVA, different stakeholders are foreseen. Each of these stakeholders will have a specific role and accordingly some associated permissions (see below in this section Policy Enforcement Service). For instance, a FP will be able to upload a VNF and upgrade it if needed. The SP will be able to select the VNFs that he is willing to deploy and use, and should not be allowed to upload/remove a given VNF from the NF Store. One of the main challenges in T-NOVA is how to administer security in a multiuser environment. To address this issue, T-NOVA will specify and develop a lightweight Role Based Access Control (RBAC) system where decisions are based on the functions a given stakeholder is allowed to perform within T-NOVA.

- The main conclusions are summarized in the following bullets:
- The different stakeholders should be authenticated before any operation on the T-NOVA system.
- The different stakeholders should be authorized to perform tasks that are associated with their roles and permissions.
- Roles are created according to their functions in T-NOVA, and stakeholders are assigned roles based on their responsibilities and qualifications.
- Roles can be reassigned or granted new permissions if needed.
- Roles and permissions should be updatable and revocable.

Functionality

The RBAC system will be offering two main functionalities as follows:

- Authentication: Authentication is the process by which the system will verify that a user of T-NOVA is exactly who he is claiming to be.
- Authorization: Authorization is the process by which a user is allowed to perform the tasks he wants to.

Architecture

The general diagram of the access control system is depicted by Figure 14.4.

To enable the T-NOVA system to provide different functionalities to the stakeholders, a mechanism for authenticating a stakeholder is required. In T-NOVA, this is performed by the authentication manager, allowing a user to register and login with username and password. In the registration case, the user has to provide the information required (username, password, email, etc.) by filling out a registration form. Finally, through the authentication process the authentication manager returns a JWT authentication token reflecting that the user is logged in.

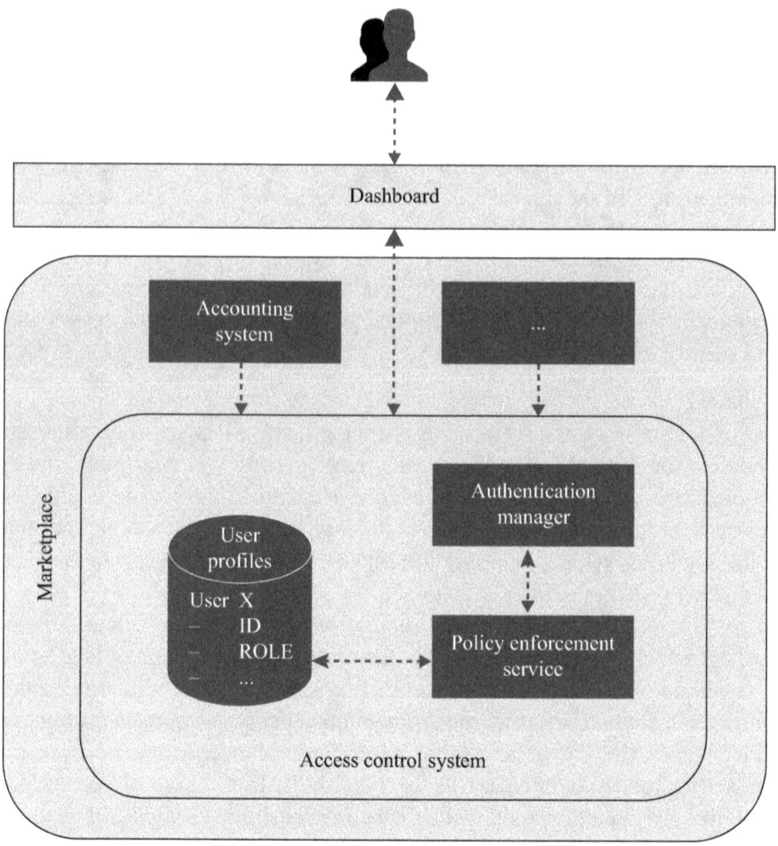

Figure 14.4 RBAC high level architecture

The Policy Enforcement service implements a RBAC mechanism that allows assigning users different roles resulting in different rights. Such an access control mechanism allows the T-NOVA system to implement functionality such as uploading a VNF or purchase a service. When a new user registers a user profile will be created containing the name and email address of the user. Furthermore, the profile will also contain the current role of the user.

The roles foreseen at this stage of the project are

- T-NOVA operator: In charge of the T-NOVA system.
- SP: It purchases several VNFs to compose a service to be sold to its final customers.
- FPs: The entities that are allowed to upload and upgrade a given VNF on the T-NOVA system.
- Customer: The entity interested in purchasing a T-NOVA service.

Interfaces

Several interfaces are foreseen to ease the communication with the other parts of the T-NOVA system. This includes the following:

- Interface to the dashboard: The T-NOVA Access Control module will provide an API to the dashboard allowing it to authenticate the T-NOVA users.
- Interface with all the other components in the marketplace: The T-NOVA Access Control System will provide an API to access the "User profiles" and features are needed to handle information, such as user permissions, personal information, and others.

14.2.5.3 Brokerage module

Towards facilitating trading between diverse actors in the NFV scene, the T-NOVA Marketplace includes an innovative brokerage module, in which VNFs by several FPs can be brokered/traded (Figure 14.5).

Functionality

Via the brokerage module API in the dashboard, the SP place their requests and requirements for the corresponding VNFs, receive offerings, and make the appropriate selections, taking into account the price and the offered SLAs. Trading policies such as long-term lease, scheduled-lease, short-term lease, or spot markets (these leasing types refer to the duration of VNFs exploitation) can be based either on fixed-price or action-based strategies.

In T-NOVA, there are several objectives in order to select an auction mechanism, which should be taken into account. The first objective is to avoid too much signaling overhead. This objective may be satisfied with the sealed-bid auction. In this auction scheme, bidders simultaneously submit sealed bids so that no bidder knows the bid of any other participant. Hence, bidders cannot change their bids after the announcement of the other bids. In the case of sealed-bid auction, the first price auction model should be implemented. Sealed-bid auction may not be truthful (truthfulness prevents market manipulation, since the bidding is performed considering the true value of the item); however, the VNFs auctions are often organized to maximize the payoff, and not to be truthful.

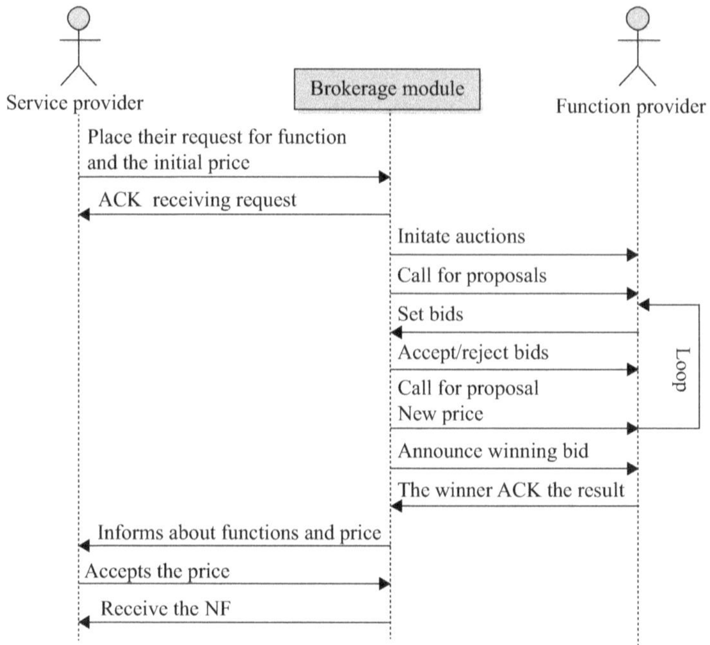

Figure 14.5 Trading process

To be precise, the implementation of the second price auction model is also possible since there is no problem in switching the payment method in an auction engine (this may be an optional feature implemented in the T-NOVA Brokerage Module), thus, making this auction truthful. The T-NOVA Brokerage Module may change the pricing rule in a flexible manner. It is a matter of implementing an extra policy in the brokerage module operation mechanism. In addition, the call price may be used to provide rational item valuation. The brokerage module will determine the proper call price for each VNF based on marketing factors. It is also possible that the bidders use their own valuation tools along with both the brokerage module, so that the former (i.e., bidders) to be able to learn the optimum call price.

In summary, the brokerage module will provide the following functionalities:

1. VNF discovery: This process is required in order the brokerage module to seek for the requested VNF.
2. Trading: This process enables the brokerage module to trade the VNFs, especially through auctions, in case that one VNF is offered by more than one FP. Figure 14.5 depicts the sequence diagram of general auction trading.

   i. The SP provides to the brokerage module the VNF request and the initial price.
   ii. The brokerage module sends an ACK that initiate auctions.
   iii. The brokerage module informs the FPs regarding the request and the initial price.
   iv. FP sends their bids for the functions (Price+SLA specification).
   v. The brokerage module solves an auction to maximize its revenue.

vi. The brokerage module informs the bid results.
vii. Depending on the type of auction, an iteration (3–6) continues until the bid winner is found.
viii. The brokerage module announces the final results.
ix. The winner acknowledges the results.
x. The brokerage module indicates the VNF's price, which is provided by the FP that won the bidding, to the SP.
xi. The SP accepts the price and SLA.
xii. The SP receives the VNF.
xiii. Price will be stored in the accounting module, and SLA agreement in the SLA management module.

Architecture

The overall architecture of the brokerage module and its interfaces is depicted in Figure 14.6.

In case the customer would like to ask for a service that is not already in the catalog, he will have the option to perform a request for a new service composition taking place. Therefore, the SP will use the brokerage module to query for specific VNFs. The process of trading between SP and FPs is then initiated according to the sequence diagram depicted in Figure 14.6.

Interfaces

The required interfaces of the brokerage module for the proper communication with the other parts of the T-NOVA system are as follows:

- Interface to the dashboard: This interface is required in order for the users of T-NOVA system (i.e., SP, FPs) to be allowed to trade. For this purpose, functionalities such as service composition/VNF request by the SP and advertise VNF/trading by the FPs are exploited.

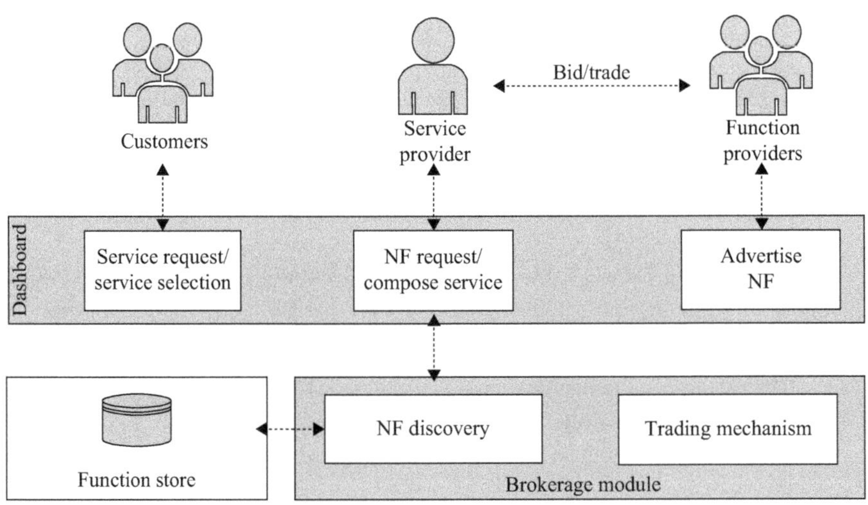

Figure 14.6 Brokerage module internal architecture

- Interface to the service selection module: With this interface, the service selection module will check if there could have been a change in the price for the VNFs as a result of the trading process. Here the SS module is an intermediate component to finally provide that information to the accounting for billing purposes.
- Interface to the SLA management module: This interface is exploited in order for the brokerage module to provide information to the SLA management module regarding the SLA agreed between SP and FPs as a result of the trading process. (The SLA management module requires such information in order to create and store the SLA contract and for SLA monitoring issues.)
- Interface to the function store: This interface is required in order for the brokerage module to retrieve information about the available VNFs for a service composition.

14.2.5.4 Business service catalog

According to the system requirements, in order to a T-NOVA user (typically the customer) to be able to easily know the services already available in the T-NOVA system, and access the description of those services, the T-NOVA Marketplace will store all this information in what we have called the "business service catalog," matching also with the approach suggested by TM Forum in its "integration framework," in which functional and nonfunctional aspects of a service based on service oriented principles are defined.

Starting from the requirements for this component, which are mainly related to the need of the catalogue to be browsable, including the service description, SLAs offered by the SPs and price for each services and SLA, we explain next its functionality, and the way the information will be stored.

Functionality

The business service catalog will be used by the SP, to store/create services and update, or delete the services. All stored services will be browsable based on criteria such as price, SLA, and other service description characteristics, by SP and Customer, which are defined in marketplace search view, through Dashboard module.

The business service catalog will be filled with the service offering information manually and offline after a service composition has taken place by the SP through the orchestrator.

Design

The business service catalog contains service offering entries; each of them is composed by: service description + SLA offer + price, according what Figure 14.7 shows.

Interfaces

- Dashboard—The business service catalog will be accessed directly and only by the dashboard module in both read and write mode (by the customer and SP respectively).
- Orchestrator—The business service catalog will push the NSD relevant fields to the orchestrator when a new service offering has been created. Once the orchestrator validates it, the availability of a service is notified to the BSC to be offered to the customer.

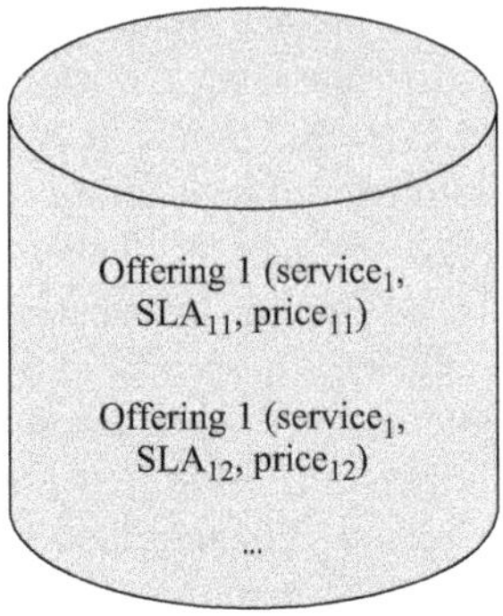

Figure 14.7 Business service catalogue

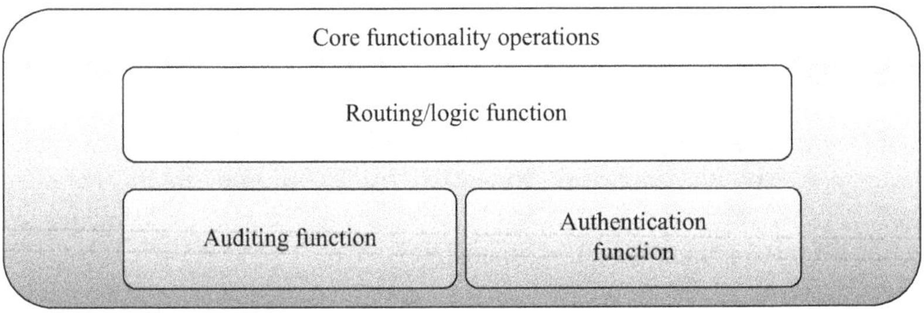

Figure 14.8 Service selection overall architecture

14.2.5.5 Service selection

Based on the requirements gathered for the Marketplace and the initial implementation steps in order to improve in modularity, it has been decided to include an intermediate component between dashboard, business service catalog and orchestrator that will be responsible to activate in the orchestrator the selected NS by the customer: the service selection module.

Functionality

The service selection module will provide the T-NOVA customer the ability to activate a NS that matches his search criteria. It will communicate with orchestrator in order to activate the service, and depending on orchestrator resources the selected NS will be provisioned, or it will be discarded [12,13]. In the case of successful activation, which would be notified by the orchestrator, the service selection module will pass the NS information to the accounting module in order to be forwarded later to the billing module. It will not take the action over the accounting in case the service is not provisioned.

Before introducing the information in the accounting module, it will make sure the information is updated and consult the brokerage module in case there has been any change during the trading process.

Design

Figure 14.8 depicts the overall architecture of the service-selection module.

The routing/logic function is responsible to forward the request to orchestrator and depending on the answer, to forward it to accounting module. The service module is based on modular architecture, in order to enhance the logic of the module, and to support additional interfaces with other modules if needed.

Interfaces

- Dashboard—Interface between the customer and the service selection where the current service selected is being configured.
- Orchestrator—It is used to notify the orchestrator about a new NS instantiation and pass the custom service configuration.
- Accounting module—It is used to create the entries in the accounting module to track every service or VNF instance created in the orchestrator for billing purposes.
- Brokerage module—This interface is used by the Service selection module to consult any change in the information regarding the price as a result of the trading process in the brokerage module before creating the accounting entry.

14.2.5.6 SLA management

SLAs represent a contractual relationship between a service consumer and a SP in order to provide a mechanism to increase trust in providers by encoding dependability commitments and ensuring the level of quality of service is maintained to an acceptable level.

In T-NOVA, there will be a SLA agreed between FP and SP and between the SP and its customers, and per each service, since the same service could have different SLA levels associated. One VNF can be offered by a FP with different flavors. Depending on the technical characteristics of the virtual infrastructure in that the VNFs will be deployed, the performance achieved for the NS will be different. The performance guarantees that a FP can offer for each VNF will be part of the negotiation through the trading mechanisms implemented by the brokerage module. Therefore, one NS can be offered with different SLA levels and different prices, being part of different offerings.

SLAs describe the service that is delivered, its properties and the obligations of each party involved. Moreover, SLAs establish that in case the guarantee is fulfilled or violated, rewards or penalties, monetary or not, can be applied, respectively. T-NOVA SLA management module will provide information for later accounting, depending on the terms and conditions gathered in the SLA and on whether this SLA has been met by all parties or not.

The requirements for the T-NOVA SLA management module, listed in Annex A, are mainly related to the need to provide mechanisms to get an agreement presented and agreed, store all the SLA agreements, to inform the orchestrator, and to know all the SLA fulfillment to inform the billing system for possible penalties.

Functionality

The SLA management module is in charge of providing mechanisms to get an agreement presented and agreed, informing the involved parties (Customer, SP and FPs), and storing the SLAs, it will later receive and will process all measurements related to the SLA from the monitoring system (in the orchestrator) and, checking if the SLAs have been fulfilled or not, will inform the accounting system for the pertinent billable items (penalties or rewarding).

A SLA basically consists of two main steps

1. Paper-signed contract, in this case, between the customer and the SP, and between the SP and FPs, including the description of the quality of service and the penalties to be applied (could also be on a web site by agreeing terms and conditions).
2. e-Contract: It is automatically negotiated between parties for each customer, depending on the demand. Always based on a paper-signed framework contract (step 1).

The SLA management module needs to be able to provide the following functionalities: publication, discovery and negotiation of SLAs requirements, in order to manage the SLA lifecycle that can be split into the following phases:

1. SLA Template Specification: For the SP (and FPs), a clear step-by-step procedure describing how to write an SLA template to provide a correct service description.
2. Publication and Discovery: Publish the provider offer and possibility for the customer to browse/compare offers.
3. Negotiation: agreement on SLA conditions between the customer and the SP and between the SP and the FPs. This could be a bargain-like transaction or simply a combo list selection of predefined choices when the customer selects a specific offering.
4. Resource Selection: Depending on the chosen SLA for every service, the SP by means of the orchestrator will map that specification to the resources that need to be assigned to the service in order to meet this SLA.
5. Monitoring and Evaluation of the SLA: Comparing all the terms of the signed SLA with the metrics provided by the monitoring system (from the orchestrator), in order to internally prevent upcoming violations.
6. Accounting: invoking the charging/billing system according to the result to inform about billable items as penalties or rewards.

Design

The information that shall be stored in the SLA management module is high level described below:

- SLA template specification: Input from the SP and FP from the dashboard
- SLA contract: (Parties involved, parameters, penalties) Input from the dashboard as the output of the SLA negotiation
- SLA fulfillment: Input from the monitoring system in the orchestrator
- Billable items: Output of the SLA management module (to be sent to the accounting module).

Interfaces

So far, several interfaces are foreseen to ease the communication with the other parts of the T-NOVA system. These are

- Interface to the dashboard: The SLA management module will provide an API to the dashboard to show the pertinent SLA information (template specification, agreement, SLA fulfillment, etc.) and to introduce the SLA templates.
- Interface to the accounting system: The SLA management module will be consulted by the accounting module about billable items as penalties or rewards when the SLA has not been achieved. Also, the accounting module will be in charge of introducing the final agreements once a purchase has occurred and to start/stop the SLA enforcement once a service has been provisioned or stopped.
- Interface to the orchestrator: The orchestrator will get the information about the terms agreed on the SLA and generate the monitoring information for each metric that will be consulted periodically by the SLA module to determine the level of fulfillment of the SLA for each service and function.

14.2.5.7 Accounting

Functionality

The accounting module in T-NOVA will be in charge of registering all the business relationships and events (subscriptions, SLA evaluations, and usage) that will be needed for billing. The accounting module will be the intermediate component between the billing module and the rest of the system.

Design

The high-level information that shall be used by the accounting module is described in Table 14.7.

Table 14.7 Accounting module information

Type	**It could be a service or a standalone VNF**
Instance ID	ID of the instance of the service (or function) in the system once it's been instantiated. It's used for interactions with the orchestrator
Client	Purchaser of the service (Customer), or function (SP)
Provider	Seller of the service (SP), or function (FP)
SLA	Id of the SLA agreement of the transaction (in order to get possible penalties to be applied)
Status	Current status or the service (or function): Running or stopped
Billing	Information on how to bill the service (or function) to the client
Dates	Date when the service (or function) was instantiated

Interfaces

- Interfaces to the SLA management module: The SLA management module will be consulted from the accounting module about billable items as penalties or rewards when the SLA has not been achieved.
- Interface to the service selection module: The Service Selection module will be in charge of creating the entries in the Accounting module. Although the orchestrator is instantiating a service, the Service Selection stores in the Accounting module all the information necessary to track the newly created instances and link them to the SLA agreement and the pricing data for a more accurate later billing.
- Interface to the orchestrator: The monitoring system in the orchestrator will use this interface to update in the accounting system the status of the different services and functions.
- Interface to the billing module: By means of this interface the billing module will get all the information it will need to issue a bill.

14.2.5.8 Billing module

The billing module in T-NOVA is in charge of generating the bills for users and providers and revenue sharing reports between the SP and the FPs at the end of a billing cycle. In order to reuse existing solutions, the billing module in T-NOVA will be an adapted version of the open source rating–charging–billing framework Cyclops.

Functionality

Cyclops allows data collection, normalization, and persistence of resource consumption data for services consumed by the customers. The framework does not implement metering itself; it assumes that the resources that are being consumed are being measured. It provides a rich set of APIs for (non-natively supported) applications to report the consumption data to Cyclops. Natively it extracts the usage values (of resources) from a few supported Infrastructure as a Service (IaaS) and Platform as a Service (PaaS) cloud platforms.

Cyclops allows custom set of meters to be defined and further allows providers to customize the rating and billing rules associated with such meters. Natively, it supports all built in meters for OpenStack and support for CloudStack is under development.

The framework generates usage data reports periodically, which has been extended to support events based usage reports generation for the T-NOVA billing scenarios. Charge data records (CDRs) are generated periodically also. This in T-NOVA is governed by the billing models, which are associated with service or VNFs instances belonging to customers.

Cyclops provides REST APIs for bill generation by specifying any desired period; hence, the user of Cyclops must determine the end of billing cycle event and use the bill generation API from Cyclops.

All the data stored within the Cyclops framework have associated timestamps, thus justifying the data storage into time-series data store. It naturally supports various data analytics and has rich APIs for data visualization.

The framework has been extended to allow revenue share computation and report generation for monetary settlements between the SP and the FP taking into account the existing revenue sharing agreement between them.

Design

Cyclops framework is inspired by micro-services design approach for distributed application development. The key micro-services in Cyclops are

1. Usage Data Record (UDR) mService
2. CDR mService
3. Billing mService.

The external applications send relevant data into Cyclops asynchronously over highly available message bus. The framework is integrated with Gatekeeper authentication and authorization service.

Interfaces

In T-NOVA, Cyclops has interfaces with the Accounting module and the marketplace dashboard. The lists of various interfaces and the functionality that are (or could be) achieved over these interfaces are described below:

- Cyclops-UDR-Accounting—It supports event-based usage reports generation for active service instances.
- Cyclops-CDR-Accounting—It supports event-based charge reports generation for active service instances and billing model details.
- Cyclops-billing-Accounting—Information flow allowing generation of revenue sharing report between SP and FP.
- Cyclops-billing-SLA—SLA violations data query from billing micro-service for a given period.
- Cyclops-billing-dashboard—Bill generation for any desired period.
- Cyclops-UDR-dashboard—Data extraction API could be used to provide rich visualization to customers.
- Cyclops-messaging-Accounting—It allows sending on billing relevant service lifecycle events to be sent to Cyclops.

At a much higher level, these interfaces can be aggregated into these two listed below:

1. Cyclops-Accounting
2. Cyclops-Dashboard.

14.3 Brokerage module

With the expanding introduction of cloud computing, the IT environment is dynamically changed into a lattice of intertwined infrastructure, platform, and application services, which are conveyed from different administration suppliers. As the quantity of cloud administration suppliers (or SPs) is increased, as well as the prerequisites of customers become unpredictable, the requirement for on-screen characters to accept a part of intermediation in the middle of suppliers/providers and consumers is getting to be more grounded. Cloud service intermediation is turning out to be progressively perceived, as a key part of cloud computing value chain.

Samples of cloud service intermediation offerings include administrations/services, towards helping to discover and think about cloud services (e.g., marketplaces/stores), to create and customize services (e.g., aPaaS—application Platform as a Service offerings), to coordinate services (e.g., iPaaS—integration Platform as a Service offerings), as well as to manage and monitor services.

Such cloud service intermediation offerings usually vary according to the types of capabilities they offer, or how these abilities are consolidated. However, they have one thing in common. This common ability is the unified characteristic that they make it less demanding, more secure, and more gainful for cloud computing adopters to explore, incorporate, consume, extend, and keep up cloud services. According to Daryl [1], this is unequivocally the quality suggestion of offerings under the general class of "Cloud Services Brokerage," a term that was authored in 2010 to refer to the emerging part of brokers in the connection of cloud computing.

14.3.1 Different roles of brokers

Gartner [14] defines "brokerage" as a model of business. This term is used to refer to "the purpose of a business that operates as an intermediary." In more detail, Gartner [15] exploits the term to denote "any type of intermediation that adds value to the consumer's use of a service." According to the same analysts, a business cannot be considered a Cloud Service Brokerage, if it does not have a "direct contractual relationship with the consumer(s) of a cloud service."

Moreover, the same analysts define a useful distinction among the terms "brokerage" and "broker," which are often alternatively used. However, they actually refer to different meanings. More specifically, according to Gartner, a broker is "a person, company or a piece of technology that delivers an instance of brokerage or, the specific application of a mechanism that performs the intermediation among consumers and providers." This analysis also states that a Broker delivers value via three primary roles: service aggregation; service integration; and service customization, whereas additional roles, such as service arbitrage, are also possible. These roles of a broker are explained below.

- Aggregation broker: This broker delivers two or more services to consumers and providers. It does not involve any integration or customization of services. Its capabilities are to support large-scale cloud provisioning, normalized discovery, access, and billing; and to support centralized management, SLAs, and security.
- Integration broker: This broker makes independent services work together for customers. It can allow process integrations, creating new value through integrated results, one-to-many, many-to-one or many-to-many. It is implemented as a PaaS with capabilities, including messaging adapters, orchestrations, and translation of event tasks. It can use policies, such as governance policy and API management, shared services for security. This broker allows cloud-to-

cloud integration, such as synchronizing between different applications, or cloud to on-premises integration, like netsuite and quickbooks synchronizing spread sheets.

- Customization broker: This broker can alter or add to the capabilities of a service to improve it. Characteristics include new functionality or new modified service. It uses the original cloud serviced enhanced, one-to-many or many-to-one capabilities to include modifications or combining services, and as a basis for implementation of new services. User interfaces and analytics messages services are typical scenarios of new and composite applications such as reports for salesforce.com. Other examples include price comparisons for bookings, business process services, and configurable processes.

With the definition of broker that Gartner defined, it basically considers any intermediation offering that increases the value of a cloud service to qualify as a cloud service broker. Any supplier of relevant services, even with the most essential intermediation abilities and a "basic" worth recommendation as now qualifies as a service broker. Some different opinions state that this definition is too comprehensive to be in any way valuable. However, Gartner states that it is a vendor-driven market research, instead of a vendor-independent assessor of best practice, and that the views are forcibly shaped by the needs of constituencies that pay for its research: distributors, system integrators, and independent software vendors (ISVs).

Alternately, in [16], cloud brokers characterizes the term of the Broker as a complex business model that offers a high value commitment in the rising cloud space. Basically, this model influences skills and abilities from every one of the three of the conventional business models; of software, consulting, and infrastructure. In Forrester's view, only integrated or aggregated services, which bring some kind of value out of the composition, may qualify as a broker, as well as an intermediary has to provide a certain complex "combined" worth recommendation so as to qualify also as broker. Forrester [17] also distinguishes three types of Cloud Brokers, as indicated by the level of the cloud stack at which they operate:

- Simple Cloud Broker—Dynamic sourcing of public IaaS services
- Full Infrastructure Broker—Dynamic sourcing across public, virtual private, and private IaaS
- SaaS broker—Unified provisioning, billing, and contract management with multiple SaaS offerings, potentially including integration of services.

In addition, the work in [18] presents the term of Broker as "an entity that manages the use, performance and delivery of cloud services, while also negotiates relationships among service providers and customers." This work also separates brokers into another three categories, according to their functionality:

- Service intermediation: A cloud broker enhances a given service, by improving some specific capability and providing value-added services to cloud consumers. The improvement can be managing access to cloud services, identity management, performance reporting, enhanced security and others.

- Service aggregation: A cloud broker combines and integrates multiple services into one or more new services. The broker provides data integration and ensures the secure data movement between the cloud consumer and multiple cloud providers.
- Service arbitrage: Service arbitrage is similar to service aggregation except that the services being aggregated are not fixed. Service arbitrage means a broker has the flexibility to choose services from multiple agencies.

14.3.2 Categorization/classification of brokerage

The classification comprises two dimensions. The first dimension concerns the type of brokerage capability concerned and includes discovery, integration, aggregation, customization, quality assurance, and optimization.

- *Discovery* deals with the provision of service that helps end users to identify and select the cloud services. For instance, through the use of marketplaces offering listings of cloud services from different providers, direct comparison of similar cloud services, ratings of cloud services, and other relevant features assisting discovery and selection.
- *Integration* is related to the provision of cloud-based software environment in order to integrate separate software systems. The integration aims at either facilitating data exchange between separate systems or realizing collaborative business processes.
- *Aggregation* concerns the provision of a cloud service that comprises multiple third-party services. An aggregate cloud service may allow users to interact with the interfaces of the third-party services directly (for instance, through a dashboard-like user interface), or indirectly, through a common interface that encapsulates the individual services and possibly adds common functionality such as authentication, billing, or SLA management across those services.
- *Customization* enables the implementation of new functionality to enrich a cloud service, by means of extension rather than modification of that service's implementation.
- *Quality assurance* is capable of ensuring that one or more cloud services obtain specific quality expectations. This can be performed by service testing, policy enforcement, SLA monitoring, and possibly by self-management mechanisms triggered to restore service quality.
- *Optimization* enables the opportunistic improvement of the consumption or provisioning of a cloud service with respect to various criteria, such as cost, functionality or performance.

The second dimension is the type of cloud service being brokered and includes the four standard cloud computing service models, that is Software as a Service (SaaS), PaaS, IaaS, and Network function virtualization as a service (NFVaaS) (Figure 14.9).

- *SaaS* concerns the provisioning of software application functionality that can be accessed through a web browser or a web API and can be paid for in a subscription-based or usage-based scheme. The granularity of the service can

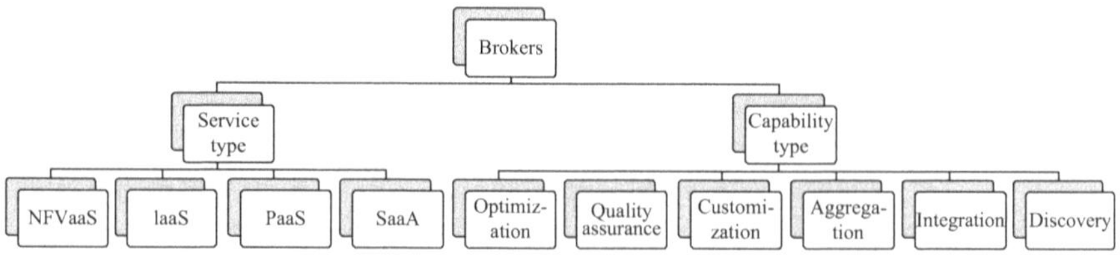

Figure 14.9 Brokers categorization based on service/capability type

range from a complete software application for customer relationship management, to a single REST/SOAP web service for obtaining stock market quotes or translating a document.

- *PaaS* concerns the cloud-based provisioning of tools and components for the development, deployment and execution of software applications, ranging from simple utility services to fully fledged development and runtime environments.
- *IaaS* concerns the provisioning of computational resources on demand following a usage-based payment scheme.
- *NFVaaS* concerns the provision of network components inside already build infrastructures.

14.3.3 Providers of brokerage modules

This section presents the brokerage providers that are classified considering the taxonomy presented above. A short description is presented for each provider.

1. Appirio [19] offers enhancements over existing SaaS offerings, such as services for contact and calendar synchronization between Salesforce customer relationship management (CRM) and Google Mail and/or Google Calendar, file management for Salesforce CRM, and more.
2. Amazon Web Services (AWS) [20] Marketplace offers discovery services for third-party SaaS and PaaS offerings. It is an online store for consumers to identify and select business applications and software infrastructure. It allows users to find, compare, and immediately deploy a SaaS or PaaS to the Amazon Elastic Compute cloud.
3. Boomi [21] aims at offering services for seamless integration between third-party SaaS offerings. Boomi comprises an integration front-end where someone can build, deploy, and manage their integration processes, and a runtime engine that executes a complete end-to-end integration process.
4. Cloudability [22] offers enhancements for managing existing SaaS, PaaS, and IaaS offerings. It enables consumers to track Key Performance Indicators for services, to create customizable cost and usage reports, to receive daily email updates on usage and cost predictions, alerts, and more.
5. CloudKick [23] offers quality assurance services for IaaS offerings. It provides a monitoring dashboard for overseeing various resources across different IaaS providers, with integrated metrics-based data collection and visualization.

6. GetApp [24] offers discovery services for SaaS. It is a free marketplace that helps consumers to identify and evaluate cloud business apps for their needs. Consumers are aided in their search with a recommendation tool, product reviews, comparison tables, and app evaluation resources.
7. Google [25] allows customization of the SaaS offerings in its Google Apps suite of services (GMail, Calendar, Drive, Docs, etc.), aggregation of third-party SaaS offerings, as well as discovery of third-party offerings through its Google Apps Marketplace.
8. Heroku [26] is the dominant platform for developing applications in the Ruby programming language. Heroku offers an add-on provider program for third-party ISVs to offer services that extend its capabilities.
9. Hojoki [27] offers aggregation services primarily for SaaS, although some IaaS services are also included. The dashboard offered by Hojoki enables consumers to launch their cloud services from within Hojoki, to keep themselves up-to-date regarding any service updates of interest, and to receive notifications in real-time of any changes in the cloud services they are using.
10. Jitterbit [28] offers integration services for SaaS through a cloud-based data and application integration platform. Jitterbit supports several integration types such as application integration, cloud/SaaS integration, ETL and data integration and business process integration.
11. Kaavo [29] offers aggregation and quality assurance services for IaaS offerings, as well as enhancements for those offerings. It provides a common API and a dashboard for managing resources across IaaS providers, offers performance monitoring, and management of service-level agreements.
12. New Relic [30] offers quality assurance services for SaaS, particularly application performance management for Ruby, PHP, Net, Java, and Python apps. Consumers can leverage New Relic's services to get real-time end-user experience monitoring for their apps and visibility across all layers of the app.
13. Rightscale [31] offers discovery, aggregation, quality assurance, and optimization services for IaaS. Through a marketplace, Rightscale allows consumers to identify and select computational resources and virtual server configuration templates, which can be deployed to different IaaS providers through a common API and UI.
14. Salesforce [32] provides not only a SaaS offering, but also a cloud application platform supporting the development of custom applications by third parties, which can be either extensions to the core CRM service by Salesforce, or used independently.
15. SnapLogic [33] offers integration services allowing users to connect any combination of Cloud, SaaS or on-premise applications and data sources. The aim of SnapLogic is to reduce vendor lock-in, providing an open and extensible integration platform for all applications and infrastructure.
16. SpotCloud [34] offers discovery and aggregation services for IaaS that assist with the identification of geographically targeted computational resources at the lowest possible cost, as well as the use of those resources through a common API.

17. StrikeIron [35] offers aggregation services for SaaS, by provisioning a common API for accessing various utility services from third-party SaaS applications. Some of the services StrikeIron offers include email verification, reverse phone lookup, postal address verification, sales tax calculation, geo IP location, and more.
18. Tapp [36] offers aggregation and quality assurance services for IaaS offerings, as well as enhancements for those offerings. Tapp introduces a management layer between a virtual infrastructure and the underlying IaaS providers, providing a common API and a dashboard for interacting with the different IaaS providers. In addition, Tapp can monitor the performance of the virtual resources across the IaaS offerings, as well as it can enhance existing IaaS offerings with migration capabilities.
19. Microsoft Azure [37] is an open cloud platform that enables developers to build, deploy, and manage applications across a global network of Microsoft-managed data centers. The platform supports development in a range of languages, tools, and frameworks. Through Windows Azure Marketplace Microsoft allows third-party SaaS offerings that are hosted on Azure or integrated with Azure to be discovered by Azure users.
20. Equinix Cloud Exchange Platform [38] is a flexible interconnection solution that provides virtualized, private direct connections that bypass the Internet, in order to provide better security and performance with a range of bandwidth options. It provides direct connectivity to several cloud providers (AWS, Google, Oracle Cloud and Microsoft Azure) and enables buyers and sellers to quickly provision to cloud connections through its portal or programmatically through APIs.
21. CoreSite Open Cloud Exchange [39] provides a portal to establish direct and secure virtual connections to cloud providers and IT SPs. It supports a wide area of Cloud Providers (AWS, Microsoft Azure, etc.) and several IT providers.

As demonstrated in Table 14.8, there are plenty brokerage modules that offer different combinations of cloud service brokerage capabilities. The majority of the brokerage SPs seem to focus on capabilities for service discovery, integration, aggregation, and customization, with a particular emphasis on SaaS services.

Only few brokerage providers enable quality assurance capabilities (New Relic, Tapp, CloudKick, Rightscale, and Kaavo, Cloud Exchange Platform). With the exception of one (New Relic), all of those offerings focus on IaaS, which happens to be the most commoditized category of cloud services today. Coverage of optimization capabilities is even sparser. Moreover, only one brokerage provider addressing this type of capability (RightScale), which also happens to focus on only one type of cloud service (IaaS).

The T-NOVA Brokerage module is the only brokering platform working with the NFVaaS concept achieving in this way benefits for the SP. More specifically, the T-NOVA SP has the ability to trade among a variety of FP's and receive the best available NFV for his service by taking into accounts the infrastructure cost and the expected performance (SLA) of the NFV.

Table 14.8 Technology ***selection***

	SaaS	PaaS	IaaS	NFVaaS
Discovery	GetApp, Google Apps, Salesforce, Windows Azure, AWS, Marketplace	Heroku, AWS Marketplace	SpotCloud, Rightscale	T-NOVA
Integration	Boomi, SnapLogic, Jitterbit	SnapLogic		T-NOVA
Aggregation	Hojoki, Google Apps, Salesforce, StrikeIron		SpotCloud, Hojoki, Tapp, Kaavo, Rightscale	T-NOVA
Customization	Appirio, Cloudability, Google Apps, Salesforce	Cloudability, Heroku	Cloudability, Tapp, Kaavo	T-NOVA
Quality assurance	New Relic		Tapp, CloudKick, Rightscale, Kaavo, Cloud Exchange Platform, OpenCloud Exchange	T-NOVA
Optimization			Rightscale	T-NOVA

14.3.4 Brokerage module architecture

The T-NOVA Marketplace has been designed as a distributed platform placed on highest layer in the overall architecture which, besides including the users front-end, comprises BSS components as billing and accounting, and innovative modules as the T-NOVA Brokerage.

Figure 14.10 depicts the high-level architecture of the T-NOVA Marketplace.

The overall architecture of the brokerage module is depicted in Figure 14.11.

It consists of five main modules that are used in various interactions, as shown in the Figure 14.11:

- *VNF Discovery Module:* This module is retrieving all the available and tradable VNFs from the NFStore.
- *Smart Filtering Module:* This module applies a smart filtering and listing to the list of the available VNFs based on the users preferences and the SLA parameters.
- *Trading Module:* This module is providing all the interaction between the SP and the FPs. The Trading module is used for requesting a new trade/offer from the SP.
- *VNF Advertise Module:* This module is responsible to advertise/return all tradable VNFs to the SPs.
- *Accepted Offers DB:* This module is responsible to store the accepted offers in order to be available for the accounting module when this required for billing purposes.

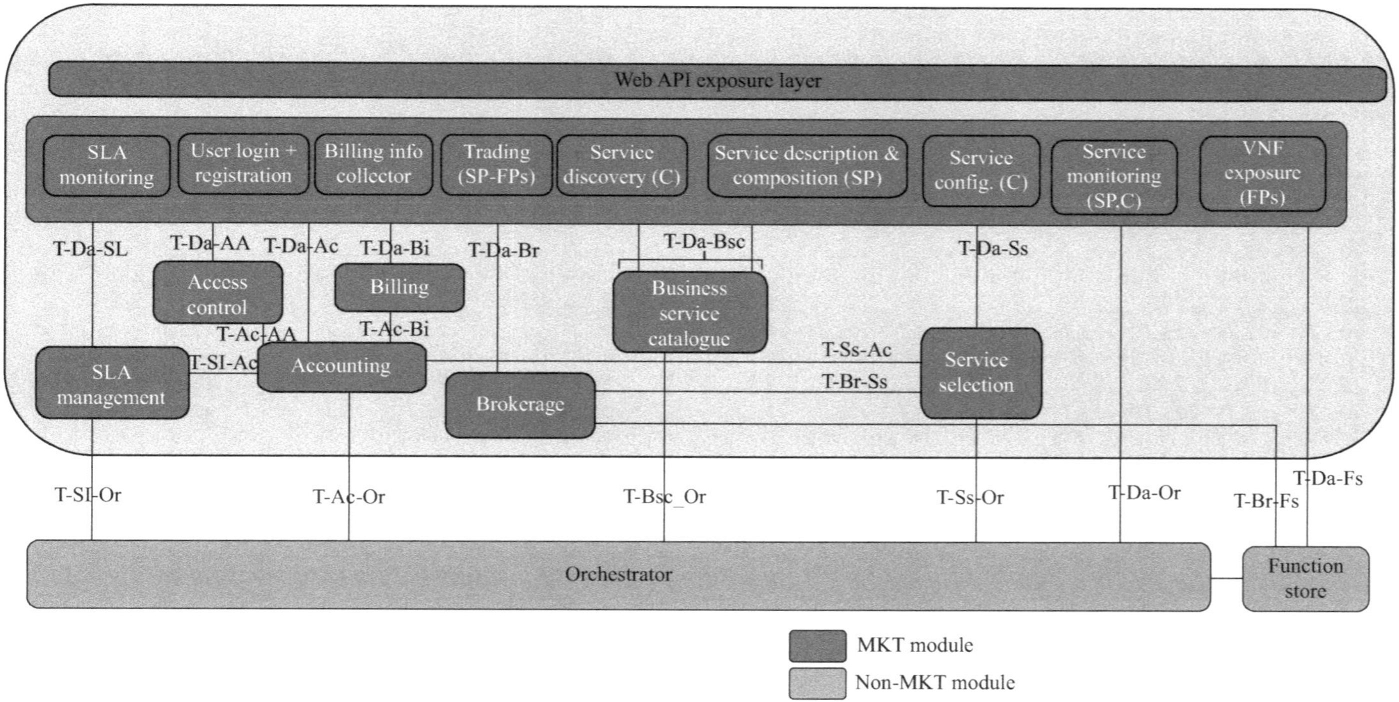

Figure 14.10 Brokerage module and its interfaces in the marketplace architecture

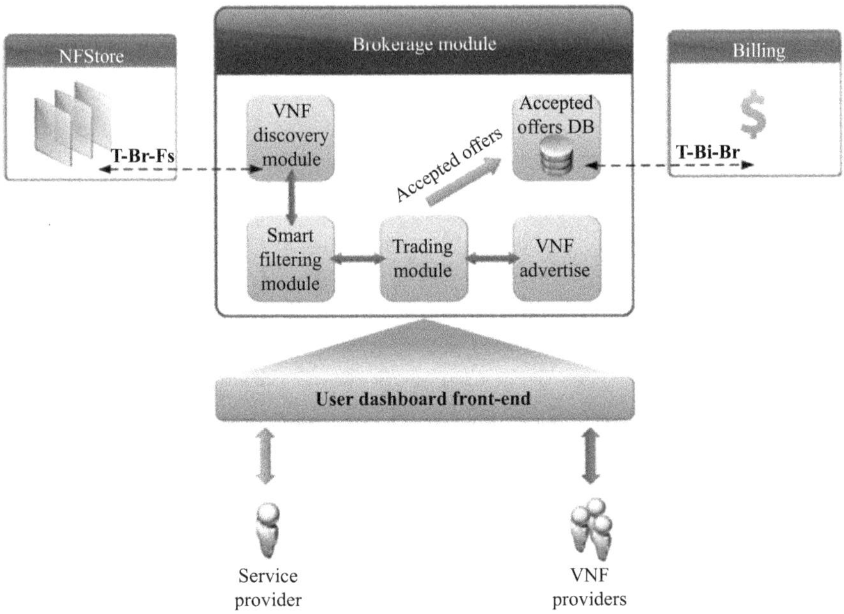

Figure 14.11 Brokerage module internal architecture

According to the proposed mechanism, the T-NOVA SP browses the offerings from the Service catalog that match his requirements. If the requested function supports Brokering/Trading the internal modules will try to fulfill the criteria set by the SP. Furthermore, the brokerage initiates the appropriate bid/trading policies according to the T-NOVA SP request inside the trading mechanism in collaboration with the NF Discovery.

The high-level architecture of the brokering module along with the interaction inside the Marketplace is depicted in Figure 14.6, all interactions are described below.

1. The SP provides to the brokerage module the VNF request and the initial price.
2. The brokerage module informs the FPs regarding the request and the initial price.
3. FP sends their bids for the functions (Price, Infrastructure cost, Setup price+SLA specification).
4. The brokerage module solves an auction to maximize its revenue based on the. Price, Setup price, Infrastructure cost, and SLA specification.
5. The brokerage module informs the bid results.
6. Depending on the type of auction, an iteration continues until the bid winner is found.
7. The brokerage module announces the final results.
8. The winner acknowledges the results.

9. The brokerage module indicates the VNF's price, which is provided by the FP that won the bidding, to the SP.
10. The SP accepts the price.
11. The SP receives the VNF.

Finally, all Setup Prices and the Price will be stored in the accounting module through the SLA management module.

14.3.5 Trading mechanism

According to the trading process (i.e., auction-based algorithm in T-NOVA), brokerage module determines the optimal allocation solution, considering the maximization of SP income. For this, the brokerage module undertakes the trading mechanism that collects bids from FPs, in order to lease the VNFs to the T-NOVA customers, through the SP. The brokerage module computes the assigning solution through this mechanism together with price and SLA per NS.

In order to calculate an Infrastructure Cost that will be used for the Trading algorithm we have used the calculation of cost based on the following Pseudo algorithm:

The vdu_cost, cpu_cost, rab_gb_cost, and the storage_gb_cost are calculated based on the price stemming from various Cloud Infrastructure providers on the internet [27].

Furthermore, when the auction-based algorithm is followed, the sellers (i.e. FPs) that are denoted as $S = \{1,2,\ldots,s\}$ lease the VNFs that denoted as $V = \{1,2,\ldots,v\}$ to $b = 1$ buyers, which is the SP. The SP is able to buy/lease xv VNFs for a specific time period ti, by reporting a price $P(b) = \{xv, ti\}$ (i.e., bid price of VNFs considering specific requirements), whereas the FPs lease yv VNFs providing a function cost fv, for a specific time ti and with a specific SLA Lv, by reporting a price $P(S) = \{fv, yv, ti, Lv\}$ (i.e., asking price of VNFs considering specific requirements). Finally, the pair (b,v) in the pseudo-code of Table 14.9 represents possible combinations of solutions, regarding "v" VNF to SP. In case that SP benefit has to be maximized, an optimization problem is formulated as follows, based on linear programming, that is, the following equation (Table 14.10):

$$\max: \sum_{s=1}^{S} (|P(b) - P(S)|) \tag{14.1}$$

In this respect and in order to facilitate competition among FPs, a novel brokerage platform is designed that will allow (i) the T-NOVA customers to search for available offerings, (ii) auctioning between the third-party function developers (FPs) and the SP, in order to find the best price for the VNFs that will be part of each T-NOVA NS.

Furthermore, while the provision of VNFs encompasses several system functionalities, VNF trading can be regarded as one part of the process that deals with the economic aspects. The trading process determines all the issues related with VNFs selling and buying (e.g., direct trading between SP and FP or via a brokerage module),

Table 14.9 Infrastructure cost calculation

```
vdu_cost = 0.03425;
cpu_cost = 0.034;
ram_gb_cost = 0.02125;
storage_gb_cost = 0.0003;
number_of_vdus = 0;
number_of_cores = 0;
number_of_ram_gb = 0;
number_of_storage_gb = 0;
for Each(vnf.vdu, function (vdu, vdu_key) {
    number_of_vdus += vdu.resource_requirements.vcpus || 0;
    number_of_ram_gb+= vdu.resource_requirements.memory || 0;
    number_of_storage_gb+= vdu.resource_requirements.storage.
    size || 0;
    number_of_vdus += 1;
});
Infrastracture.Cost = (number_of_vdus * vdu_cost) + (number_of_
cores * cpu_cost) + (number_of_ram_gb * ram_gb_cost) + (number_
of_storage_gb * storage_gb_cost);
return ( f_v, = Infrastructure.cost);
```

Table 14.10 Infrastructure cost calculation

```
Inputs: VNFs, DemandSP
Access service catalog store
Estimate the initial price per VNF
Create and advertise price-portfolio
Receive FPs offers P(S) and SP bids P(b), where P(S) = {fv, yv, ti, Lv} and P(b) = {xv, ti}
for all offers and bids do
Sort P(S) and P(b) in descending order based on price, function cost and SLA and
   create the auction-portfolio
end for
Calculate the highest valuation S[b,v] for all VNFs (i,v) □ {1, 2,..., v}
set Soptimal=S[b,v]//Random solution for algorithm initiation
for each bid P(b)do//Iteration process in order to find the best solution
if (S[b,v]) ≤ (S[b+1, v+1])//Check if the current solution is better or not to the
   neighbor solution
then save the new solution (S[b+1, v+1]) to the best found
end if
end for
return Best Solution
```

whereas pricing is a major issue that determines the value (or worth) of the VNFs to the SP and the FP.

Another issue is the competition/cooperation among function and SPs, as well as customers involved in VNF trading. Depending on the VNF trading model, the VNF access may require permission through the cooperation of SP and FP, through a payment process. To determine the optimal NF provision during the trading process, optimization and decision theory techniques can be used.

14.3.6 Dashboard integration

14.3.6.1 VNF trading

In Figure 14.12, the initial screens were the SP is able to select among the available VNFs and initiate a trade request between him and the FPs.

14.3.6.2 Trade request

In Figure 14.13, we see the pop up that is used in order to simplify the procedure of the Brokering. In the new windows, we can set new price and or Setup price. In Figure 14.13, we can see a request for a New Price for the VNF.

14.3.6.3 Pending trade request

In Figure 14.14, we can see that the SP has provided the new price and Setup Price, and the status of the VNF has changed to Pending. The pending view provides the necessary time in order the Brokering module to process the request and upon the reaction of the FP to provide the necessary answer to the SP.

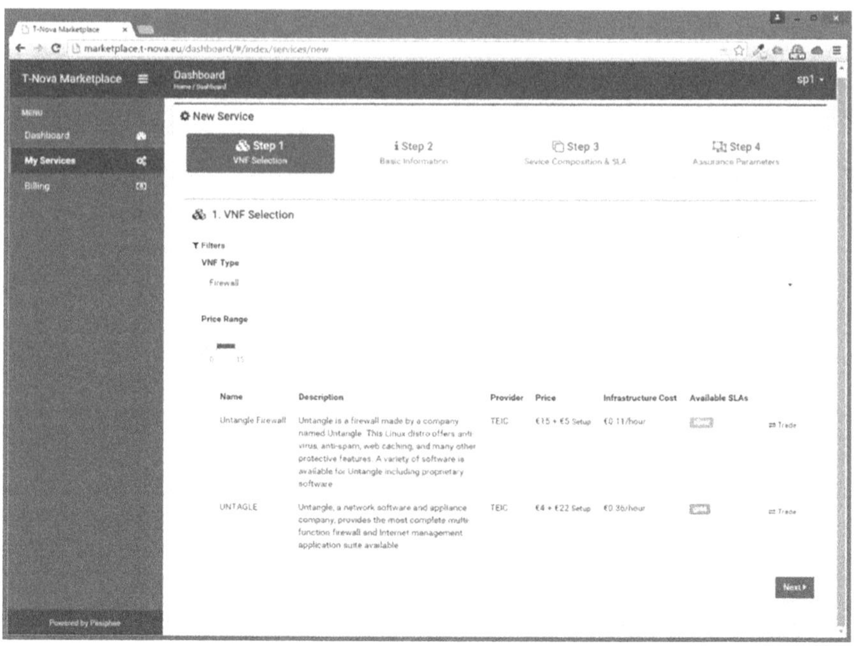

Figure 14.12 VNF trading

Figure 14.13 Trade request

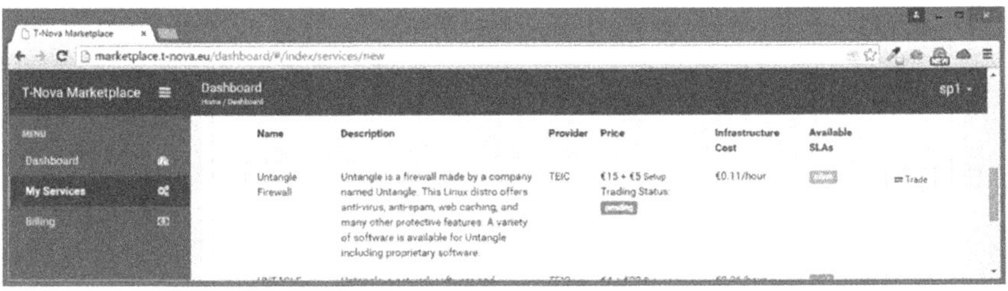

Figure 14.14 Pending trade request

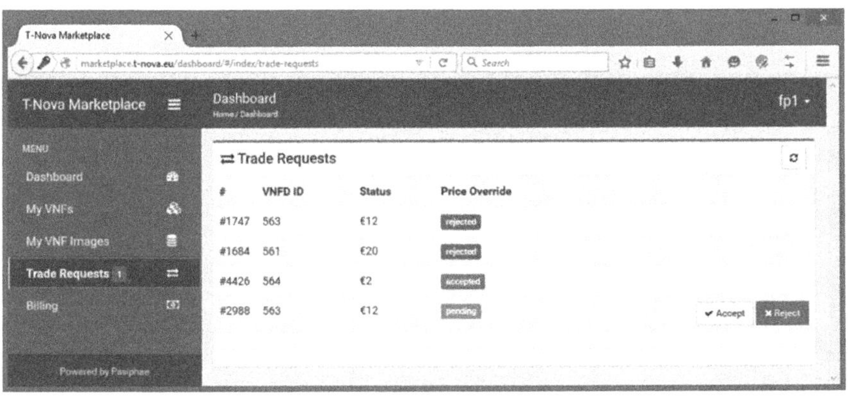

Figure 14.15 Request FP view

14.3.6.4 Request FP view

In Figure 14.15, the multiple offers from the Brokering module to the FP are depicted with the answers provided. Furthermore, we can see a pending Auction request were the FP is able to accept or reject.

14.3.6.5 Accepted trade offer

In Figure 14.16, the acceptance of new price is depicted. The colors Green for Accept or red for reject provide a quick view to the SP if multiple offers are done.

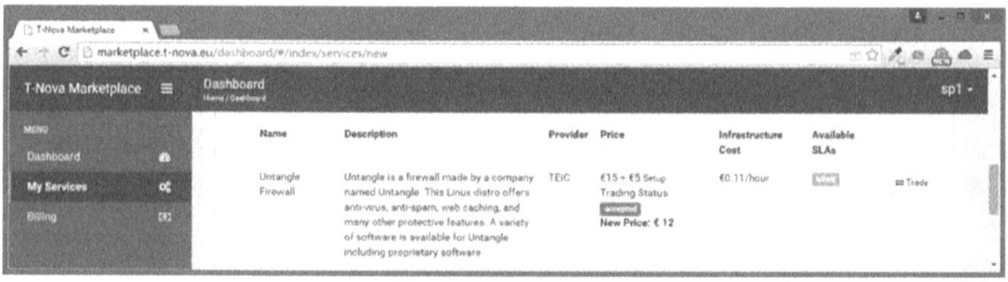

Figure 14.16 Accepted trade offer

14.4 Conclusion

This chapter provides the specification and high-level design of the NF Framework and T-NOVA Marketplace for later implementation. Analyzing the state-of-the-art, including main standardization activities, previous research projects, and commercial solutions, it has been gathered that at this stage a proper marketplace to deliver VNFs as a Service together with the Function Store as T-NOVA proposes does not exist.

The T-NOVA Marketplace has been designed to be used by three kinds of stakeholders according to the use cases analysis performed in T-NOVA [1]; therefore, a three-view dashboard will be implemented as well as and access control module that will provide AA functionalities to control their different permissions. The T-NOVA Marketplace will allow VNFs provided by a variety of software developers (FPs) to be published and traded by means of a brokerage module that will implement pricing mechanisms, for example, auctioning, when a new NS is going to be composed by a SP. T-NOVA Customers will be able to browse and select among the available NS offerings in the marketplace by means of a business–service–catalog as well as negotiate the associated SLA and price. The billing procedure contemplates not only final customers of T-NOVA NSs, but also the commercial relationship between the SP and FPs.

In relation to the specification of the NF Framework, two main tasks have been performed: the description and specification of the VNFs and the design of the NF Store. The first one includes the APIs allowing the VNF to be managed by the T-NOVA system as well as the information elements that shall be present in the VNF metadata. A key piece of data of the NF Framework is the metadata descriptor associated with the actual software implementation of the VNF. Besides the structural definition of a VNF, its behavior has been studied defining a lifecycle that is common to all the VNFs in T-NOVA. This lifecycle can be split into an inactive and active part. In relation to the active one, the lifecycle states describe the VNF when it is up and running over a virtualized execution platform. On the other hand, the inactive lifecycle states span from the software development of the VNF to its uploading into the NF Store that can be thought as the place where the VNFs are stored. The NF Store APIs provide interfaces by means of the dashboard with FPs for uploading, updating, and withdrawing the VNF software images and metadata description and interfaces with the rest of the T-NOVA system for making this information available for service orchestration.

14.5 Future work

The specification provided in this chapter has been built on the basis of the requirements at T-NOVA system level described in some relevant parts of the ETSI NFV work and TM Forum best practices. The information assembled with this process has been the critical input into a two-stage process: Stage 1 consisted on a research and design phase, where a system engineering approach was adopted to define the key functional components [3]. Stage 2 presented in this chapter has defined both the reference architecture and its functional entities and interfaces in a technology-agnostic manner to decouple the specifics of the implementations details. An additional third stage will address the details of the suitable technologies and their operation. T-NOVA project will also elaborate on the system integration and testing of all its components, for example, with the T-NOVA orchestrator and Virtualized Infrastructure Management. We do expect that system integration may detect some gaps or need of fine tuning the interface descriptions. Moreover, testing the system can identify some nonfunctional aspect that could suggest refining some part of this specification. For instance, it is difficult to figure out performance and component interaction issues with the limited experience we have with actual NFV implementation in field.

Beyond T-NOVA, we have identified some 5G research and innovation projects towards T-NOVA Marketplace and NF framework can be a very good reference to build on. These are among others,

- 5GEx [40], which aims to build a sandbox to extend software networks in a multi-domain/operator environment. Although 5GEx is not expected to implement a full marketplace layer, it should specify a northbound API for end users to access the multi-domain service catalog. T-NOVA Marketplace can be a good reference to look at, for deriving functional and non-functional requirements on business-to-customer interface.
- SONATA [41], which focuses on the implementation of an enhanced modular orchestration platform and an software development kit (SDK) to facilitate service composition by service developers. T-NOVA GUI for SPs and its interface with T-NOVA orchestrator and T-NOVA Function Store can be seen as a relevant starting point for SONATA to build the SDK for service composition.

Acknowledgment

This work was undertaken under the Information Communication Technologies (FP7-ICT-2013-11) EU FP7 T-NOVA project, which is partially funded by the European Commission under the grant 619520.

References

[1] Mastorakis G., Mavromoustakis C. X., Markakis E., Pallis E., Skourletopoulos G., Virtual network functions exploitation through a prototype resource management framework. In Telecommunications and Multimedia (TEMU), 2014 International Conference on 2014 Jul 28 (pp. 24–28). IEEE.

[2] TM Forum, "TM Forum WebSite" http://www.tmforum.org.
[3] Shiakallis O., Mavromoustakis C. X., Mastorakis G., Bourdena A., Pallis E., Traffic-based S-MAC: A novel scheduling mechanism scheduling mechanism for optimized throughput in mobile peer-to-peer systems, igi-global, 2015.
[4] "ETSI GS NFV-SWA 001" Network Functions Virtualisation (NFV) – Virtual Network Functions Architecture. Available from: http://www.etsi.org/deliver/etsi_gs/NFV-SWA/001_099/001/01.01.01_60/gs_NFV-SWA001v010101p.pdf, 2014.
[5] Mavromoustakis C. X., Mastorakis G., Bourdena A., Pallis E., Energy-efficient management using a traffic-oriented routing scheme for cognitive radio networks. International Journal of Network Management. 2015;25(6):418–434.
[6] Markakis E., *et al.*, Peer-to-peer constellations in an IMS-enabled network architecture based on interactive broadcasting. In Resource Management in Mobile Computing Environments 2014 (pp. 271–282). Springer International Publishing.
[7] Mastorakis G., *et al.*, Energy-efficient routing in cognitive radio networks. In Resource Management in Mobile Computing Environments 2014 (pp. 323–340). Springer International Publishing.
[8] Kryftis Y., Mastorakis G., Mavromoustakis C. X., Batalla J. M., Pallis E., Kormentzas G., Efficient entertainment services provision over a novel network architecture. IEEE Wireless Communications. 2016;23(1):14–21.
[9] Andreou A., Mavromoustakis C. X., Mastorakis G., Bourdena A., Pallis E., Adaptive heuristic-based P2P network connectivity and configuration for resource availability. In Resource Management in Mobile Computing Environments 2014 (pp. 221–240). Springer International Publishing.
[10] Mavromoustakis C. X., *et al.*, On the perceived quality evaluation of opportunistic mobile P2P scalable video streaming. In 2015 International Wireless Communications and Mobile Computing Conference (IWCMC) 2015 Aug 24 (pp. 1515–1519). IEEE.
[11] Markakis E., Negru D., Bruneau-Queyreix J., Pallis E., Mastorakis G., Mavromoustakis C. X., "A P2P home-box overlay for efficient content distribution," in Emerging Innovations in Wireless Networks and Broadband, pp. 199–220, igi-global, 2016.
[12] Mavromoustakis C. X., *et al.*, Dynamic cloud resource migration for efficient 3D video processing in mobile computing environments. In Novel 3D Media Technologies 2015 (pp. 119–134). Springer, New York.
[13] Bourdena A., Mavromoustakisb C. X., Kormentzasa G., Pallisc E., Mastorakisc G., Bani Yasseind M., A resource intensive traffic-aware scheme using energy-aware routing in cognitive radio networks. Future Generation Computer Systems. 2014;39(Oct):16–28.
[14] Plummer D., Lheureux B., Cantara M., Bova T., Cloud services brokerage is dominated by three primary roles. Gartner Research Note G00226509 Gartner, Inc., (2011).
[15] Plummer D., Lheureux B., Karamouzis F., Defining cloud services brokerage: taking intermediation to the next level. Gartner Research Note G00206187, Gartner, Inc., (2010).

[16] Cloud Broker—A New Business Model Paradigm by Stefan Ried. Published: August 10, 2011, Updated: September 22, 2011.
[17] Cloud Brokers Will Reshape the Cloud—Getting Ready for the Future Cloud Business Models. Forrester, September 2012.
[18] NIST, Cloud Computing Reference Architecture, September 2011. National Institute of Standards and Technology, Special Publication 500-292.
[19] http://appirio.com/.
[20] https://aws.amazon.com/.
[21] http://www.boomi.com/.
[22] https://www.cloudability.com/.
[23] https://www.crunchbase.com/organization/cloudkick.
[24] https://www.getapp.com/.
[25] http://google.com.
[26] https://www.heroku.com/.
[27] http://hojoki.com.
[28] http://www.jitterbit.com/.
[29] http://www.kaavo.com/.
[30] http://newrelic.com/.
[31] http://www.rightscale.com/.
[32] http://www.salesforce.com/eu/.
[33] http://www.snaplogic.com/.
[34] http://www.spotcloud.com/.
[35] https://www.strikeiron.com/.
[36] https://www.flexiant.com/tapp/.
[37] https://azure.microsoft.com/en-us/.
[38] http://www.equinix.com/.
[39] http://www.coresite.com/.
[40] 5GEx project – 5G Exchange – https://5g-ppp.eu/5gex/.
[41] SONATA project – Service Programing and Orchestration for Virtualized Software Networks – https://5g-ppp.eu/sonata/.

Index